T0332362

Applied Mathematics

Applied Mathematics: A Computational Approach aims to provide a basic and self-contained introduction to Applied Mathematics within a computational environment. The book is aimed at practitioners and researchers interested in modeling real-world applications and verifying the results—guiding readers from the mathematical principles involved through to the completion of the practical, computational task.

Features

- Provides a step-by-step guide to the basics of Applied Mathematics with complementary computational tools.
- Suitable for applied researchers from a wide range of STEM fields.
- Minimal pre-requisites beyond a strong grasp of calculus.

João Luís de Miranda is Professor at ESTG—Escola Superior de Tecnologia e Gestão (IPPortalegre) and a Researcher in Optimization and Process Systems Engineering (PSE) collaborating with CERENA-Centro de Recursos Naturais e Ambiente (IST/ULisboa). He has been teaching for more than 25 years in the field of mathematics (e.g., Calculus, Operations Research—OR, Management Science—MS, Numerical Methods, Quantitative Methods, Statistics) and has authored/edited several publications in Optimization, PSE, and education subjects in engineering and OR/MS contexts. João Luís de Miranda is addressing the research subjects through international cooperation in multidisciplinary frameworks, and is serving on several boards/committees at national and European levels.

Applied Mathematics
A Computational Approach

João Luís de Miranda
Portalegre Polytechnic University, *Politécnico de Portalegre*
CERENA, *Instituto Superior Técnico*, Lisboa, Portugal

CRC Press is an imprint of the
Taylor & Francis Group, an **informa** business
A CHAPMAN & HALL BOOK

Cover Image: 'Barrels Of May', by Mariana Jerónimo

First edition published 2025
by CRC Press
2385 NW Executive Center Drive, Suite 320, Boca Raton FL 33431

and by CRC Press
4 Park Square, Milton Park, Abingdon, Oxon, OX14 4RN

CRC Press is an imprint of Taylor & Francis Group, LLC

Library of Congress Cataloguing-in-Publication Data
Names: Miranda, João Luis de, author.
Title: Applied mathematics : a computational approach / João Luís de Miranda, ESTG-Escola Superior de Tecnologia e Gestão, Portugal.
Description: First edition. | Boca Raton : CRC Press, 2025. | Includes bibliographical references and index.
Identifiers: LCCN 2024011552 (print) | LCCN 2024011553 (ebook) | ISBN 9781032595245 (hardback) | ISBN 9781032610719 (paperback) | ISBN 9781003461876 (ebook)
Subjects: LCSH: Calculus--Textbooks. | Calculus--Data processing--Textbooks.
Classification: LCC QA303.2 .M57 2025 (print) | LCC QA303.2 (ebook) | DDC 515--dc23/eng/20240412
LC record available at https://lccn.loc.gov/2024011552
LC ebook record available at https://lccn.loc.gov/2024011553

ISBN: 978-1-032-59524-5 (hbk)
ISBN: 978-1-032-61071-9 (pbk)
ISBN: 978-1-003-46187-6 (ebk)

DOI: 10.1201/9781003461876

Typeset in Minion
by MPS Limited, Dehradun

Contents

Foreword

After earning my PhD in Statistics (with secondary emphases in Operations Research and Mathematics support) in 1997 from the University of Cincinnati, I have had the good fortune to coauthor eight books and over fifty academic papers in statistics, operations research, analytics, and data visualization; travel to over fifty nations; and work with hundreds (or perhaps, thousands) of intelligent, hard-working, earnest, and highly motivated colleagues across academia, industry, and the not-for-profit sector. These efforts have included establishing and organizing international Teaching Effectiveness Colloquia in Uruguay, South Africa, Colombia, India, Tanzania, Argentina, Kenya, Nepal, Cameroon, Croatia, Cuba, Estonia, Fiji, Mongolia, Moldova, Bulgaria, Tunisia, Grenada, and Sri Lanka; cofounding Statistics without Borders (https://www.statisticswithoutborders.org/), serving on the committee that founded the INFORMS Pro Bono Analytics initiative (https://connect.informs.org/probonoanalytics/home), and delivering keynote addresses to conferences in twenty-seven nations.

It has been my pleasure to know João Luís de Miranda as a colleague and trusted friend since 2011. Since I first met Dr. Miranda at the 2011 IFORS (International Federation of Operational Research Societies) Triennial Conference, I have watched him continue to grow professionally and expand his interest in and influence on making applied mathematics education more relevant and engaging for students. He has over twenty-five years of experience teaching various mathematics-oriented courses (calculus, operations research, numerical methods, quantitative methods, statistics), he has authored dozens of papers published in academic journals, and he generously gives his time and effort to initiatives designed to improve the quality of applied mathematics education we provide to our students. *Applied Mathematics: A Computational Approach* is a natural culmination of Dr. Miranda's professional efforts.

Through *Applied Mathematics,* Dr. Miranda successfully provides an easy-to-understand, self-contained introduction to the basic calculus and related mathematics topics a student must master to prepare to study STEM fields such as economics, operations research/management science, analytics, and applied statistics as well as social sciences and humanities. Dr. Miranda enhances and distinguishes his coverage of these topics through his emphasis on critical thinking and practical applications, his careful explanation of the notation, and his deliberate and thoughtful elucidation of proofs where necessary.

As student interest in STEM fields has grown over the past several years (in response to the dramatic increased in demand for these skills in the job market), the need for a book that covers basic calculus and related mathematics topics in a manner that a broader range of students and practitioners can understand has intensified. Dr. Miranda's *Applied Mathematics* has filled this need admirably. I hope you enjoy learning from *Applied Mathematics: A Computational Approach.*

Sincerely,

James J. Cochran, PhD
Professor of Statistics and the Mike &
Kathy Mouron Research Chair
The University of Alabama

Preface

IMPORTANT SKILLS FOR THE DIGITAL ERA AND THE EFFECTIVE UTILIZATION of digital systems would include high-level cognitive skills for whom automated operations and routine-basis procedures are neither being expected, nor openly proposed for deployment in the most recent years. Beyond the average years of schooling, skills complexity, and geographical factors, the prospects for education and training would include appropriate cognitive skills, e.g., active learning, complex problem solving, creativity, critical thinking, strategic thinking, along with the selection and utilization of adequate learning methods.

In that manner, a book on Applied Mathematics should overview a number of contents, covering basic notions while providing an enhanced approach to important Calculus subjects. This text is directed to post-secondary readers, or university students in the initial years, being the required concepts and tools introduced as needed. With a computational approach aiming at the understanding of new situations, creative problem-solving, and decision-making, the experimental focus is supported either in the analytical confirmation of necessary conditions or in the visual verification of sufficient conditions. It is thus envisaged the book can be effective for students of STEM fields, Economics and Management Science, Social Sciences and Humanities, and for readers interested in a computational point of view.

The text contents include Applied Mathematical tools that support judgment and decision-making, with the readers being initiated into a simple case, and facing simple issues. Then they have the opportunity to focus on more challenging situations, to select and use the appropriate procedures, and creatively solve them with the proper mathematical manipulations and computational reasoning. The importance of observation and verification, and experimental focus is illustrated through the application of diverse mathematical tools, within a typical differential calculus path that aims at integral calculus. Namely, such path includes real numbers sequences, function limits, continuity, derivation, and the main results of differential calculus; then integral calculus is introduced, by including integral sums, indefinite integrals, and definite integrals; finally, important concepts on sum series are addressed too.

Chapter 1 introduces real functions and presents in a challenging way diverse mathematical approaches directed to a simple problem, while critical thinking is promoted too. For that, a problem focusing on the cost of a cylindrical container (barrel) with counter-intuitive results is introduced. A constructive approach to the cost function is described, simplifications are discussed, and the function representations in table and graphic are followed by qualitative approaches. Such approaches are complemented with analytical developments, while the inverse rationale of giving the cost and obtaining the barrel diameter is targeted too. Finally, the main results are summarized, including the comparison of several volumes and costs.

In Chapter 2, sequences of real numbers and concepts related with sequence limits are presented, in conjunction with pertinent numerical instances. Firstly, some preliminary notions about number sequences are revisited; secondly, the main theorems about sequences are discussed, while key properties of sequence limits (e.g., uniqueness, arithmetic operations) are revisited. Important sequences are also treated, namely, geometric sequences with exponential terms, and relevant insights about the computational representation of real numbers are introduced.

The limit of a function at a point is a key concept, and it is presented in Chapter 3: the sequences study is enlarged onto the sequence limit of dependent variable, y, for a given limit of independent variable, x. Beyond the basic notions about function limits, the main properties and operations are also verified. Noting that properties and operations about function limits are analogous to those for sequence limits, the function's infinitesimal limit is focused too. While numerical examples are illustrating various types of limits (e.g., lateral limits), some remarkable limits are presented and discussed.

Chapter 4 addresses functions continuity, by treating both the continuous function at a point and in a range. Continuity is very effective in the evaluation of elementary functions limits, by allowing the point under analysis to be directly poked onto the function argument. In addition, the main properties and operations with continuous functions are revisited, together with theorems with broad specter of applications, namely, the intermediate values theorem that supports the location of function roots. Numerical instances are illustrating the referred topics, including the bisection method to obtain the cylinder diameter for a given cost.

The function derivation and its main results are described in Chapter 5. Firstly, basic notions about the derivative function are presented, along with the derivatives of elementary functions and arithmetic operations. After that, derivatives of various transcendent functions are treated, and then complemented with the derivative of inverse function. A comparison analysis is developed by inverting and deriving the cost function, and derivatives of different orders are finally addressed.

Chapter 6 addresses functions sketching, and the role of first derivative to identify the function roots and extreme points is discussed in the first place; for that, important results about differentiable functions, including both the derivative roots' theorem and the derivative average value's theorem, are addressed. After that, the second-order derivative and graphic concavity complete the study of maxima and minima. Asymptotes are also treated, either vertical asymptotes, or oblique/horizontal asymptotes; the theorem that relates the increments of two functions, as well as L'Hôpital rule for indeterminate limits are presented. Sketching procedures are described, including a table summarizing the cost function's variation, and other applications of interest are discussed.

In Chapter 7, a brief insight on integral sums is provided. Two approaches to the integral sum are focused on: the lower and the upper sum, and their convergence to the same limit. The areas quadrature is also described. For one side, by illustrating several examples of area squaring that allow direct verification, namely: (i) a rectangle; (ii) a triangle; (iii) a circle. For the other side, complementing these first steps on integral sums with the anti-derivative approach. In fact, the total variation in the area's function can be addressed by summing all the area's differentials.

Chapter 8 addresses the indefinite integral (*alias*, the primitive function); firstly, basic notions and important properties are introduced, and then immediate primitives are presented. The linearity properties of indefinite integral are discussed, namely, the integration of functions sums, and the multiplication of constant by function. General methods of integration are addressed, including integration by decomposition, by substitution, and by parts. Finally, specific integration methods for rational functions are described, also noting that both irrational functions and trigonometric functions can be transformed and treated as rational functions.

The main topics about the defined integral, including important theoretical and practical issues, are presented in Chapter 9. Initially, key results about definite integrals are focused, including the useful mean theorems, and then several applications of Barrow's rule are described. Other applications to the area calculation of positive and negative functions are detailed too, and improper integrals are also addressed. This last topic enlarges the integration tools into both unbounded domains and unbounded functions, and useful convergence criteria are defined.

Chapter 10 presents sum series, and important notions about series and some reference series are introduced first. Then, important theorems and their applications to the analysis of series convergence are presented, followed by the convergence criteria for series of non-negative terms. These topics are paving the way for the study of series with negative terms, in special alternating series, where the Leibniz criterion and related applications are focused. After that, function series, the convergence domain, and applications to both series derivation and series integration are presented. Finally, Taylor formula is revisited, and the development of functions in Taylor series is studied; namely, for the development of the cost function series. In addition, tables summarizing the main subjects described in the book are presented, including both the differential methods described until Chapter 6, and the integral and sums topics in Chapters 7–10.

With the best motivation to present computational approaches and methods that typically support mathematical applications, obviously other important topics cannot be included in this book. Namely, Ordinary Differential Equations (ODE), not only those based on function series, but also multivariate calculus, and the useful operational calculus (e.g., Laplace and Fourier Transforms) that would deserve a dedicated volume.

This book considers three main attributes, namely, important Calculus results, study of inverse mathematical applications, and making use of computational approaches (e.g., generalization/reduction, error evolution, conjugating confirmation and verification measures). In this way, the solution of numerical instances and simple examples is presented without software or digital brands; nowadays, a number of tools and solving applications are available on many digital platforms, and cloud computing is quite common. Nevertheless, presentation slides, worksheets, and other documentation can be made available for the interested reader upon request.

The materials in this book have been classroom tested for two and a half decades, including several periods of teaching mobility where invaluable feedback was received, either from STEM students or from other fields. A number of colleagues from diverse geographical places have shared their thoughts too, and I gratefully acknowledge all the precious comments and opinions.

CHAPTER 1

First Notes on Real Functions

This chapter introduces real functions, and presents in a challenging way diverse mathematical approaches directed to a simple problem, while critical thinking is promoted too. Therefore, a problem focusing on the cost of a metal container with cylindrical form (barrel) is introduced, but it conveys counter-intuitive results. A constructive approach to obtain the cost function is described, and several approximations are commented too. The function representations in table and graphic are complemented with qualitative approaches: the function's range, monotony intervals, the occurrence of maxima or minima, line curvature, and asymptotes are thus discussed. In addition, such qualitative approaches are complemented with symbolic manipulations, while the inverse rationale of defining the container cost and then obtaining the respective diameter is targeted too. Finally, the main results are summarized, with simple calculations supporting the comparison of different containers and respective costs.

1.1 INTRODUCTION

The importance of observation and verification, judgment and decision-making, and experimental focus is illustrated by a simple problem aiming at the application of several mathematical concepts. With a computational approach that promotes the understanding of new situations for creative problem-solving, the experimental focus is supported either in the verification or confirmation of mathematical-based conditions. Therefore, with the proper manipulations and computational reasoning, readers are empowered to face more challenging situations, and creatively solve them by selecting and using the appropriate procedures.

The problem instances are adjusted, they are aimed both at pre-university readers and graduation students in the initial years. More

DOI: 10.1201/9781003461876-1

difficult instances are relying on simple digital skills, e.g., by focusing on numerical instances or detailing graphs, and using mathematical applications or digital tools. In this way, such instances represent a good opportunity for mastering this type of problem, and the construction of solution strategies is crucial to adequately address the problem and related instances. Such procedures can include qualitative and quantitative approaches; different representations for real functions (e.g., analytical, graphic, table) and associated tools; verification steps and reference values for confirmation purposes; as well as comparison analysis, critical thinking, and results discussion.

In this way, the structure for the rest of this chapter follows. Initially, a simple problem that presents counter-intuitive results is challenging, and motivating for the study of real functions; second, the cost function and the constructive approach to obtain it are described; then, with the support of the table's function and related graphic, the function's main attributes are qualitatively discussed; after that, quantitative measures and developments are complementing the qualitative approaches in the prior section; from the complementary point of view, the inverse rationale of given the cost and then obtaining the barrel diameter is considered too; finally, the main results are revisited, with simple calculations and comparing volumes and costs.

1.2 A FUNCTION OF REAL NUMBERS

A simple problem that presents counter-intuitive results is challenging for additional steps in the study of real functions. Diverse approaches and mathematical tools are then applied to address in a realistic manner the cost of a metal container with cylindrical form.

The cost, C (in Euro, €), of the material used in the construction of a cylindrical metal container can be related to its diameter, d (in meters, m), according to the relation:

$$C(V, d) = 100\left(\frac{\pi}{2}d^2 + \frac{4\,V}{d}\right)$$

where V (in cubic meter, m^3) represents the volume.

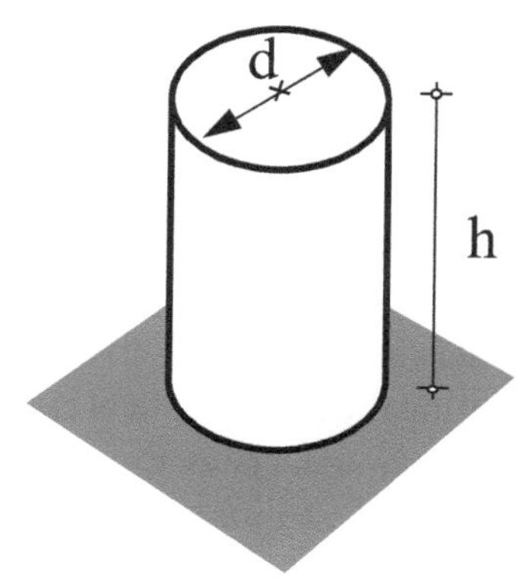

Note: It is assumed the material cost is directly proportional to the total surface area, including the areas of the base, the top, and the lateral area of a cylinder.

a. Calculate the cost for a volume of 400 liters and diameter of 1.6 meters.

b. *Ditto*, but now for a volume of 800 liters and diameter of 1.2 meters.

c. Study the cost function, C(*V*,*d*), for the given volumes and sketch it.

The formula allows the cost calculation (dependent variable, *C*), by assuming a mathematical relation with the volume and diameter (the independent variables, *V* and *d*). Or better, to evaluate the expression C(*V*,*d*) in the first member: (i) Define the values for *V* and *d*; (ii) Apply them into the mathematical expression in the second member; (iii) Obtain one and single result. Note: the inputs are real numbers, and the result is one, and only one, real number.

a. Calculate the cost for a volume of 400 liters and diameter of 1.6 meters.

$$C(V = 0.4,\ d = 1.6) = 100\left(\frac{\pi}{2} \times 1.6^2 + \frac{4 \times 0.4}{1.6}\right) = 502.12 \text{ €}$$

b. *Ditto*, but now for a volume of 800 liters and diameter of 1.2 meters.

$$C(V = 0.8,\ d = 1.2) = 100\left(\frac{\pi}{2} \times 1.2^2 + \frac{4 \times 0.8}{1.2}\right) = 492.86 \text{ €}$$

Is the Cost Formula Correct?

These results are counter-intuitive: the barrel with greater volume presents the lower cost! In addition, the related diameters are also different, with the lower diameter corresponding to the cylinder with greater volume, and vice-versa. The dimensions in use (m, m^3, €), and the involved amounts (*About five hundred Euros! Is it reasonable?*) are also issues that deserve attention, along with the procedures originating the mathematical expression for the cost function. Note the formula associates the expressions in the two sides of the equality; definitely, a more detailed approach is needed.

1.3 THE COST FUNCTION

The cost function is revisited, and the constructive approach to obtain it is described. The material utilized for the lateral, base, and top parts is combined, while approximations or simplifications in the barrel manufacturing are commented too.

- In the construction of the mathematical relation, it was assumed the total cost of material is directly proportional to its total surface area, by summing the area for the base, the top, and the lateral area of a cylinder.
- The referred assumptions benefit from simplifications, including among others, some typical issues: the barrel would correspond to a geometric cylinder, without trim losses, or material overdesign for connecting the different parts; also, metal thickness and surface treatments are not considered, and they could vary with the part at hand.

The base and top are equal (Figure 1.1c), both correspond to the circle of diameter d,

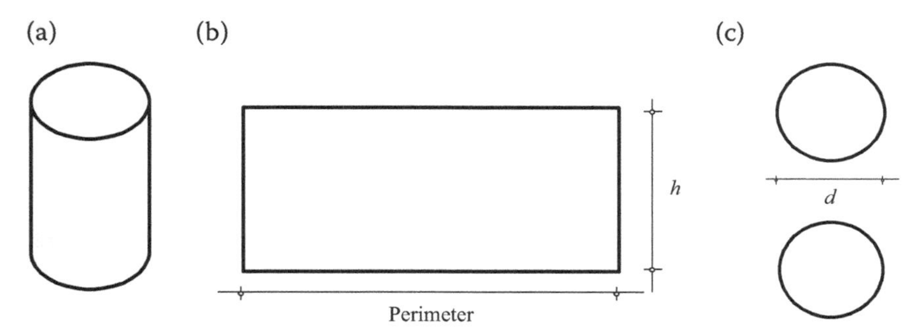

FIGURE 1.1 Views of cylinder components, from left to right: (a) Perspective view; (b) View of lateral area; (c) View from above (top and base).

$$A_{Bas} = A_{Top} = \frac{\pi}{4}d^2$$

The lateral area depends on height h and circle perimeter, $(\pi\ d)$; the lateral part is extended in the plain and corresponds to a rectangle with length equal to the circle perimeter (Figure 1.1b):

$$A_{Lat} = h\ (\pi\ d)$$

The unit cost for the material is treated as a reference value, e.g., 100, and it can be manipulated in other calculations as a percentage factor (e.g., iron or steel casts, recipient purpose):

$$c = 100\ Euro/m^2$$

The total area, A_{Total}, considers the lateral area, A_{Lat}, and two times the base area, A_{Bas},

$$A_{Total} = 2A_{Bas} + A_{Lat} = 2\left(\frac{\pi}{4}d^2\right) + h\ (\pi\ d) = \frac{\pi}{2}d^2 + h\ \pi\ d$$

When the cylinder volume is defined, the variables height h and diameter d are linked,

$$V = h\ A_{Bas} = h\left(\frac{\pi}{4}d^2\right)$$

In fact, for a given volume, when diameter increases then the height diminishes, and a larger and lower cylinder is obtained. Complementary, the height increases when the diameter diminishes, and a thinner and higher cylinder is obtained while the volume is still the same. That is,

$$h = \frac{V}{\left(\frac{\pi}{4}d^2\right)} = \frac{4\ V}{\pi\ d^2}$$

Replacing the variable height, h, in the total area relation, A_{Total}, and simplifying coefficients,

$$A_{Total} = \frac{\pi}{2}d^2 + h\pi d = \frac{\pi}{2}d^2 + \left(\frac{4\ V}{\pi\ d^2}\right)\pi d = \frac{\pi}{2}d^2 + \left(\frac{4\ V}{d}\right)$$

Then, multiplying the unit cost, c, by the total area, A_{Total},

$$C(V,\, d) = c\ A_{Total} = 100\left(\frac{\pi}{2}d^2 + \frac{4\ V}{d}\right)$$

The function expression is correct, as it wanted to be confirmed!

Consequently, further analysis is required in order to enlighten the counter-intuitive results obtained in (a) and (b). The table and graphic representations are very common when studying real functions, with useful and complementary attributes to the mathematical expression; thus, such representations of the cost function are treated in the next section.

1.4 FUNCTION REPRESENTATION IN TABLE AND GRAPHIC

For a real function, representations by table or graphic allow close observation of both function values and variation insights, as well as specific tools and qualitative approaches are developed from these useful representations. Such approaches are important, by visually providing trends, patterns, velocity, or acceleration of variations, and permitting a more complete analysis of the function at hand.

c. **Study the cost function, C(V,d), for the given volumes and sketch it.**

Thus, the function domain and its variation range (image-domain, counter-domain), the monotony intervals, the occurrence of maxima or minima points, the line curvature and inflection points, and potential asymptotes are studied with the support of the table function, in conjunction with the related graphic representation (Figure 1.2).

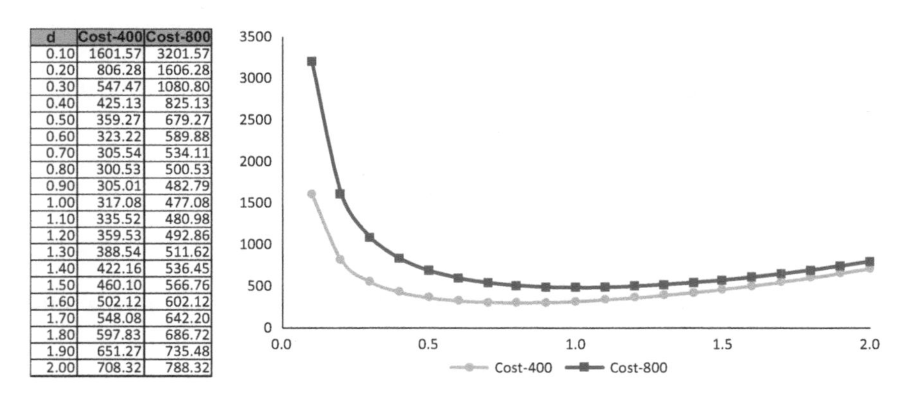

d	Cost-400	Cost-800
0.10	1601.57	3201.57
0.20	806.28	1606.28
0.30	547.47	1080.80
0.40	425.13	825.13
0.50	359.27	679.27
0.60	323.22	589.88
0.70	305.54	534.11
0.80	300.53	500.53
0.90	305.01	482.79
1.00	317.08	477.08
1.10	335.52	480.98
1.20	359.53	492.86
1.30	388.54	511.62
1.40	422.16	536.45
1.50	460.10	566.76
1.60	502.12	602.12
1.70	548.08	642.20
1.80	597.83	686.72
1.90	651.27	735.48
2.00	708.32	788.32

FIGURE 1.2 Table and graphic for the cost function with different volumes (400 liters, 800 liters).

The diameter is only assuming positive values, $d > 0$, then the costs are also positive, $C > 0$. In this way, the graphic lines in Figure 1.2 are represented only in the first quarter, and the graphic axes are not intersected, neither the vertical nor the horizontal axis.

A *set paribus* analysis follows, with one cost parameter being addressed, the diameter d, while the other parameter is kept constant, the volume V. That is, each column in the table is focused one at a time, as well as the related graphic line.

- First, by addressing the cost function for volume 800 liters when the diameter is small, the values are relatively higher; when diameter increases, the cost decreases and the lower costs occur in the diameter range of 0.9–1.1, about 477 €, alias, $C(d^{min})$; after that range, the cost is increasing with the diameter variation and the line curvature is upward.
- About the cost function for volume 400 liters, the evolution is similar to the prior one; the values are relatively higher when the diameter is small; when diameter increases, the cost decreases and the lower costs occur in the range of 0.7–0.9, about 300 €, alias, $C(d^{min})$; thereafter the cost is increasing with the diameter and the line curvature is upward too.
- By comparing the two lines for the same diameter (vertical approach), the cost for 800 liters is always greater than the corresponding cost for 400 liters. This observation is coherent with the cost expression, $C(V, d)$, but note this comparison is taking one parameter at a time (*set paribus*), while the two function arguments are altered at the same time in the original questions. In fact, in the problem statement both diameter and volume are altered, respectively, from (0.4, 1.6) to (0.8, 1.2).

In addition, the notions of increasing or decreasing function, minimum (or maximum) points, concavity upward or downward, will benefit with a more detailed description.

1.4.1 Intervals of Monotony—Increasing and Decreasing

On a given range, a function that is either increasing or decreasing is a monotonous function, with the function monotony related with its sense of variation and quantified through discrete increments (e.g., one by one), or even with the derivative concept.

For a real function $y = f(x)$, the alteration of dependent variable, y, resulting from the variation of independent variable, x, is usually measured by the function's derivative. At each point, x, the function derivative indicates the related rate of change, or the variation's velocity; for example in Figure 1.3, the derivative is taken as 0, +1, and −1, respectively, for the linear graphs (a), (b), and (c). To verify that, the increments on variable y are evaluated in face of unit increments in x.

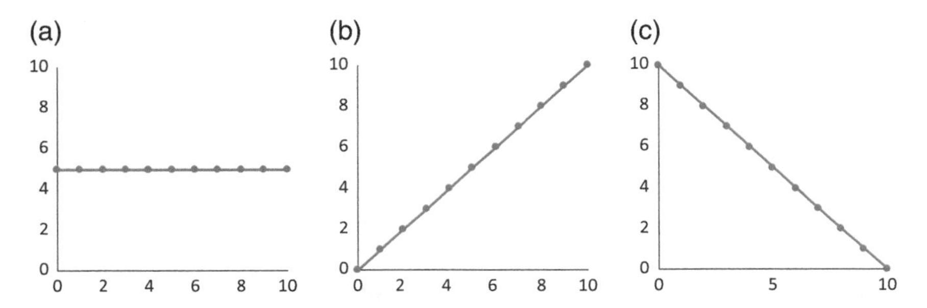

FIGURE 1.3 Graphics of linear functions. (a) Constant function, $y(x) = 5$; (b) Increasing function, $z(x) = x$; (c) Decreasing function, $w(x) = 10 - x$.

a. **Constant function, $y(x) = 5$,** with derivative zero, $y'(x) = 0$; in fact, the alteration on variable y is always zero, no matter what value x takes. When x increases one, from 0 to 1, 2, 3, … , the value of variable y still is constant and the rate of change is 0.

b. **Increasing function, $z(x) = x$,** with derivative taking a positive value, $z'(x) = 1$; when x increases successively from 0 to 1, 2, 3, … , the variable z also increases from 0 to 1, 2, 3, … ; in this way, the rate of change per unit of variable x is positive, and equal to 1.

c. **Decreasing function, $w(x) = 10 - x$,** with derivative taking a negative value, $w'(x) = -1$; at now, when x increases from 0 to 1, 2, 3, … , the variable w successively decreases one, respectively, from 10 to 9, 8, 7, … ; in this way, the rate of change for variable w is negative and equal to −1.

1.4.2 Maximum and Minimum Points

Extreme points are very important due to their real-world implications, both minima (e.g., cost, payments) and maxima (e.g., return, inflows), and they can be located by analyzing the function derivative. Namely, when the derivative is changing sign on a given range, from negative to positive, then a minimum point can be found; for the other side, when the derivative changes from positive to negative, then a maximum point can be found on that range.

From the other side, a qualitative or empirical approach can be used in simple cases, without loss of generality and subject to the necessary verification and confirmation. The occurrence of extreme points is focused now, either maxima or minima; obviously the outcome shall be coherent with the graphs observation, as follows for the quadratic functions in Figure 1.4.

a. **A minimum point** occurs at $x = 0$ for the quadratic function $y(x) = x^2$, because the square of any other point, either positive ($x > 0$) or negative ($x < 0$), will take a positive value. That is, all the power-2 of x will be greater than zero, except for point $x = 0$ that corresponds to the minimum of function $y(x) = x^2$. Then, $y(0) = 0$ is the minimum; in addition, it is the global minimum because no other point will be less than or equal to zero.

b. **A maximum point** occurs at $x = 0$ for the function $z(x) = 9 - x^2$, because this function is subtracting the prior quadratic function, $y(x) = x^2$, to the constant 9, and then the maximum of $z(x)$ will correspond to the subtraction of zero. In fact, at any other point, either positive ($x > 0$) or negative ($x < 0$), the function $y(x)$ will increase and $z(x)$ will be lower than 9. Then, $z(0) = 9$ is the function maximum, and it is the global maximum because no other point will be greater than or equal to 9.

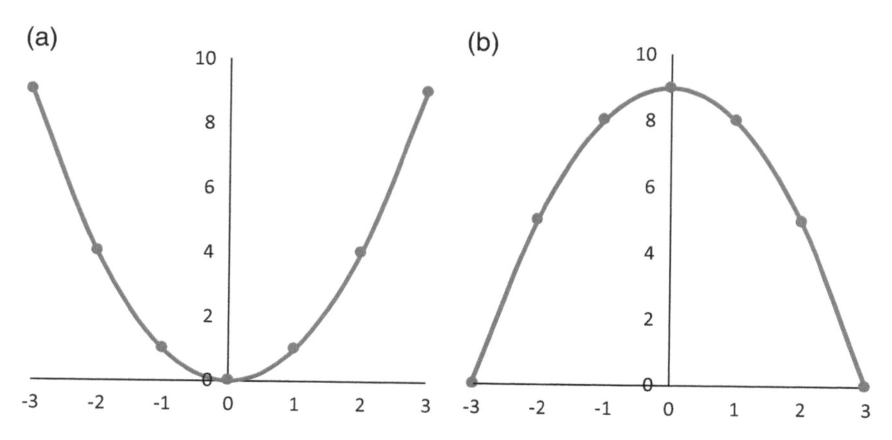

FIGURE 1.4 Monotony and extreme points of simple quadratic functions. (a) A minimum point for $y(x) = x^2$; (b) A maximum point in $z(x) = 9 - x^2$.

A derivative-based approach can be utilized to address the extreme points too, since the derivative value at each point indicates the related rate of change, that is, the variation's velocity. Once again, the increments on variable y are assessed in the context of unit increments in x; however, the derivative is varying at each point x in the two graphs of Figure 1.4.

- **By observation of the upward parabola**, in graphic (a), a minimum point is found at $y(0) = 0$; in the negative domain, $x < 0$, when x increases (from −3 to −2, −1, 0), variable y is successively decreasing (respectively, from 9 to 4, 1, 0); the rate of change varies, but it is always negative (−5, −3, −1) in the left side of the minimum.

  But immediately at the minimum's right, in the positive domain, $x > 0$, when x increases (from 0 to 1, 2, 3), then variable y is successively increasing (respectively, from 0 to 1, 4, 9); note the rate of change is varying too, but such rate is always positive (1, 3, 5) in the right side of the minimum.

  It can be later demonstrated the derivative is zero at the minimum point, $y(0) = 0$, since the related rates alter from negative to positive at that point. By contradiction, if the derivative would be negative at $x = 0$, or even at the immediate points in the right, then variable y should be decreasing and such point would not be the minimum!

- **At now observing the downward parabola** in (b), a maximum point is located at $z(0) = 9$; again in the negative domain, $x < 0$, when x increases (from −3 to −2, −1, 0), variable z is successively increasing (from 0 to 5, 8, 9), and the rate of change varies but it is always positive (5, 3, 1) in the left side of the maximum.

  But immediately at the maximum's right, in the positive domain, $x > 0$, when x increases one (from 0 to 1, 2, 3), variable z is successively decreasing (from 9 to 8, 5, 0); while the rate of change for variable z is varying too, such rate is negative (−1, −3, −5) in the maximum's right side.

  It can be also demonstrated the derivative is zero at the maximum point, $z(0) = 9$, since the related rates alter from positive to negative at that point. In fact, if the derivative would be positive at $x = 0$, or even at the immediate points in the right, then variable z should be increasing and such point would not be the maximum!

Recall that the derivative shall be zero, at both the minimum and maximum points, respectively, in both the graphs (a) and (b) of Figure 1.4. Then, another question arises:

> ***How to distinguish extreme points, if derivatives are zero in both maxima and minima?***

1.4.3 Concavity and Inflection Points

The concavity or the curvature for the function graphic can help to distinguish extreme points, to identify maxima and minima. Note that in Figure 1.4, the minimum point occurs in the upward parabola, graphic (a), while the maximum occurs in the downward parabola, graphic (b).

Thus, Figure 1.4 is revisited and the derivative variation is assessed too; the rate of change for the derivative is usually appreciated through a second-order derivative ("variation's acceleration"), by analogy with the derivative of the derivative ("velocity's rate of change").

- Again, by observing the upward parabola in graphic (a), with minimum point $y(0) = 0$, the derivative is negative for $x < 0$ (the function is decreasing), it becomes zero at the minimum ($x = 0$), and then the derivative turns positive for $x > 0$ (the function is increasing).

  The rate of change for the derivative is being positive, since the derivative is successively increasing: from negative values to zero, and then to positive values. Therefore, the second-order derivative is positive, the related graphic concavity is upward, and the extreme point is a minimum.

- And about the downward parabola in (b), with maximum at $z(0) = 9$, the derivative is positive for $x < 0$ (the function is increasing), it becomes zero at the maximum ($x = 0$), and then the derivative turns negative for $x > 0$ (the function is decreasing).

  The rate of change for the derivative is being negative, since it is successively decreasing, from positive to zero, and then to negative values. The second-order derivative is thus negative, the concavity is downward, and the extreme point is a maximum.

In summary, it is simultaneously assumed the derivatives exist and the second-order derivative maintains its sign; that is, the concavity stands and it does not alter from upward to downward, or vice-versa.

- The first-order derivative indicates the function monotony on a range: if positive, the function increases in the related range; if negative, the function decreases in that range. In case the derivative is zero, an extreme point may occur, and the second-order derivative can help to identify if a maximum or a minimum exists.
- The second-order derivative indicates the function concavity, the graphic curvature: if positive, the curvature is upward in the related range, and a minimum point exists; if it is negative, the curvature is downward, and then a maximum occurs in that range.

However, if the second-order derivative alters from positive to negative, the concavity direction is inflecting from upward to downward (or vice-versa). In such case, inflection points may occur, as in the cubic functions in Figure 1.5.

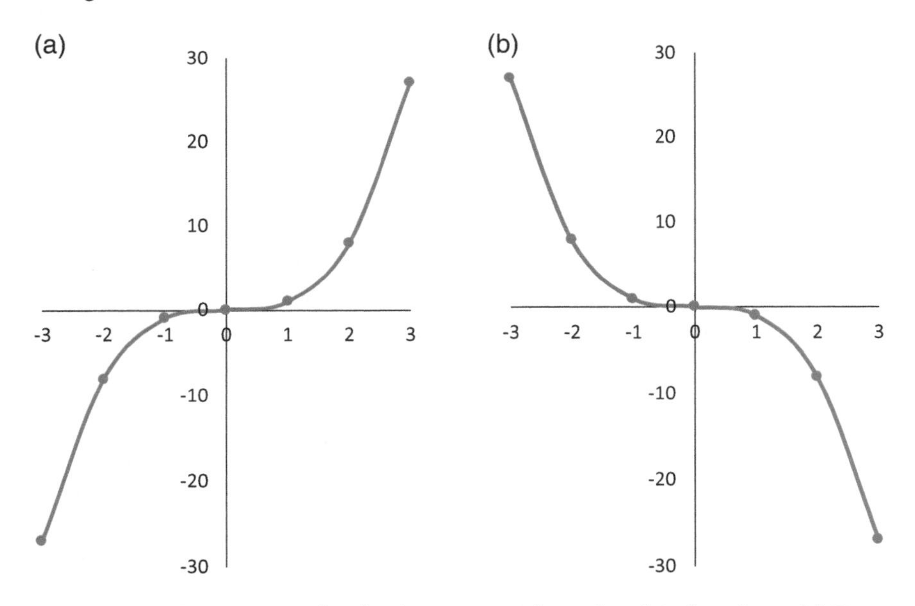

FIGURE 1.5 Concavity and inflection points of simple cubic functions. (a) From negative to positive in $y(x) = x^3$; (b) From positive to negative in $z(x) = -x^3$.

- About the cubic function $y(x) = x^3$ in graphic (a), the second-order derivative is being negative for $x < 0$ (downward concavity), and then it turns positive for $x > 0$ (upward concavity). An inflection point occurs at $y(0) = 0$, since the graphic concavity inflects at that point, from negative to positive.

It can be demonstrated the second-order derivative is zero at the inflection point, since it alters from negative to positive at that point. By contradiction, if the second derivative would be negative at $x = 0$, or even at the immediate points in the right, then the first-order derivative should be decreasing and the inflection point would not be occurring!

- Observing the cubic function $z(x) = -x^3$ in (b), the second-order derivative is positive for $x < 0$ (upward concavity), and then it turns negative for $x > 0$ (downward concavity). An inflection point occurs at $z(0) = 0$, since the graphic inflects its concavity at that point, from positive to negative.

  And again, it can be shown the second-order derivative is zero at that inflection point, since it alters from positive to negative at that point. In fact, if the second derivative would be positive at $x = 0$, or even at the immediate points in the right, then the first-order derivative should be increasing and the inflection point would not exist!

In addition, note that knowing if the function is even or odd allows us to reduce the study interval.

- The quadratic functions in Figure 1.4 are even functions, and their graphics are symmetric in relation to the vertical Y-axis. This means the even function, e.g., $y(x) = x^2$, takes the same value for the corresponding points in the positive domain ($x > 0$) and in the negative domain ($x < 0$); in fact, it is observed that

$$y(x) = y(-x)$$

- The cubic functions in Figure 1.5 are odd functions, and their graphics are symmetric in relation to the origin (0,0). It means the odd function, e.g., $y(x) = x^3$, takes symmetric values for the corresponding points in the positive and the negative domain; in fact,

$$y(x) = -y(-x)$$

The qualitative approach in this section is based on general rules, common sense, and assuming as analogous the rate of change ("variation's velocity") and the first derivative. However, the proper mathematical methods and rigorous formulations shall prevail, namely, for the elementary functions treated (linear, quadratic, and cubic functions), along with the associated continuity and derivation properties. Such properties and main results will be described later, but some introductory notes about mathematical proofs follow in the next section.

1.5 PROOFS AND MATHEMATICAL REASONING

As shown in this text, mathematical proofs and logical reasoning require special attention too. Some examples are presented in this section, and more advanced tools complement the qualitative approaches of prior section. Recall that the volume is already defined, $V = 0.4$ or $V = 0.8$, in

$$C(V, d) = 100\left(\frac{\pi}{2}d^2 + \frac{4\,V}{d}\right)$$

Therefore, the cost function can be treated as it only varies with the diameter, d, for instance:

$$\text{or } C(0.4, d) = 100\left(\frac{\pi}{2}d^2 + \frac{1.6}{d}\right), \text{ or } C(0.8, d) = 100\left(\frac{\pi}{2}d^2 + \frac{3.2}{d}\right)$$

Domain

In a realistic point of view, $d > 0$. A pragmatic approach is defining only the positive domain for diameter values, since negative values have no sense for a physical dimension, e.g., area, volume, length, height, or diameter.

- For $d = 0$, this singular point would correspond to a fading cylinder, or a void concept for the cylinder; in fact, for any finite height, h, the volume would become zero too. And also, the division in the formula's second term would not be operated, division by zero is not allowed. Thus, it is considered the following domain for variable x in the current study:

$$D =]0, \ +\infty[$$

Image-Domain and Points of Discontinuity

The function's graphic does not intersect the axes, the graphic does not show symmetry; in fact, only the first quarter is represented, because for positive diameter ($d > 0$) the cost function always takes positive values ($C > 0$).

- In addition, discontinuity points do not occur in the positive domain, $d > 0$, since the division in the formula's second term can always be operated in such domain.
- However, when the diameter diminishes and diminishes to a tiny size, the height increases more and more in a manner to obtain the cylinder volume, either 400 or 800. Thus, when the diameter is successively diminishing to a tiny scale and nears zero, the corresponding height increases without bound, and similarly the cost becomes unbounded, without any majoring (it can be later demonstrated the cost tends to infinite, and a vertical asymptote occurs at $x = 0$).

Consequently, it is considered the following image-domain, or variation range for variable y:

$$CD = [C(d^{\min}), \ +\infty[$$

Typically, equivalence results are common in mathematical proofs focusing on true statements; however, contradiction methods or even reduction to absurd can be useful to address false statements. A simple example follows from close observation of Figure 1.2.

- The graphic does not intersect the axes, and does not show symmetry or discontinuity points. Assuming only positive diameter, $d > 0$, then let it be the (false) hypothesis:

  $$C(d) = 0$$

  By applying the cost relation and manipulating successively,

  $$\Leftrightarrow 100\left(\frac{\pi}{2}d^2 + \frac{4\,V}{d}\right) = 0$$

  The factor in the parentheses shall be zero,

  $$\Leftrightarrow \frac{\pi}{2}d^2 + \frac{4\,V}{d} = 0, \quad 100 \neq 0$$

However, a contradiction is found, associated to a false equality:

$$\Leftrightarrow \frac{\pi}{2}d^2 + \frac{4\,V}{d} = 0$$

$$\Leftrightarrow \frac{\pi}{2}d^2 = -\frac{4\,V}{d}, \quad \textbf{\textit{FALSE}}$$

Since the cylinder parameters, V and d, are positive, then the first member is positive, but the second member is negative; the equality is thus false, since this relation is not possible for $d > 0$. This contradiction occurs because the initial equality, the hypothesis $C(d) = 0$, is false too.

- In case such a hypothesis would be true, then the cost should be zero at a certain diameter; in the real world, it would imply a barrel for free!

Derivatives and related results are very useful, additional notes about them allow a more rigorous approach to the function variation (monotony and extreme points; concavity and inflection points; asymptotes), and functions sketching will be detailed in Chapter 6.

Monotony and Minimum Point

The monotony ranges can be defined by studying the first-order derivative for $C(d)$:

$$C'(d) = 100\left(\frac{\pi}{2}d^2 + \frac{4\,V}{d}\right)' = 100\left[\frac{\pi}{2}(2d) + 4\,V\left(\frac{-1}{d^2}\right)\right]$$

$$= 100\left(\pi d - \frac{4\,V}{d^2}\right)$$

Then the minimum point occurs when the first derivative is zero, $C'(d) = 0$, changing from negative to positive at that point:

$$100\left(\pi d - \frac{4\,V}{d^2}\right) = 0$$

$$\pi d - \frac{4\,V}{d^2} = 0 \Leftrightarrow \pi d = \frac{4\,V}{d^2} \Leftrightarrow d^3 = \frac{4\,V}{\pi} \Leftrightarrow d = \sqrt[3]{\frac{4\,V}{\pi}}$$

In fact, by analyzing the factor inside parentheses: when d increases, the first term increases and the subtracted term decreases; then the first derivative is increasing too. In this way, the first derivative is becoming positive in the immediate points at the extreme's right; vice-versa, it is negative in the extreme's left side; thus this extreme point is a minimum and corresponds to $C(d^{\min})$.

Concavity

The graphic concavity and potential inflection points are associated with the second-order derivative too. From the prior paragraph, the first derivative is always increasing with the diameter, and the second derivative will be positive. The related calculation follows too:

$$C''(d) = 100(\pi d - 4Vd^{-2})' = 100[\pi - 4V(-2d^{-3})] = 100\left(\pi + \frac{8\,V}{d^3}\right)$$

Since the diameter has to be greater than zero, $d > 0$, then $C''(d) > 0$; therefore, the concavity is always positive (upward) in the domain, and inflection points do not occur.

Asymptotes

The function nears a vertical asymptote at point $d = 0$, and the qualitative approach in the prior section is confirmed too:

$$\lim_{d\to 0^+} C(d) = \lim_{d\to 0^+} 100\left(\frac{\pi}{2}d^2 + \frac{4\,V}{d}\right) = 0 + \infty = +\infty$$

However, oblique or horizontal asymptotes do not exist:

$$\lim_{d\to +\infty} \frac{C(d)}{d} = \lim_{d\to +\infty} \frac{100\left(\frac{\pi}{2}d^2 + \frac{4\,V}{d}\right)}{d} = \lim_{d\to +\infty} 100\left(\frac{\pi}{2}d + \frac{4\,V}{d^2}\right)$$
$$= +\infty + 0 = +\infty$$

A summary table for the variation of cost function follows, considering the relevant attributes or items, as well as the first-order derivative and the second-order derivative.

d	0	(...)	$d = \sqrt[3]{\frac{4V}{\pi}}$	(...)
$C'(d)$	*Not defined*	Negative	0	Positive
$C''(d)$	*Not defined*	Positive	Positive	Positive
$C(d)$	*Not defined*	Decreasing, upward	Minimum point	Increasing, upward

1.6 THE INVERSE RATIONALE

At now, the problem considers the inverse rationale of defining the cost, and then obtains the diameter that corresponds to the given amount. In the real world, cost is a very important economic measure that is required to stay below a given value, with many organizations and companies defining their performance indicators, not to mention staff working by quantified objectives or "target values". A common issue involves the diameter estimate for a given "target cost", e.g., obtaining the diameter for a barrel with 800 liters and cost 500 €, in this way stimulating the inverse reasoning for the current problem.

Table Analysis—By analysis of cost values in the function's table (Figure 1.6), the lowest cost occurs in the range marked by the (light line) rectangle for V = 800, and the (dark line) oval marks the initial range for V = 400. Also, the rectangle marks the costs lower than the target, C = 500,

d	Cost-400	Cost-800
0.10	1601.57	3201.57
0.20	806.28	1606.28
0.30	547.47	1080.80
0.40	425.13	825.13
0.50	359.27	679.27
0.60	323.22	589.88
0.70	305.54	534.11
0.80	300.53	500.53
0.90	305.01	482.79
1.00	317.08	477.08
1.10	335.52	480.98
1.20	359.53	492.86
1.30	388.54	511.62
1.40	422.16	536.45
1.50	460.10	566.76
1.60	502.12	602.12
1.70	548.08	642.20
1.80	597.83	686.72
1.90	651.27	735.48
2.00	708.32	788.32

FIGURE 1.6 Inverse approach targeting the cost, 500 €.

in the diameter range from 0.8 to 1.2; for the other side, the oval signals values above that target for a diameter greater than 1.6, even for $V = 400$. In addition, by careful search of the lowest costs represented in the table for the two instances under analysis:

- For V = 400 liters, the lowest cost is 300.53 €, corresponding to d = 0.8 meter. The corresponding height would be:

$$h = \frac{4\ V}{\pi\ d^2} = \frac{4 \times 0.4}{\pi\ (0.8)^2} \cong 0.795$$

- For V = 800 liters, the lowest cost is 477.08 €, corresponding to d = 1.0 meter. And the corresponding height would be:

$$h = \frac{4 \times 0.8}{\pi\ (1.0)^2} \cong 1.018$$

Now, the procedure assumes the search for the lowest cost, and then the corresponding diameter is located; note the inverse approach to the pairs, $(C, d) = (y, x)$ is related with the notion of inverse function.

Graphic Analysis—A descriptive analysis is developed, assuming the same cost—e.g., $C(d) = 500$—is corresponding to a horizontal approach in Figure 1.7. Initially, the graphic procedure started with a given diameter, and then the cost was calculated, in this way implicitly assuming a vertical approach to the pair of values, $(x, y) = (d, C)$.

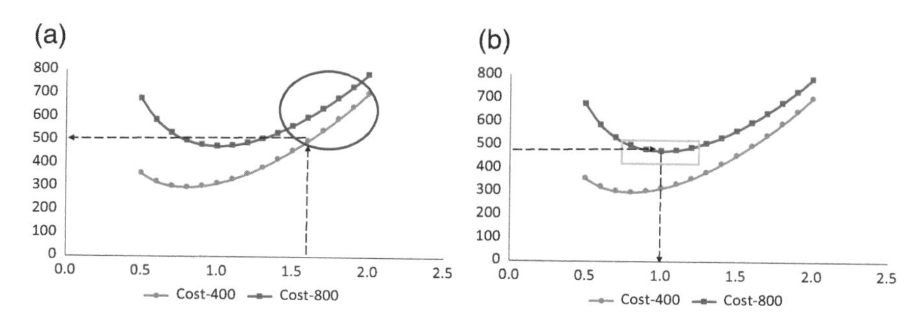

FIGURE 1.7 Cost function with target cost 500 €. (a) Direct approach, marking the diameter range with costs above the target; (b) Inverse approach, searching for the diameter range with costs below the target.

- However, when searching in the horizontal of $C = 500$, for $V = 800$, the lower values occur in the range marked by the (light line) rectangle, while the (dark line) oval range was selected for $V = 400$ in the initial statement.
- Thus, the oval's range (diameter greater than 1.6 meter) corresponds to costs above 500 €, despite the volume of only 400 liters. And the rectangle marks the diameter range (from about 0.8 to 1.2) where the cost stays below 500 € even when the volume is 800 liters.
- Additionally, the range selected in the rectangle includes the lower points for the graphic of $V = 800$, while the initial volume, $V = 400$, was assuming diameter $d = 1.6$ that is located in the increasing range and far from the lower costs.

The inverse function for the cost is the mathematical relation that indicates the diameter (dependent variable) when provided the cost (independent variable), or also $(C, d) = (y, x)$; it requires additional treatment and will be addressed later, in Chapter 5. Namely, for the target cost of 500 €, the optimal volume will be 858.4 liters, and the associated size is 1.030 meter, for equal diameter and height $(h = d)$.

1.7 DISCUSSION OF RESULTS

In this section, the main results for the cost function are revisited, with qualitative approach and simple calculations supporting the comparison of diverse volumes and costs, while additional details of interest are discussed too.

a. From the initial statement, the barrel of volume 400 liters is costing about 502 €. Thus to obtain the total volume of 800 liters, in case such volume is required, the total cost would be more than 1000 €!

   - In fact, the total cost will be about 1004 €, corresponding to the addition of a second barrel, with diameter, $d = 1.6$, and height of about $h = 0.2$, as presented in Figure 1.8.

$$h = \frac{4\,V}{\pi\,d^2} = \frac{4 \times 0.4}{\pi\,(1.6)^2} \cong 0.199$$

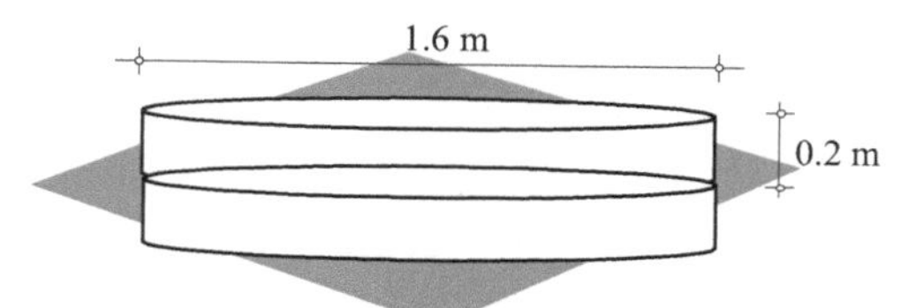

FIGURE 1.8 Two barrels of 400 liters with total cost of about 1004 €.

b. From the observation of both table and graphic, the value of about 492 € is confirmed for a barrel with a volume of 800 liters. By comparison with the previous alternative, in (a), it is noted:

- For the same amount of about 500 €, the volume can be doubled, obtaining a barrel with a volume of 800 liters rather than 400 liters.
- And for the same volume of 800 liters, the amount needed is less than half; in fact, only about 492 € is required, avoiding payment of 1004 € for two barrels of 400 liters.
- This alternative with diameter, $d = 1.2$, and height about $h = 0.7$ (Figure 1.9), will be a better option in case such design variables (diameter and height) are suitable for the barrel deployment and utilization.

$$h = \frac{4\,V}{\pi\,d^2} = \frac{4 \times 0.8}{\pi\,(1.2)^2} \cong 0.707$$

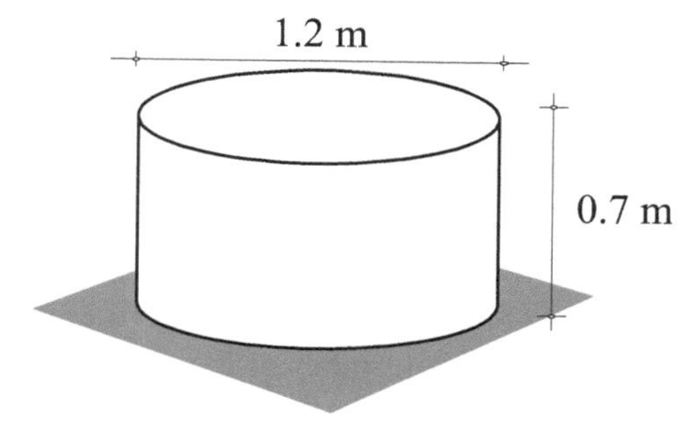

FIGURE 1.9 Barrel of 800 liters with cost of about 492 €.

c. Again from the cost analysis for volume, $V = 800$ liters, in the corresponding table (Figure 1.2 or Section 1.6), the lower amount would be about 477.08 €. However, from the cubic root expression obtained in Section 1.5, the accurate estimate for the optimum diameter is:

$$d = \sqrt[3]{\frac{4\,V}{\pi}} \cong 1.006$$

Note the height takes the same value, $h = d$ (Figure 1.10),

$$h = \frac{4\,V}{\pi\,d^2} = \frac{4 \times 0.8}{\pi\,(1.006)^2} \cong 1.006$$

and the corresponding cost is 477.06 €.

$$C\,(V = 0.8,\; d = 1.006) = 100\left(\frac{\pi}{2} \times 1.006^2 + \frac{4\; \times 0.8}{1.006}\right) = 477.06\text{ €}$$

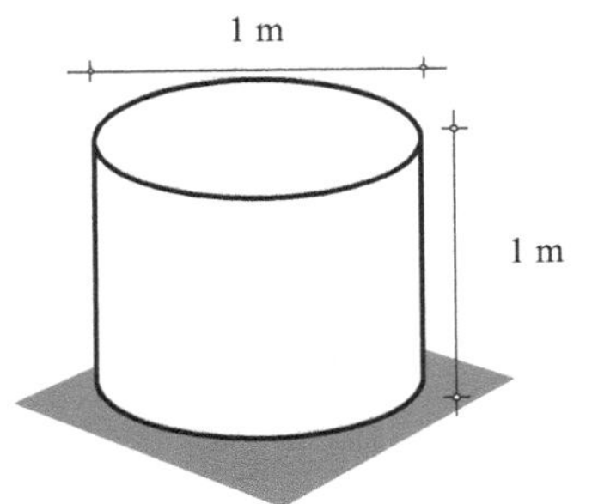

FIGURE 1.10 Barrel of 800 liters with cost 477.06 €, and diameter equal to height ($h = d \cong 1.006$).

Additionally:

- The estimate with six decimal digits indicates a diameter of about $d = 1.006159$, and the corresponding amount (with six decimal places too) is 477.061682 €. Such estimate with more digits is also of great interest, in case a very large number of containers is to be produced.

  For example, in the agro-food industry, in case one million juice cans are produced, the sixth rounding digit will induce an error until 1 € while an inaccuracy in the third decimal place (e.g., one-tenth of a cent) will enlarge the error bound until 1000 €.

- For the instance at hand, the difference between the lower amount obtained from the table (477.08 €) and the function minimum (477.06 €) implies a deviation of two cents, 0.02 €; therefore, the error bound for one million cans would be about 20000 €!

Due to the error propagation that occurs when large quantities are multiplied, decimal amounts are also presented with ten or more decimal digits in business and finance sectors, namely, for accounting purposes or currency exchanges.

1.8 CONCLUDING REMARKS

Important notions about real functions are introduced, recalling the mathematical conditions for extreme points, either maxima or minima, inflection points, as well as functions monotony and concavity are crucial for many applications. Real functions are key for many developments in exact sciences, technological developments, engineering tools, and in economics and enterprise sciences, e.g., either to minimize costs or to maximize resources utility for a given amount.

The importance of closely observing, verifying and experimenting, is illustrated through the cost of a cylindrical container, which motivates the application of Calculus important topics. In addition, it is also verified that the barrel's diameter and height are equal ($h = d$) for the minimum cost. This type of relations can be obtained from more advanced approaches, namely, by adequately studying the inverse function and derivation topics (Chapter 5). In a "pull approach" along the main differential subjects, the proposed path considers:

- Sketching functions in Chapter 6, including roots location and axis intersections, monotony and extreme points (maxima and minima), concavity and inflection points, together with the diverse asymptotes; thus, the treatment of indeterminate limits is required, along with important theorems about differentiable functions that allow the location of derivative roots (e.g., for the function's extreme points).
- Derivation properties and operations are previously presented in Chapter 5, for both elementary and transcendent functions, together with the inverse function and higher order derivatives.

- Since differentiable functions also satisfy the continuity attributes, the properties and operations of continuous functions are focused in Chapter 4. Namely, the important results about extreme points in a closed range, about intermediate values, or about limits calculation. In the latter case, by directly evaluating the point at hand in the continuous function, since it presents equal image and lateral limits at that point.
- Before that, function limits, their properties and operations, are addressed in Chapter 3. Through an experimental approach that conjugates analytical treatments and numerical instances, the limit of dependent variable, y, is associated with the limit of independent variable x. That is, by analogy with the sequence limits, the function limit is approached as a sequence of function values that converges to point b, when the sequence of real numbers, x, tends to point a.

In this way, limits of number sequences are studied in the next chapter, once again conjugating theoretical and experimental approaches, with graphics and tables representation, aiming at:

- Initially, revisiting notions and concepts about number sequences, as well as reminding the main theorems and properties of sequence limits.
- In addition, geometric sequences are studied, namely, the sequences with exponential terms, $X_n = a^n$, and relevant insights about numerical approximations are introduced.

CHAPTER 2

Sequences of Real Numbers

This chapter addresses sequences of real numbers; the key concepts are presented and then related with sequence limits, while numerical instances allow verification procedures. First, some preliminary notions about number sequences are revisited, reminding basic concepts; second, the main theorems about sequences, along with the properties of sequence limits (e.g., uniqueness, arithmetic operations), are revisited too. Important sequences are studied too, namely, the geometric sequences with exponential terms, in particular for base-2 that supports the binary set {0,1}. An experimental approach utilizing base-10 is described, and relevant insights about approximations and the computational representation of real numbers are introduced.

2.1 INTRODUCTION

The study of real numbers sequence is crucial to the study of function limits, and many properties and results about the two types of limits are similar. It is also a good opportunity to encourage rigorous habits, and promote systematic verification and control practices, namely, checking numerical tables, comparing graphics and the related ranges, or experimenting with some kind of approximation or simplification.

The main properties of convergent sequences are based on simple reasoning about real numbers, and contradiction procedures are very useful to show a number of results. Infinitesimal sequences are converging to zero and they are of great interest: in particular, to construct sequences that converge for a given point, in the preferred mode (e.g., decreasing, increasing, oscillating).

In general, the monotony and boundedness attributes are instrumental to understanding a convergent sequence; their mathematical proofs are commonly based on, respectively, superiority relations,

DOI: 10.1201/9781003461876-2

bounds existence, and neighborhood radius. In addition, the associated definitions and procedures are inviting numerical verification, and even double verification. For example, consider the well-known sequence,

$$X_n = \left(1 + \frac{1}{n}\right)^n,$$

with a significant number of terms presented in the following table and graphic (Figure 2.1).

n	Xn
1	2.000000
10	2.593742
15	2.632879
20	2.653298
25	2.665836
30	2.674319
40	2.685064
50	2.691588
100	2.704814
200	2.711517
300	2.713765
400	2.714892
500	2.715569
1000	2.716924
10000	2.718146
100000	2.718268

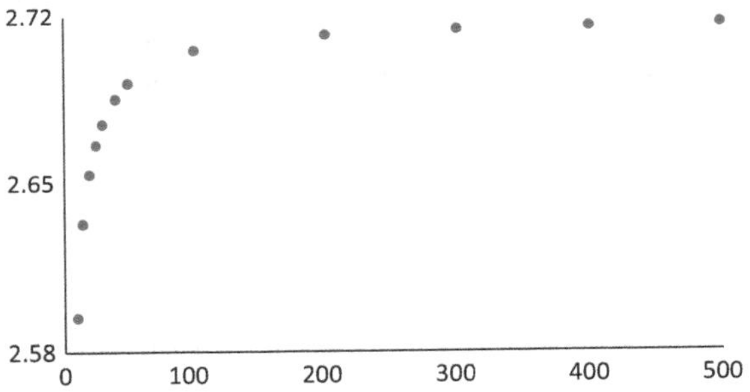

FIGURE 2.1 Table and graphic for sequence X_n.

The convergence proofs can be difficult and complex; however, from the numerical table and graphic (Figure 2.1), it can be observed that:

- The sequence X_n is strictly increasing;
- The sequence X_n is upper bounded, in fact, the sequence values are all less than 3, or for accuracy purposes, less than 2.72.

The sequence is convergent: it is both upper bounded and increasingly monotonous, then the sequence converges to the minimum of its majors, the sequence supremum. And it is well known the sequence limit is the natural base, *e*, or the Neper number:

$$\lim X_n = \lim\left(1 + \frac{1}{n}\right)^n = e \cong 2.718282$$

In this way, preliminary notions about sequences of real numbers are presented in the next section; after that, sequence limits are addressed, with both analytical developments and numerical instances; from the other point of view, the main theorems about sequences are discussed, while important properties of sequence limits are revisited; in addition, important sequences are also studied, namely, geometric sequences with diverse bases a are treated; and finally, an experimental approach that utilizes base-10 is detailed, while important insights about numerical approximations are introduced. In addition, by promoting the recognition of patterns and structures in the sequence limits presented here, it will facilitate the study of function limits to be developed later on.

2.2 PRELIMINARY NOTIONS

In this section, preliminary notions about sequence of real numbers are presented, including the subsequence of a sequence, monotony, bounding a sequence, or even the convergent sequence.

Sequence of Real Numbers—A sequence, S_n, is the entire map of N to R,

$$\begin{aligned} N &\rightarrow R \\ n &\rightarrow S_n \end{aligned}$$

Where: n—the term order; and S_n—value for the term of nth order.

A subsequence of a sequence is any sequence that is obtained from another sequence by deleting some of its terms.

Example

Let a sequence be $X_n = 1/n$.

Two subsequences are directly obtained, namely, the subsequence with only even terms, and the subsequence of odd terms:

$$\begin{aligned} X_{2n} &= \frac{1}{2n} \Rightarrow \frac{1}{2}, \frac{1}{4}, \frac{1}{6}, \frac{1}{8}, \ldots, \frac{1}{2n-2}, \frac{1}{2n}, \ldots \\ X_{2n-1} &= \frac{1}{2n-1} \Rightarrow \frac{1}{1}, \frac{1}{3}, \frac{1}{5}, \frac{1}{7}, \ldots, \frac{1}{2n-3}, \frac{1}{2n-1}, \ldots \end{aligned}$$

Monotonous Sequence—A sequence, S_n, is said to be monotonous if it is increasing or decreasing, in the broad (*latus*) or in strict sense:

$$\text{Increasing:} \quad \begin{cases} S_{n+1} \geq S_n, & \textit{latus} \text{ sense} \\ S_{n+1} > S_n, & \textit{strict} \text{ sense} \end{cases}$$

$$\text{Decreasing:} \quad \begin{cases} S_{n+1} \leq S_n, & \textit{latus} \text{ sense} \\ S_{n+1} < S_n, & \textit{strict} \text{ sense} \end{cases}$$

Example

Again, let a sequence be $X_n = 1/n$.

The first values are listed and the associated graphic follows in Figure 2.2.

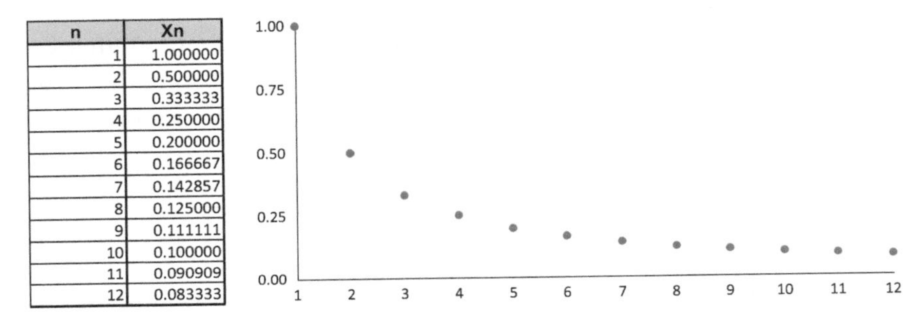

n	Xn
1	1.000000
2	0.500000
3	0.333333
4	0.250000
5	0.200000
6	0.166667
7	0.142857
8	0.125000
9	0.111111
10	0.100000
11	0.090909
12	0.083333

FIGURE 2.2 Table and graphic for sequence $X_n = 1/n$.

The sequence is decreasing, by close observation of both graphic and table; the corresponding proof is initiated by assuming the term of order $(n+1)$ is lower than the prior term, the nth term:

$$X_{n+1} < X_n$$

$$\frac{1}{n+1} < \frac{1}{n} \Leftrightarrow n < n+1 \Leftrightarrow 0 < 1, \textbf{\textit{TRUE}}$$

The true relation in the final step confirms the first relation, the hypothesis that indicates the sequence is decreasing.

- Else, in case a false relation would be obtained, e.g., $1 < 0$, then the decreasing relation (hypothesis) would be false too, and monotony could not be confirmed.

Bounded Sequence—A sequence, S_n, is said to be bounded if the set of its terms admits a bounding number, B, both minoring and majoring the sequence:

$$-B < S_n < B$$

$$\exists\, B \in R^+, \forall\, n \in N \underset{n}{\Rightarrow} |S_n| < B$$

Example

Let it be again, $X_n = 1/n$.

A positive number, B, is searched in a way that:

$$\left|\frac{1}{n}\right| < B \Leftrightarrow -B < \frac{1}{n} < B \Leftrightarrow \frac{1}{n} < B$$

Since the sequence starts with 1 and it is strictly decreasing, as observed in the prior developments, let it be $B = 2$, and

$$\frac{1}{n} < 2 \Leftrightarrow \frac{1}{2} < n \Leftrightarrow n > \frac{1}{2}, \ \textbf{\textit{TRUE}}$$

Again, the true relation in the final step is confirming the hypothesis that indicates the sequence boundedness, $|X_n| < 2$.

- Else, in case a false relation would be obtained in the last step, e.g., $n < 1/2$, then the hypothesis would be false, and such bound B could not be confirmed.

2.3 LIMIT AND CONVERGENCE OF A SEQUENCE

In this section, sequence limits are addressed, while analytical concepts related with convergent sequence or infinitely large sequence are introduced, and illustrated through numerical examples.

Convergent Sequence—A sequence S_n is convergent to a (or the real number a is the sequence limit) when, for every positive number r, there exists a corresponding number p, such that,

$$n > p \underset{n}{\Rightarrow} |S_n - a| < r$$

Or also, all the sequence terms after order p are located in a neighborhood of point, a, and radius, r; in this way, all those terms belong to the open interval, $]a\text{-}r, a\text{+}r[$,

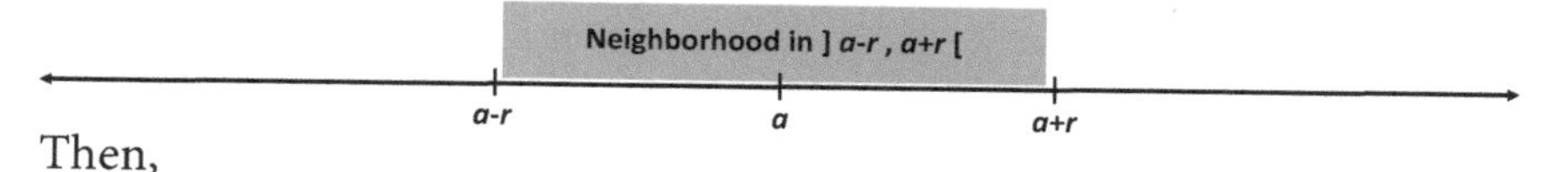

Then,

$$\lim S_n = a, \quad \text{or} \quad S_n \to a$$

A relation between n and r, $n(r)$, is thus foreseen, and the flowchart in Figure 2.3 outlines the pathway to obtain it for a sequence of real numbers, X_n, converging to the limit, a.

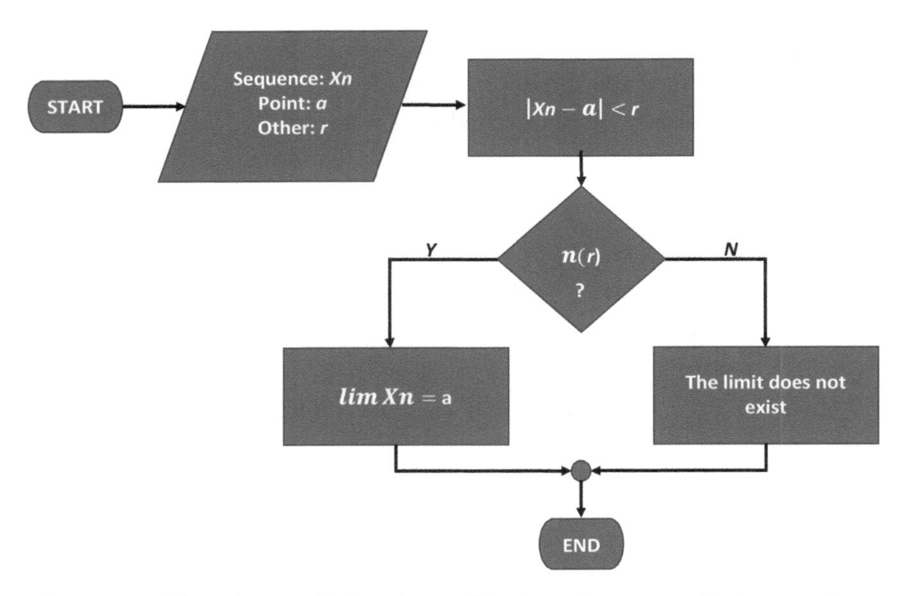

FIGURE 2.3 Flowchart outlining the verification of sequence limit on point a.

Example

Let it be, again, $X_n = 1/n$, and assuming this sequence is converging to zero, that is, 0 is the limit for the sequence X_n if all its terms are in the neighborhood r of 0:

$$|X_n - 0| < r$$

All the sequence terms after order p are located in a neighborhood of point, 0, and radius, r; in this way, all those terms belong to the open interval,]-r, +r[,

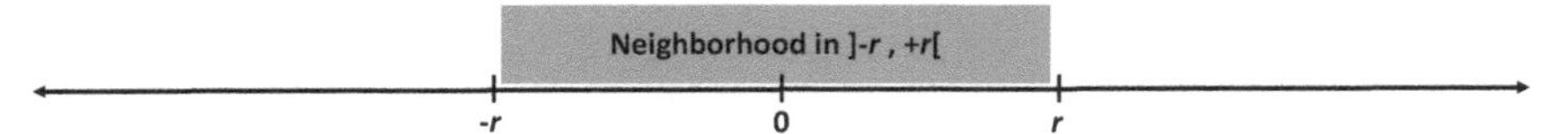

Substituting the relation for X_n, and neglecting the null term, as well as the negative branch for the inequality (the sequence terms are all positive):

$$\left|\frac{1}{n}\right| < r \Leftrightarrow -r < \frac{1}{n} < r \Leftrightarrow \frac{1}{n} < r$$

Finally, a relation between n and r, $n(r)$, is obtained:

$$n > \frac{1}{r}$$

For each neighborhood radius, r, by evaluating the relation, $p = p(r)$, and the integer, $n = \text{INT}(p) + 1$, the table with the related nth terms that satisfy such relation follows:

r	p	n = INT(p)+1	Xn	|Xn - 0|
0.1	10.00	**11**	0.090909	0.090909
0.01	100.00	**101**	0.009901	0.009901
0.001	1000.00	**1001**	0.000999	0.000999

***Sequence Converging to* 0**—For a given term of order p, any term of order n greater than p belongs to the neighborhood of 0, that is, such terms of sequence X_n satisfy the inequality:

$$n > p \underset{n}{\Rightarrow} |X_n - 0| < r$$

In other words, the limit for sequence X_n is 0, or also, X_n tends to 0:

$$\lim X_n = 0, \quad \text{or } X_n \to 0$$

The sequence $X_n = 1/n$ is converging to 0, it is also called infinitely small or infinitesimal sequence. Such infinitesimal sequences are very important to construct converging sequences to a certain point, either by larger values ("right side") or lower values ("left side"); the infinitely small sequences are key for additional developments, as presented in the next two examples.

Example

It also can be shown that sequence, $Y_n = 2 + 1/n$, is decreasing, lower bounded, and converging by larger values to the limit, 2, as presented in Figure 2.4.

n	Yn
1	3.000000
2	2.500000
3	2.333333
4	2.250000
5	2.200000
6	2.166667
7	2.142857
8	2.125000
9	2.111111
10	2.100000
11	2.090909
12	2.083333

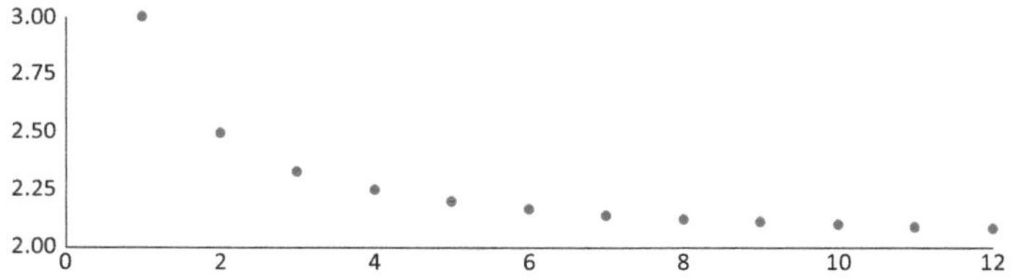

FIGURE 2.4 Table and graphic for sequence $Y_n = 2 + 1/n$.

The sequence Y_n at hand is:

- A decreasing sequence, with $Y_{n+1} < Y_n$;
- A bounded sequence, with values ranging between the first term $Y_1 = 3$, while Y_n never reaches the lower bound, 2;
- The decreasing and bounded sequence Y_n is thus convergent, and the associated limit is 2.

  The related proofs would follow a very similar path to the previous study for sequence X_n; the addition of a constant, 2, does not change the key attributes (monotony, boundedness, convergence) of the infinitesimal term, $X_n = 1/n$.

Example
In the opposite sense, let the sequence converging to 2 by lower values be, $Y_n = 2 - 1/n$; it is obtained by subtracting the infinitesimal sequence X_n to the same constant, 2, and the related table and graphic are presented below (Figure 2.5).

- At now, this new sequence Y_n is increasing, and the proof is initiated by hypothetically assuming the term of order $(n+1)$ is greater than the nth term:

$$Y_{n+1} > Y_n$$

$$2 - \frac{1}{n+1} > 2 - \frac{1}{n} \Leftrightarrow -\frac{1}{n+1} > -\frac{1}{n} \Leftrightarrow \frac{1}{n+1} < \frac{1}{n} \Leftrightarrow n < n + 1$$
$$\Leftrightarrow 0 < 1, \; \textbf{\textit{TRUE}}$$

n	Yn
1	1.000000
2	1.500000
3	1.666667
4	1.750000
5	1.800000
6	1.833333
7	1.857143
8	1.875000
9	1.888889
10	1.900000
11	1.909091
12	1.916667

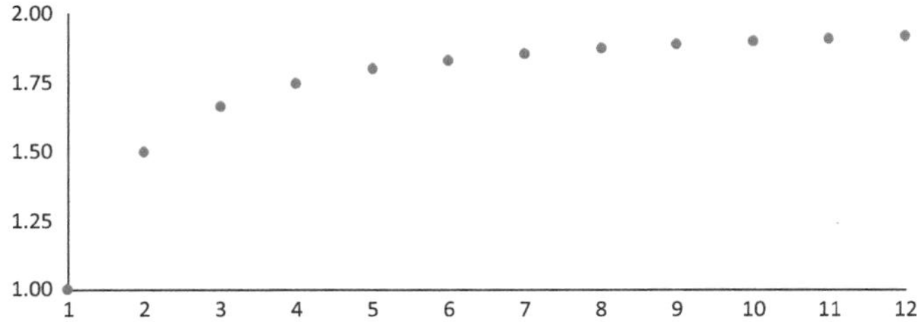

FIGURE 2.5 Table and graphic for sequence $Y_n = 2 - 1/n$.

Once again, the true relation in the final step confirms the hypothesis in this proof, which assumed Y_n as increasing sequence, $Y_{n+1} > Y_n$, is correct too.

- In addition, the monotonous sequence Y_n is limited, with values ranging above the first term, $Y_1 = 1$, while Y_n never reaches the upper bound in 2:

$$\left|2 - \frac{1}{n}\right| < B \Leftrightarrow -B < 2 - \frac{1}{n} < B \Leftrightarrow 2 - \frac{1}{n} < B$$

Namely, assuming $B = 2$,

$$2 - \frac{1}{n} < 2 \Leftrightarrow -\frac{1}{n} < 0 \Leftrightarrow -1 < 0, \textbf{\textit{TRUE}}$$

- The increasing and bounded sequence Y_n is thus convergent; the associated limit is 2, as shown in the following steps:

$$|Y_n - 2| < r$$

By substituting the expression of Y_n, canceling the limit term, 2, as well as neglecting the negative branch for the inequality (since the sequence terms are all positive), it follows:

$$\left|\left(2 - \frac{1}{n}\right) - 2\right| < r \Leftrightarrow \left|\frac{1}{n}\right| < r \Leftrightarrow -r < \frac{1}{n} < r \Leftrightarrow \frac{1}{n} < r$$

And a relation between n and r, $n(r)$, is obtained once again:

$$n > \frac{1}{r}$$

For each neighborhood radius, r, evaluating again the relation, $p = p(r)$ and the integer $n = \text{INT}(p) + 1$, then the numerical table with the related nth terms follows:

r	p	n = INT(p)+1	Yn	|Yn - 2|
0.1	10.00	**11**	1.909091	0.090909
0.01	100.00	**101**	1.990099	0.009901
0.001	1000.00	**1001**	1.999001	0.000999

It is observed that for a given term of order p, any term of order n greater than p belongs to the neighborhood of 2; that is, such terms of sequence Y_n satisfy the inequality:

$$n > p \underset{n}{\Rightarrow} |Y_n - 2| < r$$

In other words, the limit of Y_n is 2, or also, the sequence Y_n approaches 2:

$$\lim Y_n = 2, \text{ or } Y_n \to 2$$

Infinitely Large Sequence—A sequence X_n tends to infinite if, for every positive number M, there is a corresponding number p, such that,

$$n > p \underset{n}{\Rightarrow} |X_n| > M$$

Then,

$$\lim X_n = \infty, \quad \text{or} \quad X_n \to \infty$$

A relation between n and M, $n(M)$, is thus foreseen, and the flowchart in Figure 2.6 outlines the pathway to obtain it.

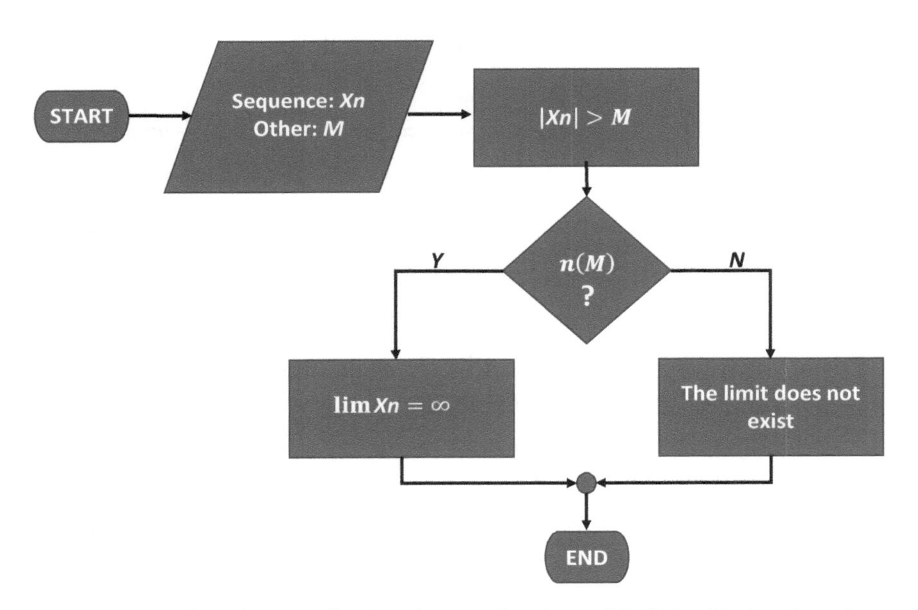

FIGURE 2.6 Flowchart outlining the verification of infinite limit of a given sequence.

Example
Let it be $Z_n = n$; the corresponding table and graphic are also presented (Figure 2.7):

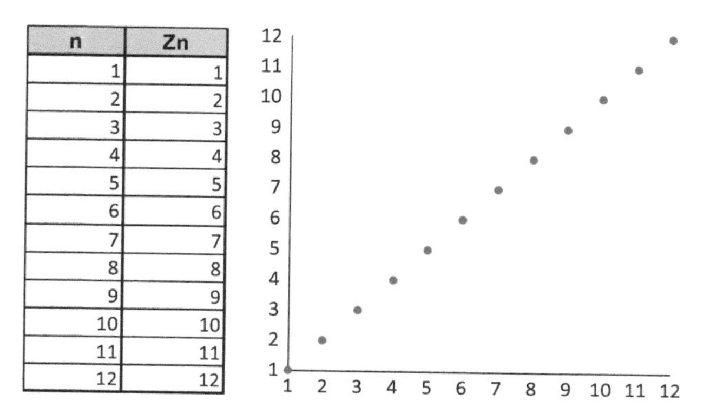

n	Zn
1	1
2	2
3	3
4	4
5	5
6	6
7	7
8	8
9	9
10	10
11	11
12	12

FIGURE 2.7 Table and graphic for sequence $Z_n = n$.

The sequence Z_n is linearly increasing, and the monotony proof is straight:

$$Z_{n+1} > Z_n \Leftrightarrow n + 1 > n \Leftrightarrow 1 > 0, \ \textbf{\textit{TRUE}}$$

The linear sequence Z_n is increasing without an upper bound, it is unbounded, and a majoring value M cannot be defined; it directly follows that:

$$|Z_n| > M$$

By substituting Z_n with the related expression, and by neglecting the negative branch in the inequality because all the sequence terms are positive, then:

$$|n| > M \Leftrightarrow n < -M \vee n > M$$

Finally, a relation between n and the majoring M, $n(M)$, is obtained:

$$n > M$$

By evaluating the relation, $p = p(M)$, and also the integer, $n = \text{INT}(p) + 1$, the table with numerical values for the majoring, M, and the related nth *term* follows:

M	p	n = INT(p)+1	Zn	|Zn| - M
1000.0	1000.0	**1001**	1001.0	1.0
10000.0	10000.0	**10001**	10001.0	1.0
100000.0	100000.0	**100001**	100001.0	1.0

It is observed that for a given term of order p, any term of order n greater than p is larger than the majoring M; that is, such terms of sequence Z_n satisfy the inequality:

$$n > p \underset{n}{\Rightarrow} |Z_n| > M$$

In other words, the sequence Z_n is unbounded, it increases without majoring; the sequence Z_n is an infinitely large sequence, or also, Z_n tends to infinite:

$$\lim Z_n = \infty, \text{ or } Z_n \to \infty$$

The Limit on the Operation Corresponds to the Operation on Limits

When treating sequence limits involving arithmetic operations, e.g., addition, subtraction, multiplication, via the basic properties of real numbers and arithmetic operations, they result in the arithmetic operation on the limits, assuming such limits exist and operations are possible. Namely: the limit of summing (subtracting) two sequences corresponds to the sum (subtraction) of the sequence limits; the limit of multiplying (dividing) two sequences corresponds to the multiplication (division, if not by zero) of limits; and the approach also applies for the limit of a power, or the corresponding root (the root sequence would be non-negative, if power p is even).

$$\lim(S_n \pm T_n) = \lim S_n \pm \lim T_n \qquad \lim(k.S_n) = k.\lim S_n$$

$$\lim(S_n.T_n) = \lim S_n.\lim T_n \qquad \lim(S_n/T_n) = \frac{\lim S_n}{\lim T_n}$$

$$\lim \sqrt[p]{S_n} = \sqrt[p]{\lim T_n}$$

2.4 THEOREMS ABOUT SEQUENCES

In this section, the main theorems about sequences are discussed, and important properties of sequence limits (uniqueness, the constant sequence, trapped sequence) are revisited too.

If a Sequence Is Convergent for a Given Limit, That Limit Is Unique (Uniqueness)

The contradiction approach is very useful for uniqueness proofs, typically it is assumed a second limit would exist, and it would be different from the previous one.

- Suppose that $b \neq a$; by definition, after a certain order, $n = \text{INT}(p) + 1$, the S_n terms would be within the neighborhood of limit, b. However, they cannot be located in such neighborhood of b, since S_n is converging to limit, a, and the contradiction occurs.
- Then, the second limit b cannot exist unless it would take the same value, a. The limit is thus unique, and it takes one single value, $\lim S_n = a$.

The limit of a constant sequence is the constant itself, that is, $S_n = k \Rightarrow \lim S_n = k$

The limit of a constant sequence directly results from the convergence definition, as follows:

$$n > p \underset{n}{\Rightarrow} |S_n - k| < r$$

By substitution of the sequence term,

$$|k - k| < r \Leftrightarrow 0 < r, \textit{TRUE}$$

If a Sequence Is Convergent, Then Any of Its Subsequences Tend toward the Same Limit

Let the sequence be $X_n = 1/n$, with limit 0; this theorem indicates the two prior subsequences, the subsequence with even terms and the subsequence of odd terms, are both converging to the same limit, 0. In fact, for each radius, r, all the terms from order $n = \text{INT}(p) + 1$ are in the neighborhood of zero, and this includes both even and odd terms; thus, the even-subsequence tends to 0, and the odd-subsequence approaches 0 too.

$$\frac{1}{1}, \frac{1}{3}, \frac{1}{5}, \frac{1}{7}, \ldots, \frac{1}{2n-3}, \frac{1}{2n-1}, \ldots \to 0$$
$$\frac{1}{2}, \frac{1}{4}, \frac{1}{6}, \frac{1}{8}, \ldots, \frac{1}{2n-2}, \frac{1}{2n}, \ldots \to 0$$

Through a similar reasoning, the general result can be induced too: **If the sequence is converging to limit, *a*, all of its subsequences are converging to the same limit.**

Every Monotonous and Bounded Sequence Is Convergent

The following two instances illustrate both the cases, a decreasing and lower bounded sequence and an increasing and upper bounded sequence.

- Let it be again, $X_n = 1/n$; this sequence is decreasing and bounded, as shown in the prior section; for point 0, the greatest of sequence minors, the point $b = 0 + r$ is not a minor; thus, there is at least a term of order p in the interval $]0, 0 + r[$, and all the terms $n = \text{INT}(p) + 1$, are within that interval, $]0, r[$, because the sequence is decreasing. Therefore, the sequence is converging to 0, or also, X_n tends to the greatest of its minors (the sequence infimum).
- And let it be, $Y_n = 2 - 1/n$; the sequence Y_n is increasing and bounded, as shown before; for point 2, the lowest of its majors, the point $b = 2 - r$ is not a major; thus, there is at least a term of order p in the interval $]2 - r, 2[$, and all the terms $n = \text{INT}(p) + 1$, are within that interval, because the sequence is increasing. Therefore, Y_n is approaching 2, or also, the sequence tends to the lowest of its majors (the sequence supremum).

The general result can be induced too, including these two cases, and then stating that every monotonous (decreasing, or increasing) and bounded sequence (minored, or majored) is convergent; namely, to the infimum or its terms, or to the supremum or its terms.

If the Convergent Sequence S_n Is, from a Certain Order, $S_n \geq 0$, Then $\lim S_n \geq 0$

The proof by contradiction is very useful, typically it develops well in cases or results that are analogous to common sense.

- For the convergent sequence of positive terms, $S_n \geq 0$, let the negative limit be $a < 0$; by definition, after a certain order, $n = \text{INT}(p) + 1$, all the sequence terms are within the neighborhood of point a, that would occur in the negative domain. However, the sequence cannot take negative values, and a contradiction occurs; such contradiction is waved in case the limit takes a non-negative value, $a \geq 0$.

***If the Convergent Sequences* $\mathbf{S_n}$ *and* $\mathbf{T_n}$ *Are, from a Certain Order,* $S_n \geq T_n$*, Then lim* $S_n \geq$ *lim* T_n**

The current result directly follows from the last one, thus defining the difference between the two convergent sequences:

$$S_n \geq T_n \Leftrightarrow S_n - T_n \geq 0$$

The difference-sequence, $U_n = S_n - T_n$, will be non-negative, and so it will be the related limit,

$$U_n \geq 0 \Rightarrow \lim U_n \geq 0$$

The difference between the two limits will be non-negative too, and then

$$\lim S_n - \lim T_n \geq 0 \Leftrightarrow \lim S_n \geq \lim T_n$$

***If Sequences* $\mathbf{S_n}$ *and* $\mathbf{T_n}$ *Are Convergent for the Same Limit, lim* $S_n =$ *lim* $T_n = a$*, and If from a Certain Order,* $S_n \leq U_n \leq T_n$*, Then lim* $U_n = a$**

The proof by contradiction can be useful in this case too, the result is seen as common sense.

- For the sequences, $S_n \leq U_n \leq T_n$, consider the limit $b \neq a$; by the limit definition, after a certain order, $n = \text{INT}(p) + 1$, the U_n terms would be within the neighborhood of limit, b. However, both the sequences S_n and T_n cannot be located in such neighborhood since they are converging to limit, a, and the contradiction occurs.

- Such contradiction is waved in case sequence U_n takes the same limit, a, similarly as sequences S_n and T_n; in this manner, for all terms after a certain order,

$$a - r < S_n \leq T_n < a + r$$

$$a - r < S_n \leq U_n \leq T_n < a + r$$

The last relation indicates the terms of sequence, U_n, also are in the neighborhood of point a, in this way confirming the limit is the same, $\lim U_n = a$.

2.5 STUDY OF IMPORTANT SEQUENCES

Some sequences are very important due to the number of related applications, namely the sequence of exponential terms, a^n, that corresponds to a geometric sequence. Thus, geometric sequences with diverse bases (e.g., greater than one; less than one, but positive; negative, but greater than −1; or even base less than −1) are focused in this section.

- Let the sequence be $X_n = a^n$, and consider the results summary:

$$a > 1 \Rightarrow \lim a^n = \infty, \quad \text{or} \quad a^n \to \infty$$

$$a = 1 \Rightarrow \lim a^n = 1, \quad \text{or} \quad a^n \to 1$$

$$0 < a < 1 \Rightarrow \lim a^n = 0, \quad \text{or} \quad a^n \to 0$$

$$a = 0 \Rightarrow \lim a^n = 0, \quad \text{or} \quad a^n \to 0$$

$$-1 < a < 0 \Rightarrow \lim a^n = 0, \quad \text{or} \quad a^n \to 0, \text{ convergent oscillating}$$

$$a < -1 \Rightarrow \text{Divergent oscillating}$$

Case $a > 1$

- Let it be, $Z_n = 2^n$, with the corresponding table and graphic of exponential values in Figure 2.8; note the exponential growth is so fast that only nine terms are represented, even when the vertical scale is about 400 times larger than the horizontal scale.

  The sequence is strongly increasing, from the first term with value 2; the exponential presents positive values that increase faster and faster, thus it does not accept any upper bound. Assuming sequence Z_n is unbounded, increasing above any majoring value, M:

$$|Z_n| > M$$

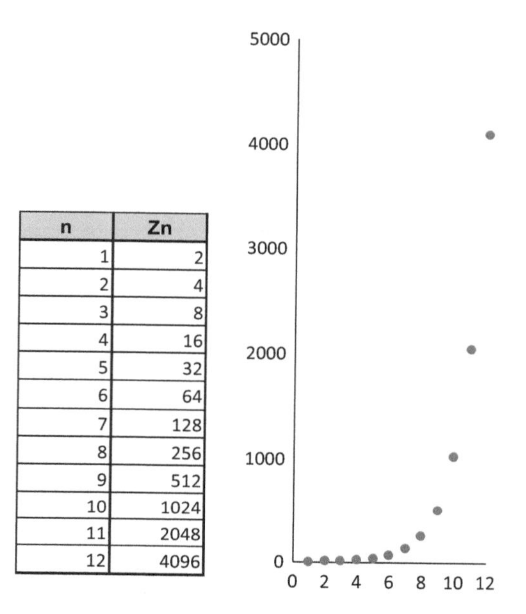

n	Zn
1	2
2	4
3	8
4	16
5	32
6	64
7	128
8	256
9	512
10	1024
11	2048
12	4096

FIGURE 2.8 Table and graphic for sequence $Z_n = 2^n$.

In this way,

$$|2^n| > M \Leftrightarrow 2^n > M \vee 2^n < -M \Leftrightarrow 2^n > M$$

The exponential of base-2 is suitable for manipulating the inequality:

$$2^n > 2^{\log_2 M}$$

The exponential is increasing with the associated exponent, the relation between the exponents on both sides of the inequality shall be satisfied too:

$$n > \log_2 M$$

That is, by manipulating the logarithm base,

$$n > \frac{\log_{10} M}{\log_{10} 2} \Leftrightarrow n > \frac{\ln M}{\ln 2}$$

For each majoring, M, evaluating again the relation, $p = p(M)$, and the integer $n = \text{INT}(p) + 1$, the numerical table with the related nth terms follows too:

M	p	n = INT(p)+1	Zn	|Zn| - M
1000.0	9.97	**10**	1024.0	24.00
1000000.0	19.93	**20**	1048576.0	48576.00
1000000000.0	29.90	**30**	1073741824.0	73741824.00

Through the same reasoning, it can be shown that similar relations apply for any base a greater than one, $a > 1$. Namely, showing that:

$$|a^n| > M \Leftrightarrow a^n > M$$

As well as treating both the inequality members as exponentials with the same base, $a > 1$,

$$a^n > a^{\log_a M}$$

The inequality also applies for the exponents in the first and second member, and a relation between n and the majoring M, $n(M)$, is obtained once again:

$$n > \log_a M$$

By manipulating the logarithm base,

$$n > \frac{\log_{10} M}{\log_{10} a} \Leftrightarrow n > \frac{\ln M}{\ln a}$$

Assuming $a > 1$, as described above, for any term nth term of order greater than p,

$$n > p \underset{n}{\Rightarrow} |a^n| > M$$

Then,

$$\lim a^n = \infty, \quad \text{or} \quad a^n \to \infty$$

***Case* 0 < *a* < 1**

- At now, the positive base a is lower than 1, that is, $0 < a < 1$; the sequence $X_n = (1/2)^n$ is studied, and the associated table and graphic follow (Figure 2.9).

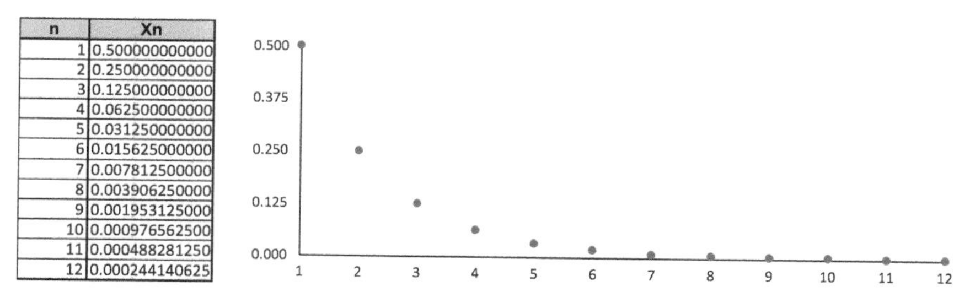

n	Xn
1	0.500000000000
2	0.250000000000
3	0.125000000000
4	0.062500000000
5	0.031250000000
6	0.015625000000
7	0.007812500000
8	0.003906250000
9	0.001953125000
10	0.000976562500
11	0.000488281250
12	0.000244140625

FIGURE 2.9 Table and graphic for sequence $X_n = (1/2)^n$.

The sequence is decreasing, from the very first term with value $1/2 = 0.5$; the exponential presents only positive values, thus it never reaches 0. Let 0 be the limit of sequence X_n:

$$|X_n - 0| < r$$

By substituting the expression for X_n, and neglecting both the null term and the negative branch in the inequality (since the exponential terms are all positive), it results:

$$\left|\left(\frac{1}{2}\right)^n - 0\right| < r \Leftrightarrow -r < \left(\frac{1}{2}\right)^n < r \Leftrightarrow \left(\frac{1}{2}\right)^n < r$$

The exponential of base-2 is preferred for manipulating the inequality, because exponentials with base greater than 1 are strictly increasing functions; then,

$$2^{-n} < r \Leftrightarrow 2^{-n} < 2^{\log_2 r}$$

The exponential of base-2 increases when the exponent increases, and thus the relation between the exponents on both sides of the inequality shall be satisfied too:

$$-n < \log_2 r \Leftrightarrow n > -\log_2 r$$

By changing logarithm to the decimal base, 10, or even to the so-called natural base, *e*,

$$n > -\frac{\log_{10} r}{\log_{10} 2} \Leftrightarrow n > -\frac{\ln r}{\ln 2}$$

For each neighborhood radius, *r*, by calculating the relation, $p = p(r)$, and the integer $n = \text{INT}(p) + 1$, the table with the associated *n*th terms follows too:

r	p	n = INT(p)+1	Xn	|Xn - 0|
0.1	3.32	4	0.062500	0.062500
0.01	6.64	7	0.007813	0.007813
0.001	9.97	10	0.000977	0.000977

Similar relations for any positive base *a* lower than one, $0 < a < 1$, can be shown. In fact, assuming that $a = 1/b$, with *b* positive and greater than one, $b > 1$:

$$\left| \left(\frac{1}{b}\right)^n - 0 \right| < r \Leftrightarrow \left(\frac{1}{b}\right)^n < r$$

The comparison between two exponentials of same base, $b > 1$, corresponds to the comparison of their exponents:

$$b^{-n} < r \Leftrightarrow b^{-n} < b^{\log_b r}$$

Thus,

$$-n < \log_b r \Leftrightarrow n > -\log_b r$$

Or also, changing to base-10, or the natural base, *e*, that are commonly used:

$$n > -\frac{\log_{10} r}{\log_{10} b} \Leftrightarrow n > -\frac{\ln r}{\ln b}$$

For the base a under analysis, recall the positive and inverse value, $b = (1/a) > 1$, corresponds to the range $0 < a < 1$; then, for any nth term of order greater than p, the inequality holds:

$$n > p \underset{n}{\Rightarrow} |a^n - 0| < r$$

Consequently, when the positive base is lower than one, $0 < a < 1$, the limit of sequence a^n is 0, or a^n tends to 0:

$$\lim a^n = 0, \quad \text{or} \quad a^n \to 0$$

Case $-1 < a < 0$

A similar situation occurs when addressing the base a in the range $-1 < a < 0$; in such case, the symmetric approach arises when multiplying by (-1) the inequality terms and inverting its sense; then it is obtained $1 > -a > 0$, and such case is already analyzed as convergent sequence.

- An illustrative example follows in the table, and graphic too, for $a = -1/2$ (Figure 2.10). The sequence generated with such negative base is convergent to 0; however, it shall be noted the oscillatory behavior for this sequence, with even terms being positive, while odd terms are presenting negative values.

n	Xn
1	-0.500000000000
2	0.250000000000
3	-0.125000000000
4	0.062500000000
5	-0.031250000000
6	0.015625000000
7	-0.007812500000
8	0.003906250000
9	-0.001953125000
10	0.000976562500
11	-0.000488281250
12	0.000244140625

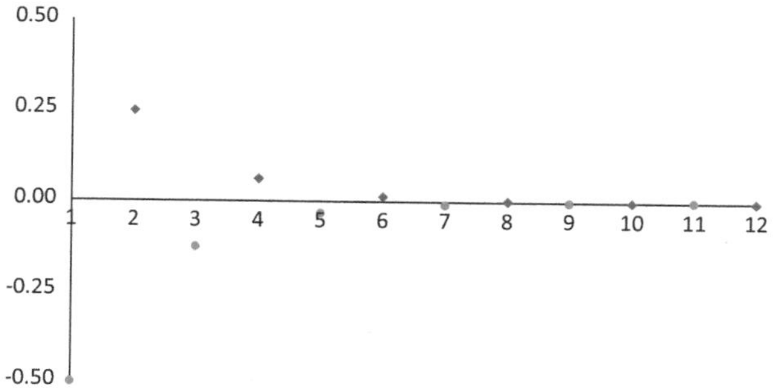

FIGURE 2.10 Table and graphic for sequence $X_n = (-1/2)^n$.

Case $a < -1$

Note the oscillatory behavior too when addressing the exponential sequence with base $a < -1$; once again, the symmetric approach arises when multiplying by (-1) the inequality terms and inverting its sense;

n	Zn
1	-2
2	4
3	-8
4	16
5	-32
6	64
7	-128
8	256
9	-512
10	1024
11	-2048
12	4096

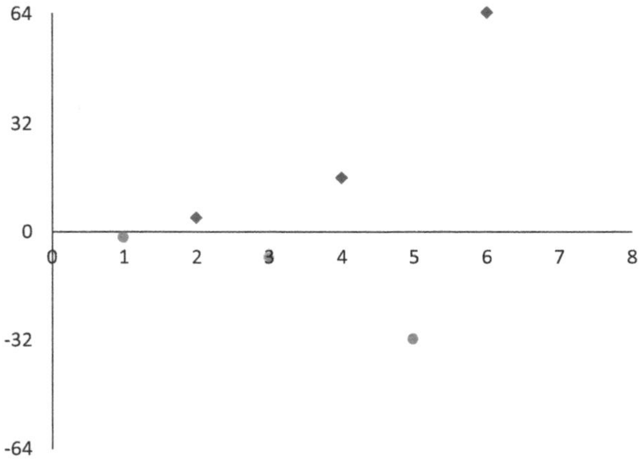

FIGURE 2.11 Table and graphic for sequence $Z_n = (-2)^n$.

it results, $-a > 1$, and such case is already studied as an unbounded sequence.

- An example follows in Figure 2.11, for $a = -2$. This sequence is oscillatory divergent, since the even terms are positive and increase without upper bound, while odd terms are negative and decrease without lower bound. Thus, the subsequence with even terms would tend to a positive infinite, while the subsequence with odd terms would tend to a negative infinite, and the sequence would present two different limits! Due to the uniqueness property, the limit for this sequence does not exist.

2.6 NOTES ON NUMBERS COMPUTATION

In this section, an experimental approach to the sequence that tends to the optimum diameter for the barrel instance (V = 800 liters) is presented. The oscillating geometric sequence utilizes negative base, (−1/10), and important insights about numerical approximations and numbers rounding are introduced. After that, the sequence $X_n = (1 + 1/n)^n$ is revisited, and additional notes about numbers computation (e.g., overflow, underflow) are presented.

2.6.1 A Sequence for the Optimum Diameter

The optimum diameter was obtained (Chapter 1) by the cubic root formula, $d = \sqrt[3]{\frac{3.2}{\pi}} \cong 1.006$.

The numerical approximation to the optimum diameter is rounding the third decimal digit, that corresponds to the millimeters range; however, laboratory procedures and industry practice (e.g., aeronautics, automotive) can be very requiring with the associated procedures, including quality assurance, very thin cuts, or material losses. Usually, the approximation by rounding the third decimal digit is related with an error within the fourth digit, majored by $0.0005 = 5 \times 10^{-4}$. In fact, the usual approach considers adding 1 to the third digit, in case the fourth digit is greater than or equal to 5; and keeping equal the third digit, by truncating the fourth digit (and other digits in the right side) in case it ranges between 0 and 4.

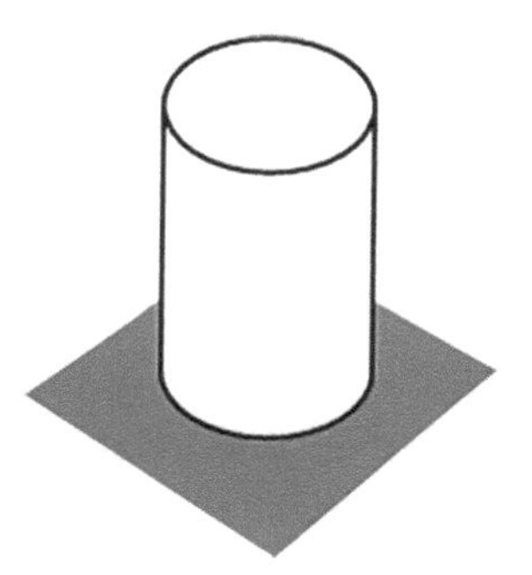

At now, let the oscillating sequence be

$$X_n = \sqrt[3]{\frac{3.2}{\pi}} + \left(-\frac{1}{10}\right)^n.$$

This sequence presents a constant in the first term, which is added to an exponential with negative-basis, $a = -1/10$, in the second term.

The absolute value of the exponential base, $a = 1/10$, is lower than 1, then the oscillating exponential is converging to zero. The table and graphic for sequence X_n follow in Figure 2.12.

n	Xn
1	0.90615919832087
2	1.01615919832087
3	1.00515919832087
4	1.00625919832087
5	1.00614919832087
6	1.00616019832087
7	1.00615909832087
8	1.00615920832087
9	1.00615919732087
10	1.00615919842087

FIGURE 2.12 Table and graphic for sequence X_n.

The sequence is converging, and it approaches the limit, $\sqrt[3]{\frac{3.2}{\pi}}$, at step, $1/10 = 0.1$, although it is oscillating around that limit. The convergence proof for sequence X_n follows too:

$$\left| X_n - \sqrt[3]{\frac{3.2}{\pi}} \right| < r$$

By substituting the expression for X_n, and canceling the cubic root terms,

$$\left|\left[\sqrt[3]{\frac{3.2}{\pi}}+\left(-\frac{1}{10}\right)^n\right]-\sqrt[3]{\frac{3.2}{\pi}}\right|<r\Leftrightarrow\left|\left(-\frac{1}{10}\right)^n\right|<r$$

Treating the absolute value, and neglecting the negative branch for the inequality since all the exponential terms are positive, it results:

$$-r<\left(\frac{1}{10}\right)^n<r\Leftrightarrow\left(\frac{1}{10}\right)^n<r$$

At now, the base-10 exponential is preferred for manipulating the inequality, because exponentials with base greater than 1 are strictly increasing functions; then:

$$10^{-n}<r\Leftrightarrow 10^{-n}<10^{\log_{10}r}$$

The exponential is increasing when the associated exponent increases too, then the relation between the exponents on both sides of the inequality shall be satisfied too:

$$-n<\log_{10}r\Leftrightarrow n>-\log_{10}r$$

Or also,

$$n>-\ln(r)/\ln 10$$

Calculating again the relation, $p=p(r)$, and also $n=\text{INT}(p)+1$, then a table with numerical values for radius, r, and the associated nth terms follows:

r	p	n = INT(p)+1	Xn	|Xn - a|
0.0005000000	3.30	**4**	1.006259198321	0.000100000000
0.0000005000	6.30	**7**	1.006159098321	0.000000100000
0.0000000005	9.30	**10**	1.006159198421	0.000000000100

Note the number of terms n in need of decimal rounding and the nth order digit are coincident, since the exponential of base-10 is utilized. Namely, by rounding the third digit the deviation is thus related with the fourth digit ($n=4$, $r=0.0005$), and by rounding the sixth digit the deviation is related with the seventh digit ($n=7$, $r=0.0000005$).

2.6.2 Additional Notes on Computational Calculations

Let the sequence again be $X_n = (1 + 1/n)^n$, but consider the different expressions,

$$X_n = \left(\frac{n+1}{n}\right)^n = \frac{(n+1)^n}{n^n}$$

A limited number of terms are presented in Figure 2.13, because the computation of integer numbers is upper bounded; the largest integer number, $(n+1)^n$, indicated in the table is about $7.7 \times 10^{+300}$, and neither additional terms are calculated nor presented in the graphic. A common machine will suffer overflow for integer numbers greater than 10^{+308}, and underflow for positive fractional numbers lower than 10^{-308}. Indeed, a machine working in double precision (64 bits), is limited in its ability to accommodate the binary (0, 1) strings that represent real numbers.

n	n+1	(n+1)^n	Xn
1	2	2.0000000000000000E+00	2.0000000000000000E+00
10	11	2.5937424601000000E+10	2.5937424601000000E+00
15	16	1.1529215046068500E+18	2.6328787177279200E+00
20	21	2.7821842944695200E+26	2.6532977051444200E+00
25	26	2.3677383000796800E+35	2.6658363314874200E+00
30	31	5.5061852034591100E+44	2.6743187758703000E+00
40	41	3.2460430015432000E+64	2.6850638383899700E+00
50	51	2.3906104021463700E+85	2.6915880290736000E+00
100	101	2.7048138294215200E+200	2.7048138294215200E+00
140	141	7.7745591440660400E+300	2.7086368139211600E+00
150	151	#NÚM!	#NÚM!
300	301	#NÚM!	#NÚM!
400	401	#NÚM!	#NÚM!

FIGURE 2.13 Table and graphic for the updated expression of X_n.

Returning to the initial expression, the number of terms can be largely extended (Figure 2.14). However, the second term inside parentheses is an infinitesimal sequence, $I_n = 1/n$, and the relative weight is decreasing and decreasing when adding to the first term, 1:

$$X_n = \left(1 + \frac{1}{n}\right)^n$$

In that manner, at a certain point, the second term is added to 1 but this value is not altered, and all the powers of 1 also result 1, as presented in Figure 2.14 (graphic with logarithmic X-axis).

n	Xn
100000	2.718268237197530
1000000	2.718280469156430
10000000	2.718281693980370
100000000	2.718281786395800
1000000000	2.718282030814510
10000000000	2.718282053234790
100000000000	2.718282053357110
1000000000000	2.718523496037240
10000000000000	2.716110034086900
100000000000000	2.716110034087020
1000000000000000	3.035035206549260
10000000000000000	1.000000000000000
100000000000000000	1.000000000000000
1000000000000000000	1.000000000000000

FIGURE 2.14 Table and graphic for the extended calculation of sequence X_n.

In this instance, while the machine was able to represent the infinitesimal term until about 10^{-14}, or even 10^{-15}, it was not able to operate it and add to 1 for values lower than 10^{-16}. Thus the machine precision of such 64-bit machine is about 10^{-14}–10^{-15}.

2.7 CONCLUDING REMARKS

Important notions about sequence limits, with the main purpose of supporting additional developments about function limits, are introduced.

- Thus, preliminary notions about sequences of real numbers are presented, including monotony, bounded sequence, or convergent sequence; terminology and typical Greek symbols are adjusted in order to promote understanding, as well as graphics and their axis are tuned to improve the perception about the ranges at hand.
- Sequence limits are illustrated, including the main concepts of convergence or infinitely large sequences; the verification of convergence radius is a necessary condition to be satisfied in all the numerical examples, otherwise the occurrence of counter-example would support the contradiction procedure.
- The main theorems about sequences are discussed too, while important properties of sequence limits (uniqueness, the constant sequence, operations) are revisited.
- Important sequences are also studied, namely, the geometric sequences with exponential terms, a^n; diverse bases a are focused

on, in particular base-2 and base-2^{-1}, while the associated binary instances are treated.

A rigorous confirmation, or double verification, would consider the comparison between the term, n = INT(p) + 1 and the prior term, INT(p); they shall be located, respectively, inside and outside the neighborhood radius, r.

- And finally, a sequence that tends to the optimum diameter for the cost instance is presented; an experimental approach that utilizes base-10 is detailed, and relevant insights about the computational representation of real numbers are introduced.

These concepts and results about limits of real numbers sequence support the study of function limits. By analogy, dependent variable y can be viewed as a sequence of function values that converges to point b, when the numbers sequence associated with independent variable x converges to point a. Next chapter about function limits also includes:

- Basics about the limit of a function either at a point or at infinite, as well as the properties of infinitesimal limits and lateral limits at a point;
- Important theorems about function limits, together with the related operations rules;
- Some remarkable function limits, and associated discussion.

CHAPTER 3

Limit of a Function

The limit of a function at a point is a very important concept, due to its relevant applications in many scientific areas and technologic fields. In the Mathematics fields, many developments are supported in the function limit, such as function continuity, derivatives and related applications, integrals, sum series, just to name a few. In this chapter, beyond the basic notions about the function limit at a point, and also in the infinite, the main properties and operations are addressed too. The infinitesimal limit is also treated, it occurs when the function is infinitely small and tends to zero, and the related properties and operations are similar to those of infinitesimal sequences. Finally, some remarkable limits are presented, while several numerical instances are illustrating the various cases, namely, the lateral limits at both the left and right side of the point under analysis.

3.1 INTRODUCTION

The function limit is a key notion that supports other important concepts, such as continuity at a point, which occurs when the function limit at the point a under analysis equals the function image at such point. It is again a good occasion to reinforce rigorous habits, promoting tables verification and graphics comparison, or experimenting with infinitesimal limits.

The function limit can be perceived as enlarging the sequence concept onto a second convergent variable, while the composite function limit can be seen as three convergent variables in cascade. The current approach is based on the study of sequences of real numbers, by assuming the variable x is converging to a point a, $x \to a$, and the dependent variable $y(x)$ is converging to the corresponding value b, $y \to b$. That is, the sequence of real numbers X_n for independent variable, x, is originating a sequence Y_n for dependent variable, y; then the properties and main results concerning the functions limit are treated through an analogous approach, such as focusing sequence, Y_n.

DOI: 10.1201/9781003461876-3

- By analogy with the operating rules for limits of real number sequences, similar rules (e.g., sum and subtraction, multiplication and division, powers and roots) are also applying for the calculation of functions limits, obviously assuming the independent variable in such calculations approaches the same point a, $x \to a$.
- Thus, the recognition of similar properties and operations as those presented in sequence limits may help to develop systematic practices for function limits.

Note the existence of limit at a certain point, $x = a$, even when the function is not defined at that point. In particular, at $x = 0$, both the sine function and linear function, x, tend to zero; but the related ratio of limits becomes 1, even when the well-known function is not defined at point 0 (Figure 3.1):

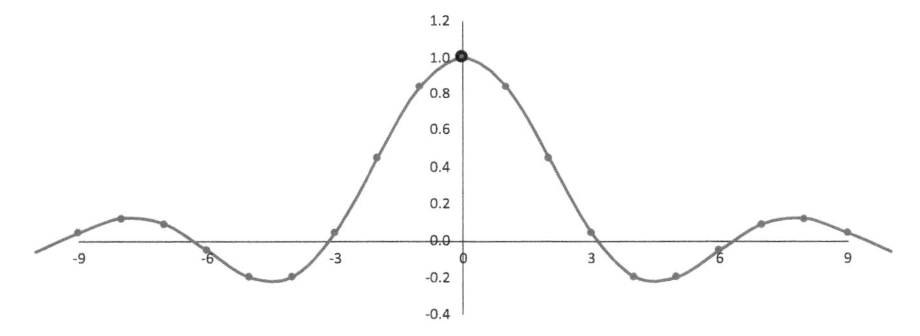

FIGURE 3.1 Graphic of function, $y(x) = \sin(x)/x$, which is not defined at $x = 0$.

$$\lim_{x \to 0} \frac{\sin(x)}{x} = 1$$

The structure for the rest of this chapter is as follows: first, basic notions about the limit of a function are addressed, either at a point or at infinite, and the properties of infinitesimal limits are described too; second, lateral limits are focused, and both the left side and the right side limits at a point are detailed; after that, important theorems about function limits are revisited, together with the related operations rules; and finally, some remarkable limits are presented too, because these well-known limits are widely applied in other developments.

3.2 NOTIONS ABOUT FUNCTION LIMITS

In this section, basic notions about the limit of a function at a point are addressed, along with the limit of a function at infinite when the independent variable, x, increases without any bound or majoring. From other side, the infinite limit of a function is focused when the dependent variable, y, increases without any bound, while the infinitesimal limit $I(x)$ occurs at such point that the function is infinitely small and tends to zero. The properties of infinitesimal limits are described too, due to their importance in a number of applications.

Limit of a Function at a Point

Since $f(x)$ is a real function of a real variable, it is said that $\lim_{x \to a} f(x) = b$ if, and only if, to any sequence X_n of terms from the $f(x)$ domain approaching finite point, a, for values different from a, there corresponds a sequence of images $f(X_n)$ that tends to b.

$$X_n \to a \Rightarrow Y_n \to b, \text{ or}$$

$$\lim X_n = a \Rightarrow \lim Y_n = b$$

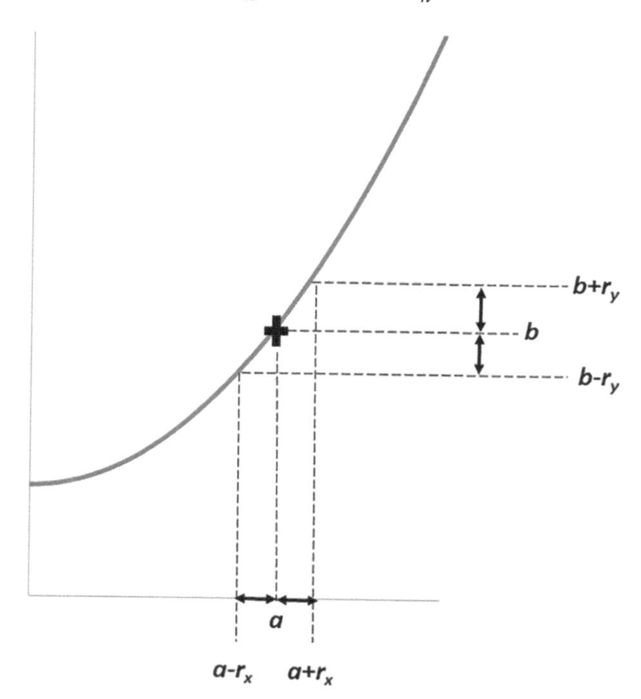

FIGURE 3.2 Function limit at point a, and related neighborhoods.

In this manner, the X-neighborhood radius, r_x, and the Y-neighborhood radius, r_y, are connected (Figure 3.2):

$$|X_n - a| < r_x \Rightarrow |Y_n - b| < r_y$$

Therefore, a relation between r_x and r_y is foreseen, $r_x(r_y)$, and the flowchart in Figure 3.3 outlines the pathway to obtain it.

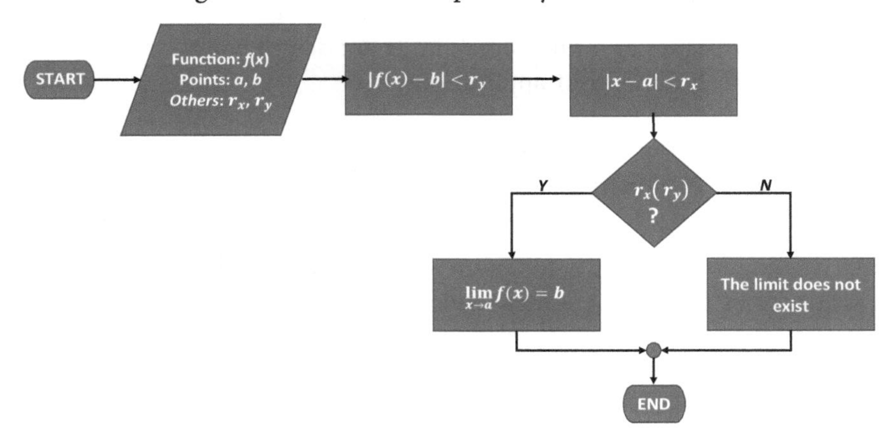

FIGURE 3.3 Flowchart outlining the verification of function limit at point a.

Note that point a may not even belong to the domain of the function, but it can never be an isolated point. For instance, $x = 0$ does not belong to the domain of function, $y(x) = \sin(x)/x$, as in Figure 3.1, but the limit at point 0 exists and it is 1 at that point.

Example

Let it be $y(x) = -2x + 3$, and the sequence $X_n = 1-(1/2)^n$.

The sequence X_n corresponds to the sum of a constant, 1, with an infinitely small sequence of oscillating exponential; thus sequence X_n approaches 1, and the X_n terms satisfy the inequality:

$$X_n \to 1 \Rightarrow |X_n - 1| < r_x$$

Assuming sequence $Y(X_n)$ is also approaching point 1, that is, 1 is the limit of sequence Y_n if its terms are in the neighborhood r_y of 1:

$$|Y_n - 1| < r_y$$

For both the sequences X_n and Y_n on the respective limits, the two neighborhood radius, r_x and r_y, are thus connected (Figure 3.4):

n	Xn	Y
1	0.500000	2.000000
2	0.750000	1.500000
3	0.875000	1.250000
4	0.937500	1.125000
5	0.968750	1.062500
6	0.984375	1.031250
7	0.992188	1.015625
8	0.996094	1.007813
9	0.998047	1.003906
10	0.999023	1.001953

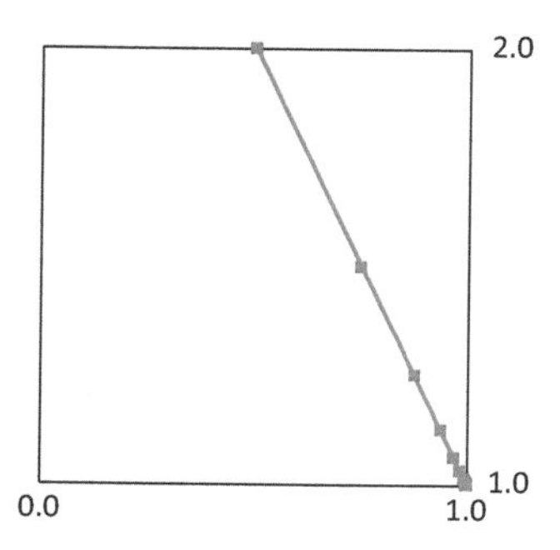

FIGURE 3.4 Table and graphic for limit of function, $y(x) = -2x + 3$, at point $x = 1$.

$$X_n \to 1 \Rightarrow Y_n \to 1, \text{ or}$$

$$|X_n - 1| < r_x \Rightarrow |Y_n - 1| < r_y$$

Substituting the relation for Y_n, and manipulating the terms in the absolute value:

$$| Y_n - 1| < r_y$$

$$|(-2x + 3) - 1| < r_y \Leftrightarrow |-2x + 2| < r_y \Leftrightarrow |2x - 2| < r_y \Leftrightarrow 2|x - 1| < r_y$$

$$\Leftrightarrow |x - 1| < \frac{r_y}{2}$$

That is,

$$-\frac{r_y}{2} < x - 1 < \frac{r_y}{2}$$

Noting that,

$$-r_x < x - 1 < r_x$$

then the neighborhood, r_x, is selected in such a way to satisfy

$$-\frac{r_y}{2} < -r_x < x - 1 < r_x < \frac{r_y}{2}$$

Finally, a relation between r_x and r_y, $r_x(r_y)$, is obtained: $r_x < \frac{r_y}{2}$.

For each tiny value in the Y-neighborhood, r_y, the related X-neighborhood is obtained from relation, $r_x(r_y)$; after that, by evaluating both the typical relation, $p = p(r_x)$ and the integer, $n = \text{INT}(p) + 1$, the table with numerical values that allow the verification of such convergence relations follows:

Ry	Rx	p	n=INT(p)+1	Xn	Y	ABS (Yn-1)
0.1	0.050000	4.32	5	0.968750	1.062500	0.062500
0.01	0.005000	7.64	8	0.996094	1.007813	0.007813
0.001	0.000500	10.97	11	0.999512	1.000977	0.000977

Limit of Function at Infinite

The function $f(x)$ converges to the finite point, b, as x tends to infinite,

$$\lim_{x \to +\infty} f(x) = b$$

if, and only if, to any sequence X_n of terms belonging to the domain of $f(x)$ that tends to infinite, there corresponds a sequence of images $f(X_n)$ that converges to b.

$$X_n \to +\infty \Rightarrow Y_n \to b, \text{ or}$$

$$\lim X_n = +\infty \Rightarrow \lim Y_n = b$$

A relation between M and r_y, $M(r_y)$, is thus foreseen, and the flowchart in Figure 3.5 outlines the pathway to obtain it.

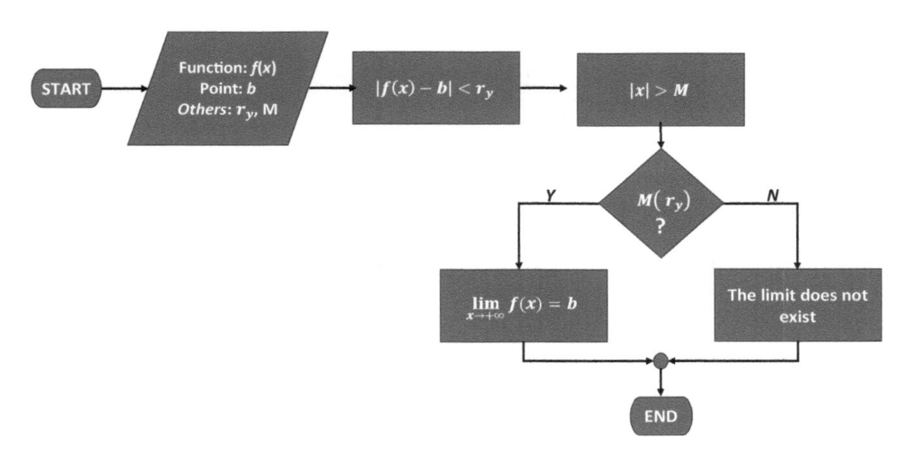

FIGURE 3.5 Flowchart outlining the verification of function limit at infinite.

$$|X_n| > M \Rightarrow |Y_n - b| < r_y$$

Example

At now, let it be $y(x) = 4 + 1/x$, and $X_n = n$

The limit for sequence X_n is infinite, it is increasing infinitely without any upper bound or major; the sequence X_n thus tends to infinite, and the X_n terms satisfy the inequality:

$$X_n \to +\infty \Rightarrow |X_n| > M$$

Assuming the sequence $Y(X_n)$ is converging to 4, that is, 4 is the limit of Y_n if its terms all are within the neighborhood, r_y, of 4:

$$|Y_n - 4| < r_y$$

The two neighborhood relations, for both the sequences X_n and Y_n on the associated limits, M and r_y, are thus connected (Figure 3.6):

$$X_n \to +\infty \Rightarrow Y_n \to 4$$

$$|X_n| > M \Rightarrow |Y_n - 4| < r_y$$

n	Xn	Y
1000	1000.00	4.001000
2000	2000.00	4.000500
3000	3000.00	4.000333
4000	4000.00	4.000250
5000	5000.00	4.000200
6000	6000.00	4.000167
7000	7000.00	4.000143
8000	8000.00	4.000125
9000	9000.00	4.000111
10000	10000.00	4.000100
11000	11000.00	4.000091
12000	12000.00	4.000083

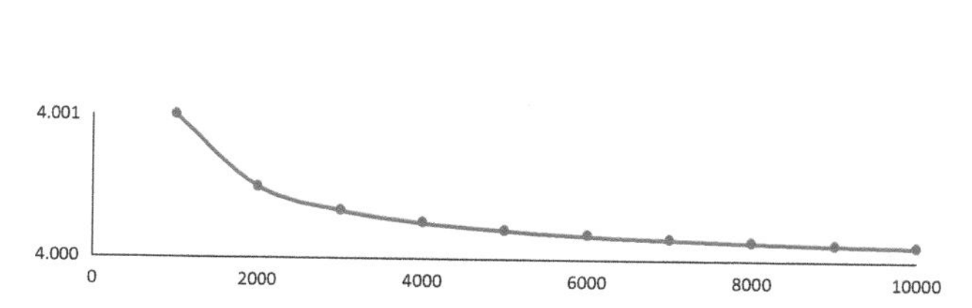

FIGURE 3.6 Table and graphic for limit of function, $y(x) = 4+1/x$, at infinite.

Substituting the relation for Y_n, cancelling the constants and neglecting the negative branch for the inequality (the sequence terms are all positive):

$$|Y_n - 4| < r_y$$

$$\left|\left(4 + \frac{1}{x}\right) - 4\right| < r_y \Leftrightarrow \left|\frac{1}{x}\right| < r_y$$

$$\frac{1}{r_y} < |x|, \text{ or also } |x| > \frac{1}{r_y}$$

Also noting that,

$$|X_n| > M$$

then the major, M, is selected in such a way to satisfy

$$|X_n| > M > \frac{1}{r_y}$$

Finally, a relation between M and r_y, $M(r_y)$, is thus obtained:

$$M > \frac{1}{r_y}$$

For each tiny radius in the Y-neighborhood, r_y, the related infinitely large terms, M, are obtained from that relation, $M(r_y)$; thereafter, by obtaining both the integer part in the typical relation, $p = p(\mathrm{M})$ and $n = \mathrm{INT}(p) + 1$, the numerical table allowing the verification of such relations follows:

Ry	M(Ry)	p	n=INT(p)+1	Xn	Y	ABS (Y-4)
0.1	10.0	10.0	**11**	11.0	4.090909	0.090909
0.01	100.0	100.0	**101**	101.0	4.009901	0.009901
0.001	1000.0	1000.0	**1001**	1001.0	4.000999	0.000999

Infinite Limit of a Function

The function $f(x)$ tends to infinite as x approaches a,

$$\lim_{x \to a} f(x) = +\infty$$

if, and only if, to any sequence X_n of terms belonging to the domain of $f(x)$ that approaches finite point, a, there corresponds a sequence of images $f(X_n)$ that tends to infinite (infinitely large).

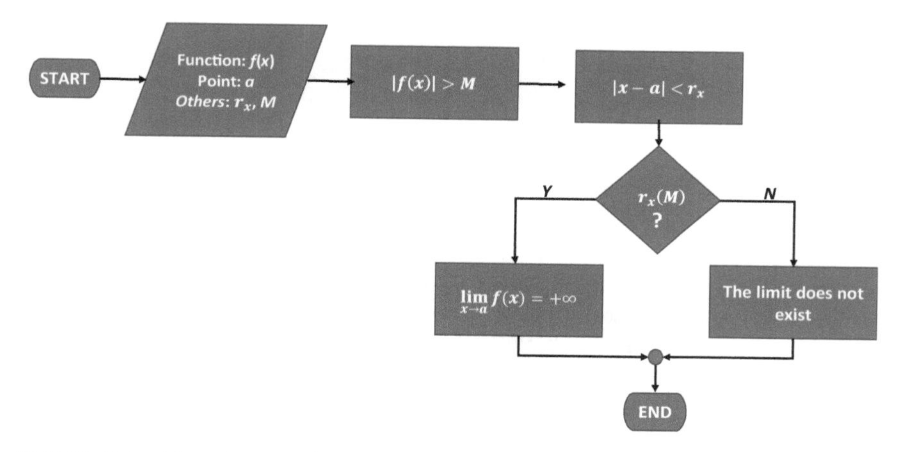

FIGURE 3.7 Flowchart outlining the verification of infinite limit for a function.

$$X_n \to a \Rightarrow Y_n \to +\infty, \text{ or}$$

$$\lim X_n = a \Rightarrow \lim Y_n = +\infty$$

A relation between r_x and M, $r_x(M)$, is thus foreseen, and the flowchart in Figure 3.7 outlines the pathway to obtain it.

$$|X_n - a| < r_x \Rightarrow |Y_n| > M$$

Example

And now, consider $y(x) = 1/(x-3)^2$, and $X_n = 3-1/n$.

The limit of sequence X_n is 3, because it corresponds to a constant, 3, minus an infinitesimal sequence; thus X_n approaches 3, and the X_n terms satisfy the inequality:

$$X_n \to 3 \Rightarrow |X_n - 3| < r_x$$

Assuming that sequence $Y(X_n)$ tends to $+\infty$, that is, Y_n is an infinitely large sequence if its terms are larger than M:

$$|Y_n| > M$$

The two neighborhood relations, for both the sequences X_n and Y_n on the associated limits, are thus connected (Figure 3.8):

n	Xn	Y
1000	2.999000	1000000.000000
2000	2.999500	3999999.999997
3000	2.999667	9000000.000002
4000	2.999750	16000000.000018
5000	2.999800	25000000.000006
6000	2.999833	35999999.999912
7000	2.999857	49000000.000098
8000	2.999875	63999999.999844
9000	2.999889	81000000.000234
10000	2.999900	99999999.999578

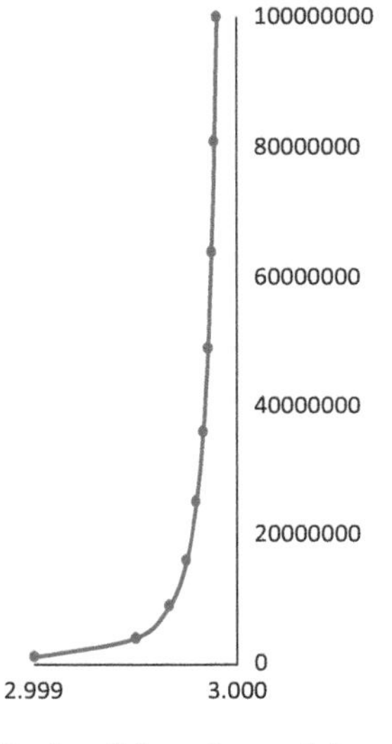

FIGURE 3.8 Table and graphic for infinite limit of function, $y(x) = 1/(x-3)^2$, at point $x = 3$.

$$X_n \to 3 \Rightarrow Y_n \to +\infty$$

$$|X_n - 3| < r_x \Rightarrow |Y_n| > M$$

By substituting the relation for Y_n, and manipulating the absolute value and square root inequalities:

$$|\,Y_n| > M$$

$$\left|\frac{1}{(x-3)^2}\right| > M \Leftrightarrow |(x-3)^2| < \frac{1}{M} \Leftrightarrow |x-3| < \sqrt{\frac{1}{M}}$$

That is,

$$-\sqrt{\frac{1}{M}} < x - 3 < \sqrt{\frac{1}{M}}$$

Also noting that,

$$-r_x < x - 3 < r_x$$

then the X-neighborhood at point 3 is selected in such a way to satisfy,

$$-\sqrt{\frac{1}{M}} < -r_x < x - 3 < r_x < \sqrt{\frac{1}{M}}$$

Finally, a relation between r_x and M, $r_x(M)$, is obtained:

$$r_x < \sqrt{\frac{1}{M}}$$

For each infinitely large value, M, the related X-neighborhood is obtained from relation, $r_x(M)$; after that, by evaluating both the sequence relation, $p = p(r_x)$, and integer $n = \text{INT}(p) + 1$, the numerical table that allows verification follows too:

M	Rx	P	n=INT(p)+1	Xn	Y	ABS(Y) - M
1000.0	0.031623	31.62	**32**	2.968750	1024.00	24.00
10000.0	0.010000	100.00	**101**	2.990099	10201.00	201.00
100000.0	0.003162	316.23	**317**	2.996845	100489.00	489.00

Infinitesimal Limit of a Function

The function $f(x)$ is an infinitesimal (infinitely small)

$$\lim_{x \to a} f(x) = 0$$

if, and only if, to any sequence X_n of terms belonging to the domain of $f(x)$ that approaches a, finite or infinite, there corresponds a sequence of images $f(X_n)$ that tends to zero.

$$X_n \to a \Rightarrow Y_n \to 0$$

$$\lim X_n = a \Rightarrow \lim Y_n = 0$$

$$|X_n - a| < r_x \Rightarrow |Y_n - 0| < r_y$$

Example

And now, consider $y(x) = (x-3)^2$, and $X_n = 3-1/n$.

The limit of sequence X_n is again 3, it is the same sequence as in the prior example; in fact, X_n approaches 3, and the X_n terms satisfy the inequality too:

$$X_n \rightarrow 3 \Rightarrow |X_n - 3| < r_x$$

Assuming the sequence $Y(X_n)$ is converging to zero, that is, zero is the limit for the sequence Y_n if its terms are all in the neighborhood r_y of 0:

$$|Y_n - 0| < r_y$$

The two neighborhood relations are thus connected, that is, the sequences X_n and Y_n on the associated limits and the neighborhood parameters, r_x and r_y (Figure 3.9):

$$X_n \rightarrow 3 \Rightarrow Y_n \rightarrow 0$$

$$|X_n - 3| < r_x \Rightarrow |Y_n - 0| < r_y$$

n	Xn	Y
10	2.900000	0.01000000
20	2.950000	0.00250000
100	2.990000	0.00010000
200	2.995000	0.00002500
300	2.996667	0.00001111
400	2.997500	0.00000625
500	2.998000	0.00000400
600	2.998333	0.00000278
700	2.998571	0.00000204
800	2.998750	0.00000156
900	2.998889	0.00000123
1000	2.999000	0.00000100

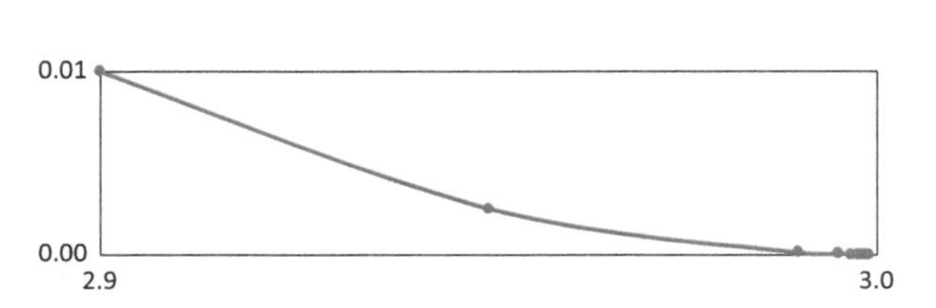

FIGURE 3.9 Table and graphic for infinitesimal limit of function, $y(x) = (x-3)^2$, at point 3.

By substituting the relation for Y_n, neglecting the null term, and manipulating the absolute value and the square root, it follows:

$$|Y_n - 0| < r_y$$

$$|(x - 3)^2 - 0| < r_y \Leftrightarrow |(x - 3)^2| < r_y \Leftrightarrow |x - 3| < \sqrt{r_y}$$

That is,

$$-\sqrt{r_y} < x - 3 < \sqrt{r_y}$$

Also noting that,

$$-r_x < x - 3 < r_x$$

then the neighborhood at point 3 is selected in such a way to satisfy

$$-\sqrt{r_y} < -r_x < x - 3 < r_x < \sqrt{r_y}$$

Finally, a relation between r_x and r_y, $r_x(r_y)$, is obtained:

$$r_x < \sqrt{r_y}$$

Again, for each tiny Y_n neighborhood, r_y, the related X_n neighborhood is obtained from this relation, $r_x(r_y)$; then, by evaluating both the sequence relation, $p = p(r_x)$, and the integer $n = \text{INT}(p) + 1$, the table that allows the convergence's verification follows too:

Ry	Rx	p	n=INT(p)+1	Xn	Y	ABS (Yn-0)
0.1	0.316228	3.16	**4**	2.750000	0.062500	0.062500
0.01	0.100000	10.00	**11**	2.909091	0.008264	0.008264
0.001	0.031623	31.62	**32**	2.968750	0.000977	0.000977

3.3 LATERAL LIMITS AT A POINT—THE EXTENDED COST FUNCTION

The lateral limits at a point are focused in this section, and both the left-side limit and the right-side limit are detailed. In particular, the referred lateral limits are detailed for the (negative) point where the extended cost function is zero: the variable independent, x, is allowed to take any value except x = 0, being such values either positive or negative. Since the calculated root is negative, there is no physical or realistic meaning for such root.

Giving the volume, V = 800 liters, the cost function depends only on diameter d,

$$C(1, d) = C(d) = 100\left(\frac{\pi}{2}d^2 + \frac{3.2}{d}\right)$$

The calculation of the function's root assumes the extension to the negative domain, as follows:

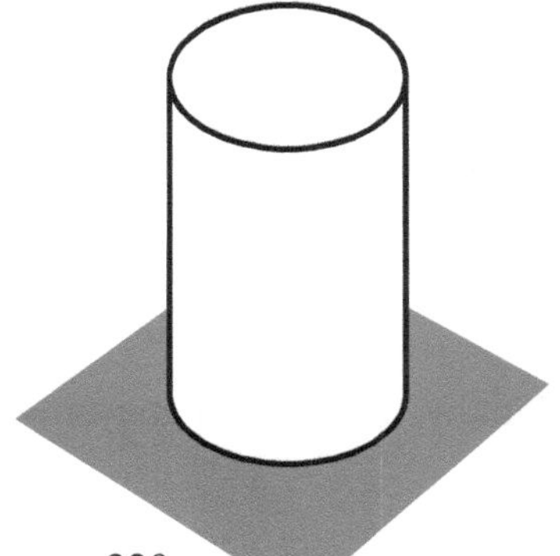

$$y(x) = 100\left(\frac{\pi}{2}x^2 + \frac{3.2}{x}\right), \forall\, x \neq 0$$

The root on the extended domain for variable, x,

$$50\pi x^2 + \frac{320}{x} = 0 <=> 50\pi x^2 = -\frac{320}{x} \Leftrightarrow x^3 = -\frac{320}{50\pi}$$

However, such root does not present a physical meaning, since negative sizes are not realistic:

$$x = -\sqrt[3]{\frac{32}{5\pi}} \approx -1.267681\ldots$$

Left-Side Limit—It is said that $\lim_{x \to a^-} f(x) = b$ if, and only if, to any sequence X_n that approaches a, for values smaller than a, there corresponds a sequence of images $f(X_n)$ that tends to b. In this way, b is the function's limit when x approaches point a, by values to the left of a (Figure 3.10).

Right-Side Limit—It is also said that $\lim_{x \to a^+} f(x) = b$ if, and only if, to any sequence X_n that approaches a, for domain values greater than a, there corresponds a sequence of images $f(X_n)$ that tends to b. In this way, b is the function's limit when x approaches a, by values to the right of point a (Figure 3.11).

The Equality of Lateral Limits—In addition, function $f(x)$ has limit at point a, if and only if, the two lateral limits (to the left side, and to the right side) exist and are equal:

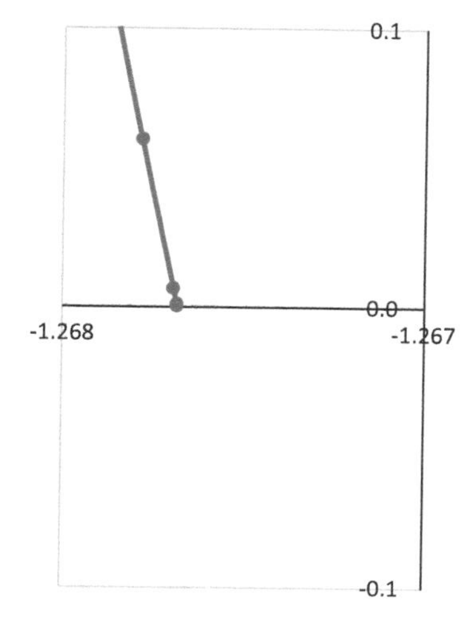

n	Xn	Y
1	-1.367681153510	59.852918046
2	-1.277681153510	5.973929640
3	-1.268681153510	0.597380794
4	-1.267781153510	0.059738067
5	-1.267691153510	0.005973807
6	-1.267682153510	0.000597381
7	-1.267681253510	0.000059738
8	-1.267681163510	0.000005974
9	-1.267681154510	0.000000597
10	-1.267681153610	0.000000060

FIGURE 3.10 Table and graphic for lateral limit: the left side of function root.

n	Xn	Y
1	-1.167681153510	-59.872589694
2	-1.257681153510	-5.973931595
3	-1.266681153510	-0.597380794
4	-1.267581153510	-0.059738067
5	-1.267671153510	-0.005973807
6	-1.267680153510	-0.000597381
7	-1.267681053510	-0.000059738
8	-1.267681143510	-0.000005974
9	-1.267681152510	-0.000000597
10	-1.267681153410	-0.000000060

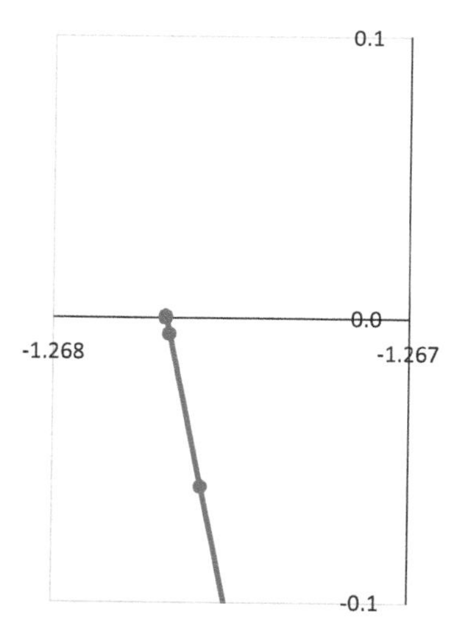

FIGURE 3.11 Table and graphic for lateral limit: the right side of function root.

$$\lim_{x \to a} f(x) = b \Leftrightarrow \lim_{x \to a^-} f(x) = \lim_{x \to a^+} f(x) = b$$

Note that, at a given point, a:

- The function limit is the common value of these two lateral limits;
- If the two lateral limits are different, the function has no limit at that point.

Example

At now, consider again, $y(x) = 50\pi x^2 + \frac{320}{x}$, and $x \to -\sqrt[3]{32/5\pi}$.

The limit of sequence X_n is again the cubic root, $a = -\sqrt[3]{32/5\pi}$, it is the same sequence as in prior instances; in fact, when X_n approaches point a, the X_n terms satisfy the inequality:

$$X_n \to -\sqrt[3]{32/5\pi} \Rightarrow |X_n - (-\sqrt[3]{32/5\pi})| < r_x$$

Assuming sequence $Y(X_n)$ is converging to zero, that is, zero is the limit for sequence Y_n if its terms are all in the neighborhood r_y of 0:

$$|Y_n - 0| < r_y$$

The two neighborhood relations are thus connected, that is, the converging sequences X_n and Y_n and the neighborhood parameters, r_x and r_y, shall satisfy (Figure 3.12):

$$X_n \to -\sqrt[3]{32/5\pi} \Rightarrow Y_n \to 0$$

n	Xn	Y
3	-1.392681153510	74.892875473
6	-1.252056153510	-9.334551548
9	-1.269634278510	1.166760043
12	-1.267437012885	-0.145844892
15	-1.267711671088	0.018230611
18	-1.267677338813	-0.002278826
21	-1.267681630347	0.000284853
24	-1.267681093905	-0.000035607
27	-1.267681160960	0.000004451
30	-1.267681152578	-0.000000556

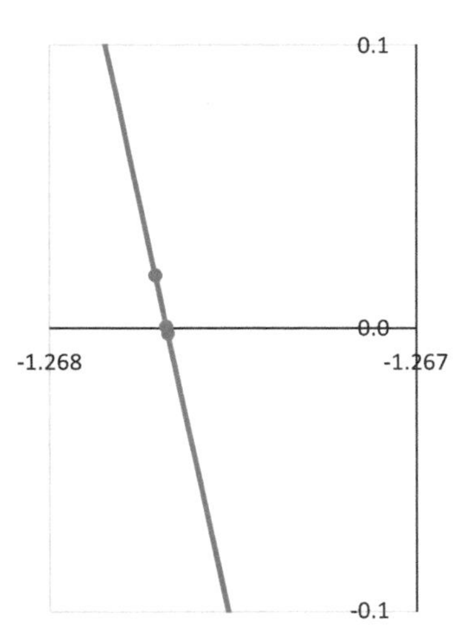

FIGURE 3.12 Table and graphic for equal lateral limits, in both sides of function root.

$$|x - (-\sqrt[3]{32/5\pi})| < r_x \Rightarrow \left|\left(50\pi x^2 + \frac{320}{x}\right) - 0\right| < r_y$$

A variable substitution is useful, namely, making $h = x - (-\sqrt[3]{32/5\pi})$ or $h = x - a$; then parameter h tends to 0 when variable x approaches point $a = -\sqrt[3]{32/5\pi}$:

$$x \to a \Rightarrow h \to 0$$

$$x = h + a = h - \sqrt[3]{32/5\pi}$$

In this way, replacing variable x by parameter h, and updating the prior inequalities,

$$|h| < r_x \Rightarrow \left|50\pi\,(h + a)^2 + \frac{320}{(h + a)} - 0\right| < r_y$$

Manipulating the absolute value in the second inequality and grouping the similar terms,

$$\left|50\pi\,(h + a)^2 + \frac{320}{(h + a)}\right| < r_y \Leftrightarrow \left|\frac{50\pi\,(h + a)^3 + 320}{(h + a)}\right| < r_y$$

$$\Leftrightarrow \left|\frac{50\pi\,(h^3 + 3h^2a + 3ha^2 + a^3) + 320}{(h + a)}\right| < r_y,$$

By substituting the cubic root expression, $a = -\sqrt[3]{32/5\pi}$,

$$\left|\frac{50\pi\,h^3 - 150\pi\,h^2\sqrt[3]{32/5\pi} + 150\pi\,h(\sqrt[3]{32/5\pi})^2 - 50\pi\,(\sqrt[3]{32/5\pi})^3 + 320}{h - \sqrt[3]{32/5\pi}}\right| < r_y$$

Noting the two last terms in numerator cancel,

$$-50\pi\,(\sqrt[3]{32/5\pi})^3 + 320 = -50\pi\frac{32}{5\pi} + 320 = -10 \times 32 + 320 = 0$$

$$\left| \frac{50\pi\ h^3 - 150\pi\ h^2 \sqrt[3]{\frac{32}{5\pi}} + 150\pi\ h\left(\sqrt[3]{\frac{32}{5\pi}}\right)^2}{h - \sqrt[3]{\frac{32}{5\pi}}} \right| < r_y$$

$$\Leftrightarrow \left| \frac{h^3 - 3\ h^2 \sqrt[3]{\frac{32}{5\pi}} + 3\ h\left(\sqrt[3]{\frac{32}{5\pi}}\right)^2}{h - \sqrt[3]{\frac{32}{5\pi}}} \right| < \frac{r_y}{50\pi}$$

and factoring parameter h,

$$\left| h.\frac{h^2 - 3\ h\ \sqrt[3]{32/5\pi} + 3(\sqrt[3]{32/5\pi})^2}{h - \sqrt[3]{32/5\pi}} \right| < \frac{r_y}{50\pi}$$

For every positive h, the second-order polynomial in numerator does not present real roots, and it takes only positive values; when parameter h is decreasing and decreasing to zero, the associated terms can be neglected in comparison with the constant terms, and

$$\left| h.\frac{0 - 0 + 3(\sqrt[3]{32/5\pi})^2}{0 - \sqrt[3]{32/5\pi}} \right| < \frac{r_y}{50\pi} \Leftrightarrow \left| h.\frac{3(\sqrt[3]{32/5\pi})^2}{\sqrt[3]{32/5\pi}} \right| < \frac{r_y}{50\pi}$$

That is,

$$|3h\ \sqrt[3]{32/5\pi}| < \frac{r_y}{50\pi} \Leftrightarrow |h| < \frac{r_y}{150\pi\ \sqrt[3]{32/5\pi}}$$

Thus, the neighborhood at point a is selected in such a way to satisfy,

$$|h| < r_x < \frac{r_y}{150 \times 2\ \sqrt[3]{\frac{4\pi^3}{5\pi}}} < \frac{r_y}{300\ \sqrt[3]{\frac{4}{5}}\ \sqrt[3]{\pi^2}}$$

Finally, a relation between r_x and r_y is obtained:

$$r_x < \frac{r_y}{300\ \sqrt[3]{\pi^2}}$$

For each tiny value for the Y-neighborhood, r_y, the related X-neighborhood is obtained from that relation, $r_x(r_y)$; thereafter, by obtaining both the sequence relation, $p = p(r_x)$, and the integer, $n = \text{INT}(p) + 1$, the table allowing the verification of such convergence relations follows:

Ry	Rx	p	n=INT(p)+1	Xn	Y	ABS (Yn-0)
0.1	0.000155	12.65	**13**	−1.267803	0.072922	0.072922
0.01	0.000016	15.97	**16**	−1.267666	−0.009115	0.009115
0.001	0.000002	19.30	**20**	−1.267680	−0.000570	0.000570
0.0001	0.000000	22.62	**23**	−1.267681	0.000071	0.000071

3.4 PROPERTIES OF FUNCTION LIMITS

In this section, important theorems about function limits are revisited, together with the operations rules on function limits. It includes the limit of trapped functions, the non-negative limit for a non-negative function, or the superiority relation between two function limits, in an analogous manner with the properties for sequence limits. In fact, function limits present properties and rules similar to those for the limits of sequences, at now assuming the dependent variable, y, is being treated as a sequence of real numbers.

These properties are analogous to those for limits of real number sequences, namely, they are based on arithmetic operations and subsequent results of sequences converging to a point, and then generalized to the entire domain. Despite the importance of rigorous habits and associated proofs, the computational approach in this text makes the formal manipulations being neglected for this topic.

Operations on Function Limits—Typically, the function limit of the operation corresponds to the operation on the function limits (e.g., sum and subtraction, multiplication and division):

$$\lim_{x \to a} (f + g)(x) = \lim_{x \to a} f(x) + \lim_{x \to a} g(x)$$

$$\lim_{x \to a} (f \,.\, g)(x) = \lim_{x \to a} f(x) \,.\, \lim_{x \to a} g(x)$$

$$\lim_{x \to a} (f/g)(x) = \frac{\lim_{x \to a} f(x)}{\lim_{x \to a} g(x)}$$

- Note the arithmetic operations on limits of functions, at a given point a, follow similar rules to the operations on limits of real number sequences. Once again, it is assumed all the operations are possible, that is, all the limits exist and the limit on the quotient's denominator is not zero.

- The limit of power function corresponds to the power of limit; in the same way, the limit of the root function (rational power) can be obtained through the root of the limit; of course, if the root operation is possible, because if the root index is even, the root object must be positive.

$$\lim_{x \to a} [f(x)]^p = \left[\lim_{x \to a} f(x)\right]^p, p \in N$$

$$\lim_{x \to a} \sqrt[p]{f(x)} = \sqrt[p]{\lim_{x \to a} f(x)}$$

If a Function Has a Limit at a Point, That Limit Is Unique (Uniqueness)
The uniqueness property also holds for function limits, and the contradiction approach is very useful once more; in fact, when assuming a second limit different from the previous one could exist, by manipulation a false expression will occur.

The Limit of a Constant Function Is the Constant Itself, That Is, $\lim_{x \to a} (k) = k$
The limit of a constant function results from the limit definition, it always holds when the constant k is directly substituted in the expression: $|Y_n - k| = 0 < r_y$, for every positive radius, r_y.

If the Function Is Monotonously Increasing and Upper-Bounded, Then It Converges to the Supreme of Its Values

$$\left.\begin{array}{l} f(x) \leq M \\ \text{increasing } f(x) \end{array}\right\} \Rightarrow \lim_{x \to a} f(x) = b \leq M$$

The convergence of a monotonous and bounded function, by analogy with the sequence of a bounded and monotonous variable, $Y_n = f(X_n)$, also assumes the opposite: if the function is monotonous decreasing and bounded from below, then it converges to the infimum of its terms.

The Limit of a Non-negative Function Is Also Non-negative

$$\left.\begin{array}{l} f(x) \geq 0, \quad x \to a \\ \lim\limits_{x \to a} f(x) = b \end{array}\right\} \Rightarrow b \geq 0$$

The non-negative property also holds for the limit of non-negative functions, and the contradiction approach works well once again; in fact, when assuming a negative limit could exist, $b < 0$, for the non-negative variable, $Y_n = f(X_n) \geq 0$, by manipulation a false expression will occur since the function at hand cannot take negative values.

Superiority Relationship between Limits

$$f(x) \geq g(x), \ x \to a \Rightarrow \lim_{x \to a} f(x) \geq \lim_{x \to a} g(x)$$

The current result directly follows from the last one; thus, by defining the difference between the two convergent functions at point a, a non-negative function occurs, $h(x) = f(x) - g(x) \geq 0$; in that way, this difference function and the dependent variable will be non-negative, $Y_n \geq 0$; and so it will be the related limit, as well as the superiority relationship holds on.

3.4.1 The Limit of Trapped Functions - A Numerical Conjecture

$$\left.\begin{array}{l} u(x) \leq f(x) \leq v(x) \\ \lim\limits_{x \to a} u(x) = \lim\limits_{x \to a} v(x) = b \end{array}\right\} \Rightarrow \lim_{x \to a} f(x) = b$$

This property for function limits is similar to those for the limits of sequences, at now assuming the dependent variables on the functions at hand are treated as sequences of real numbers, while the sequence of independent variable, X_n, approaches point a.

Example

For positive values in the zero neighborhood, $x > 0$, the following estimates are assumed:

$$x - \frac{x^3}{6} < \sin(x) < x$$

More advanced approaches will be useful to demonstrate these superiority relations (e.g., Chapter 10), however the table and graphic in Figure 3.13 allow such numerical conjecture. In addition, when x approaches zero,

$$\lim_{x \to 0}(x) = \lim_{x \to 0}(x - x^3/3) = 0$$

because the limit of sum and multiplication corresponds to the sum and multiplication of the associated limits. All the function terms also are in the neighborhood of zero, and then

$$\lim_{x \to 0} \sin(x) = 0$$

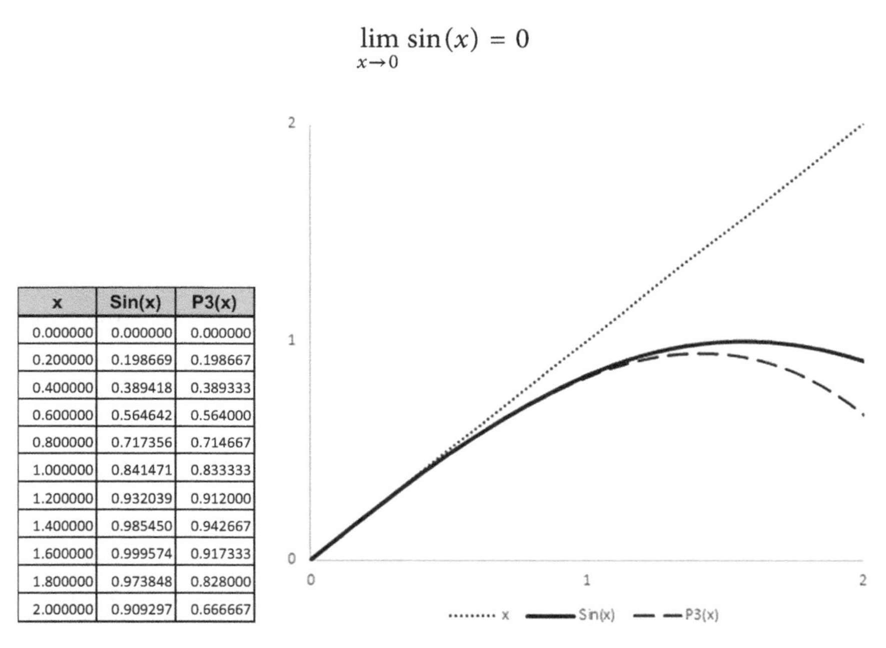

x	Sin(x)	P3(x)
0.000000	0.000000	0.000000
0.200000	0.198669	0.198667
0.400000	0.389418	0.389333
0.600000	0.564642	0.564000
0.800000	0.717356	0.714667
1.000000	0.841471	0.833333
1.200000	0.932039	0.912000
1.400000	0.985450	0.942667
1.600000	0.999574	0.917333
1.800000	0.973848	0.828000
2.000000	0.909297	0.666667

FIGURE 3.13 The trigonometric function, $f(x) = \sin(x)$, trapped in-between the linear function, $u(x) = x$, and third-order polynomial, $P_3(x) = x - x^3/3$.

3.5 REMARKABLE LIMITS

In this section, some remarkable limits at point zero are presented, since the related results are widely applied in a number of different developments. Namely, for some functions that are not defined in that point, but the lateral limits exist and take the same value, the unit 1. In fact, the limit of sine function at zero tends to be equal to x, and the related ratio becomes, 1, even if their function is not defined at that point, $x = 0$.

Other remarkable limits are indicated too; such results are widely applied in a number of different developments. Namely,

$$\lim_{x \to 0} \frac{e^x - 1}{x} = 1$$

$$\lim_{x \to 0} \frac{\ln |x + 1|}{x} = 1$$

3.5.1 The Limit of Function, $y(x) = \sin(x)/x$, When x Approaches Zero

Let it be $y(x) = \sin(x)/x$, and the sequence $X_n = -1/\text{n}$. The table and graphic of both the function and the negative sequence X_n follow in Figure 3.14.

The sequence X_n corresponds to an infinitely small sequence with negative terms; thus sequence X_n approaches zero, that is, the X_n terms satisfy the inequality:

$$\lim X_n = 0 \Rightarrow |X_n - 0| < r_x$$

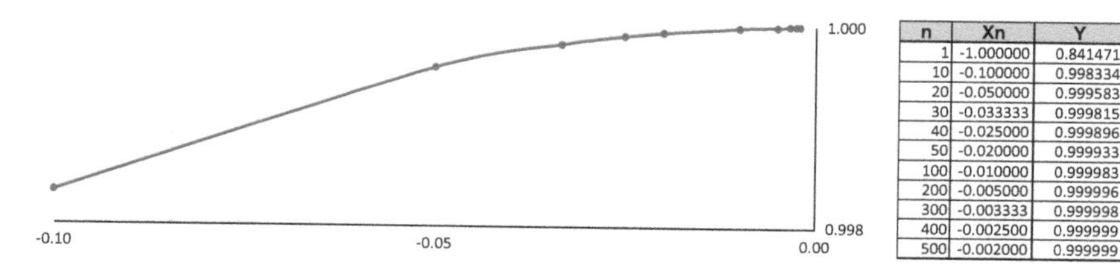

n	Xn	Y
1	-1.000000	0.841471
10	-0.100000	0.998334
20	-0.050000	0.999583
30	-0.033333	0.999815
40	-0.025000	0.999896
50	-0.020000	0.999933
100	-0.010000	0.999983
200	-0.005000	0.999996
300	-0.003333	0.999998
400	-0.002500	0.999999
500	-0.002000	0.999999

FIGURE 3.14 Lateral limit of function, $y(x) = \sin(x)/x$, at the left side of $x = 0$.

Assuming that sequence $Y(X_n)$ is converging to 1, that is, 1 is the limit of sequence Y_n if its terms are in the neighborhood r_y of 1:

$$\left| \frac{\sin(x)}{x} - 1 \right| < r_y$$

The two neighborhood relations, for both the sequences X_n and Y_n on the associated limits, are thus connected:

$$X_n \to 0 \Rightarrow \frac{\sin(x)}{x} \to 1$$

$$|X_n - 0| < r_x \Rightarrow \left|\frac{\sin(x)}{x} - 1\right| < r_y$$

Similarly, let the sequence be $X_n = 1/\text{n}$; the table and graphic of both the function and the positive sequence X_n follow too (Figure 3.15).

n	Xn	Y
1	1.000000	0.841471
10	0.100000	0.998334
20	0.050000	0.999583
30	0.033333	0.999815
40	0.025000	0.999896
50	0.020000	0.999933
100	0.010000	0.999983
200	0.005000	0.999996
300	0.003333	0.999998
400	0.002500	0.999999
500	0.002000	0.999999

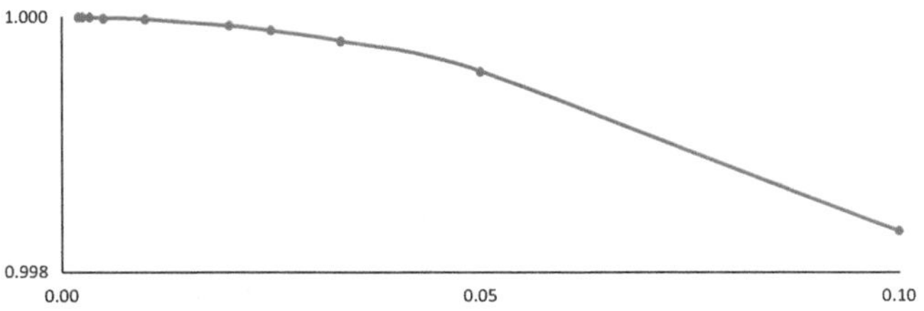

1.000
0.998
0.00
0.05
0.10

FIGURE 3.15 Lateral limit of function, $y(x) = \sin(x)/x$, at the right side of $x = 0$.

For positive values in the zero neighborhood, $x > 0$, the following relations (section 3.4.1) are assumed:

$$x - \frac{x^3}{6} < \sin(x) < x$$

$$x\left(1 - \frac{x^2}{6}\right) < \sin(x) < x$$

Dividing all terms by x,

$$1 - \frac{x^2}{6} < \frac{\sin(x)}{x} < 1$$

Subtracting 1 to all the terms,

$$-\frac{x^2}{6} < \frac{\sin(x)}{x} - 1 < 0$$

The following relation thus applies too,

$$-\frac{x^2}{6} < \frac{\sin(x)}{x} - 1 < 0 < \frac{x^2}{6}$$

Then the absolute value is verified,

$$\left|\frac{\sin(x)}{x} - 1\right| < \frac{x^2}{6}$$

And also assuming that

$$\left|\frac{\sin(x)}{x} - 1\right| < \frac{x^2}{6} < r_y$$

Or better,

$$\frac{x^2}{6} < r_y \Leftrightarrow x^2 < 6r_y \Leftrightarrow x < \sqrt{6r_y}$$

Also noting that,

$$-r_x < x - 0 < r_x$$

then the neighborhood, r_x, is selected in such a way to satisfy

$$-\sqrt{6r_y} < -r_x < x < r_x < \sqrt{6r_y}$$

Finally, a relation between r_x and r_y, $r_x(r_y)$, is obtained:

$$r_x < \sqrt{6r_y}$$

Once again, for each infinitesimal value on the Y-neighborhood, r_y, the related X-neighborhood is obtained from the relation, $r_x(r_y)$; then, by evaluating both the sequence relation, $p = p(r_x)$, and the integer, $n = \text{INT}(p) + 1$, the table that allows the verification of such relations follows:

Ry	Rx	p	n=INT(p)+1	Xn	Y	ABS (Yn-1)
0.1	0.774597	1.29	2	0.500000	0.958851	0.041149
0.01	0.244949	4.08	5	0.200000	0.993347	0.006653
0.001	0.077460	12.91	13	0.076923	0.999014	0.000986

3.6 CONCLUDING REMARKS

A generalization approach is initiated with the limit of a function at a point, which is based on the numbers sequence of dependent variable, $y(x)$; and the limit of a function at infinite is treated when the independent variable, x, increases without upper bound. From other side, the infinite limit of a function occurs when the dependent variable, y, increases without bound, while the infinitesimal limit $I(x)$ occurs at when the function is infinitely small and tends to zero. The properties of the infinitely small are referred too, due to the relevant number of applications.

- Diverse limits for the cost function are discussed too, searching for a relation between the convergence radius of both the variables, the independent variable x and the dependent variable y: that is, $r_x(r_y)$. In particular, the lateral limits are treated at the point where the cost function is zero: the function's root is negative, and no physical or real meaning for the negative diameter is expected; the calculations were difficult, but the continuity attributes allow the root location.
- Important theorems about function limits are revisited, so as the operations rules. By analogy, the properties and rules for function limits are similar to those for numbers sequence, but now the sequence of dependent variable values, $Y(Xn)$, is focused.
- Some remarkable limits at point $x = 0$ are presented, since the related results are widely applied in a number of different developments. In particular, the limit of the sine function at zero tends to be equal as limit of x, and the related ratio becomes 1, even if the ratio is not defined at that point. This remarkable limit can be also verified graphically or numerically, and the trapped theorem is recalled to frame the limit at zero of function, $y(x) = \sin(x)/x$ —L'Hôpital Rule (Chapter 6) cannot be used due to circular reasoning.

Note the equality of lateral limits and the function image at a given point corresponds to the function continuity in such point. Thus, limits calculation involving continuous functions can be solved in a straightforward manner, by directly substituting the point at hand in the function argument, since the function image and the function limit are equal at that point. In this sense, the next chapter addresses continuity, the associated properties and operations about continuous functions, and it treats important results both about extreme points and intermediate values in a closed range.

CHAPTER 4

Continuity

This chapter addresses functions continuity, including the continuity of a function at a point, a, by treating both the continuity on the left side and the right side of that point, as well as continuity in a range. Continuity is very important in limits evaluation for elementary functions; in this case, do substitute the point under analysis at the function argument, and the limit calculation follows directly. In addition, the main properties and operations with continuous functions in a closed interval are focused on, along with important results with a broad spectrum of applications; namely the intermediate values corollary for location of equations roots. Numerical instances illustrate the referred topics about functions continuity, including the bisection method to locate the barrel diameter for a given cost.

4.1 INTRODUCTION

The support of continuity and related properties is instrumental to the study of real functions, either to analyze the existence of extreme points or to locate roots of continuous functions. The operations with continuous functions are based in the operations about limits, because continuity at a point requires the equality of lateral limits and the existing image at that point. However, in such cases, remarkable limits occur in points of discontinuity, as presented in the next section.

Continuity topics are also adequate to strength verification and control practices, including graphics comparison in both open and closed intervals, the computational estimates and associated errors, or by experimenting iterative procedures such as the bisection method.

In general, function continuity and associated attributes are supporting the study of function derivatives, and symbolic manipulations with two different variables or working parameters (or the independent variable, x; or the alteration or increment parameter, h) are useful.

- Therefore, the continuity definition and related procedures are again inviting instances verification: the procedure assumes the

DOI: 10.1201/9781003461876-4

continuity proofs using definition, and then confirming the result via the operating rules for continuous functions.

Obviously, the necessary condition requires the referred verification; however, new advances are not expected from such procedure, due to the circular reasoning involved.

- The recognition of binomial structures in the continuity proofs of power functions, typically drives polynomials with coefficients obtained by combinatory (or the Pascal Triangle), and similar tasks will occur in derivation procedures.

In another point of view, the practical approach to the Weierstrass theorem of extreme values will be enlarged for advanced Optimization approaches (e.g., multivariable functions). And the rest of this chapter is as follows: the continuity of a function at a point is treated in the initial step; then, continuity in a range is analyzed, by both considering the interior points and the lateral limits in the range border points; in addition, the main properties about continuous functions are addressed, together with the typical operations with continuous functions; after that, important results about continuous functions on a closed range are addressed, namely the Weierstrass theorem and Bolzano theorem, while roots location of continuous function is discussed too; after that, the important application of obtaining equation roots is detailed.

4.2 CONTINUITY AT A POINT

The continuity of a function at a point is treated in this section, it considers both the continuity on the left side and the continuity on the right side of the point under study. In addition, continuity assumes a very important role in the calculation of limits for elementary function, by directly substituting the point at hand in the function argument to obtain the final result.

Continuity of a Function at a Point

A function is said to be continuous at point, $x = a$, on its domain, if and only if,

$$\lim_{x \to a} f(x) = f(a)$$

Or also,

$$\lim_{x \to a^-} f(x) = \lim_{x \to a^+} f(x) = f(a)$$

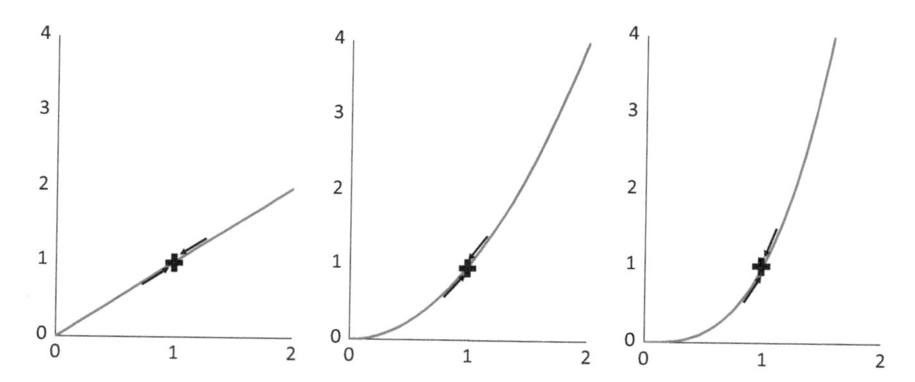

FIGURE 4.1 Continuous functions at point $x = 1$.

If a function is simultaneously left and right continuous at point, a, then it is continuous at that point. The graphics in Figure 4.1 illustrate the continuity of several power functions at $x = 1$, namely, by representing from left to right: (a) $\lim_{x \to 1} x = 1$; (b) $\lim_{x \to 1} x^2 = 1$; and (c) $\lim_{x \to 1} x^3 = 1$.

In other words, since the continuous function takes the image value at the point at hand, then the infinitesimal difference at that point is zero,

$$\lim_{x \to a} [y(x) - y(a)] = 0 \Leftrightarrow \lim_{x \to a} \Delta y(x) = 0$$

and the flowchart in Figure 4.2 outlines a pathway to verify and confirm it.

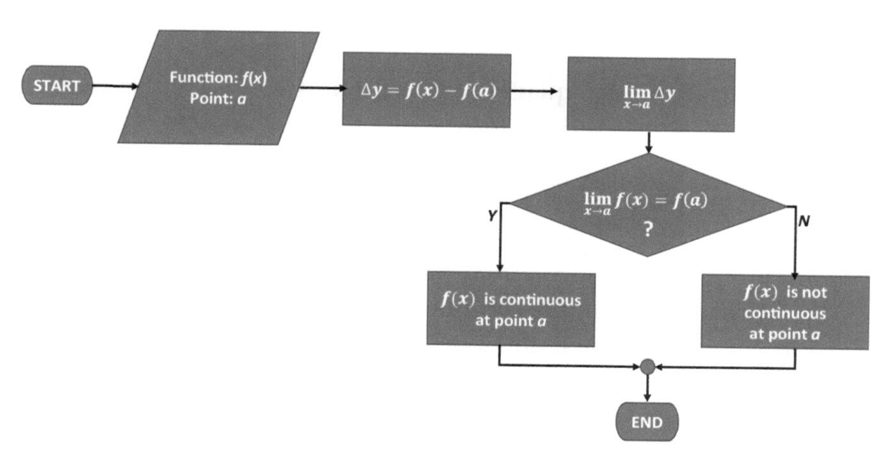

FIGURE 4.2 Flowchart outlining the verification of function continuity at point a.

Examples—Power Functions

- Let it be the **linear function** $y(x) = x$; then the relation $\lim_{x \to a} \Delta y(x) = 0$ takes the form,

$$\lim_{x \to a} y(x) = y(a)$$

In other words, the limits for both the variables are connected:

$$x \to a \Rightarrow y(x) \to y(a)$$

$$|x - a| < r_x \Rightarrow |y(x) - y(a)| < r_y$$

By substituting the function $y(x)$ in the right side's relation,

$$|x - a| < r_y$$

The neighborhood r_x is defined in such a way that

$$|x - a| < r_x < r_y$$

The obtained relation $r_x(r_y)$ is confirming the function's continuity at any point a, that is,

$$\lim_{x \to a} (x - a) = 0 \Leftrightarrow \lim_{x \to a} x = a, \quad \forall\, a$$

- And let it be the **quadratic function,** $y(x) = x^2$; the relation $\lim_{x \to a} \Delta y(x) = 0$ is verified too,

$$\lim_{x \to a} (x^2 - a^2) = \lim_{x \to a} (x - a)(x + a) = 0$$

since the limit in the first factor results zero, as shown in the prior example about the continuity of $y(x) = x$. Then, the continuity for $y(x) = x^2$ at any point, a, is confirmed:

$$\lim_{x \to a} x^2 = a^2$$

- And now, let it be **the general case of nth-order power,** $y(x) = x^n$; by mathematical induction, the continuity of such power function considers:

$$\lim_{x \to a} x^n = a^n$$

By multiplying both the equality members by the same factor, $\lim_{x \to a} x$,

$$\lim_{x \to a} x^n . \lim_{x \to a} x = a^n . \lim_{x \to a} x$$

Since the function $y(x) = x$ is continuous at any point, a, the limit in second member thus takes the referred point a,

$$\lim_{x \to a} (x^n . x) = a^n . a$$

$$\lim_{x \to a} x^{n+1} = a^{n+1}, \forall\, a$$

The graphics in Figure 4.1 illustrate the continuity of several power functions at point 1; in fact, the infinitesimal difference at that point results zero, the function limit and the image are equal, and then there exists concordance with the continuity definition at point 1.

For the other side, recall the existence and equality of the two lateral limits, (on left side, by lower values; and on right side, by greater values) at point a, and also the image function, $f(a)$, $\lim_{x \to a^-} f(x) = f(a) = \lim_{x \to a^+} f(x)$. Some kind of discontinuity will occur in case one of these three parts is missing, either the left lateral limit, or the right lateral limit, or even the function's image at point a, as follows in the next paragraphs.

Continuity Left

A function is said to be continuous to the left of a point $x = a$ in its domain, if and only if the limit to the left of that point exists and is equal to the corresponding function's image:

$$\lim_{x \to a^-} f(x) = f(a)$$

For instance, in Figure 4.3, the function is continuous on the left side of point, 1, with $\lim_{x \to 1^-} y(x) = y(1)$; however, the limit on the right side, $\lim_{x \to 1^+} y(x) = 1$, while the function's image is 2 at that point, $y(1) = 2$; then the infinitesimal difference Δy is not zero, in fact it results -1, and the function is continuous at the left side of point 1 and discontinuous on the right side.

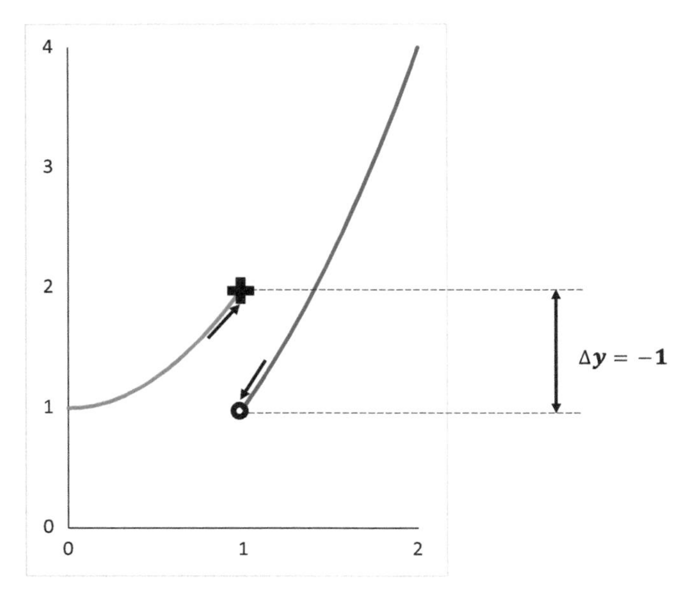

FIGURE 4.3 Continuous function on the left side of $x = 1$.

Continuity Right

A function is said to be continuous to the right of a point $x = a$ in its domain, if and only if the limit to the right of that point exists and is equal to the corresponding function's image:

$$\lim_{x \to a^+} f(x) = f(a)$$

Now, in Figure 4.4, the function is continuous on the right side of point, 2, with $\lim_{x \to 2^+} y(x) = y(2)$; however, the limit on the left side, $\lim_{x \to 2^-} y(x) = 2$, while the function's image is 3 at that point, $y(2) = 3$; again, the infinitesimal difference Δy is not zero, it becomes +1, and the function is continuous on the right side of point 2 and discontinuous on the left side.

Discontinuity in a point where the limit exists

If a function is not continuous at a given point $x = a$ on its domain, it is said to be discontinuous at that point. As presented in Figure 4.5, even in case the limits exist, the discontinuity occurs because the function cannot be evaluated at that point. Namely, the lateral limits on both right side and left side exist, and they take the same value, 1; however, the function images at point 0 cannot be calculated for the well-known ratios:

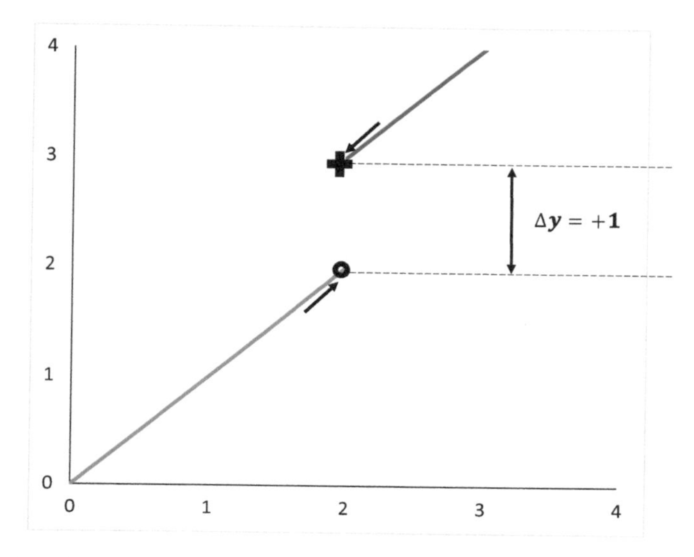

FIGURE 4.4 Continuous function on the right side of $x = 2$.

$$\lim_{x \to 0^-} \frac{\sin(x)}{x} = \lim_{x \to 0^+} \frac{\sin(x)}{x} = 1$$

$$\lim_{x \to 0^-} \frac{e^x - 1}{x} = \lim_{x \to 0^+} \frac{e^x - 1}{x} = 1; \text{ and also } \lim_{x \to 0^-} \frac{\ln|x+1|}{x} = \lim_{x \to 0^+} \frac{\ln|x+1|}{x} = 1$$

The corresponding graphics in the neighborhood of $x = 0$ follow too (Figure 4.5):

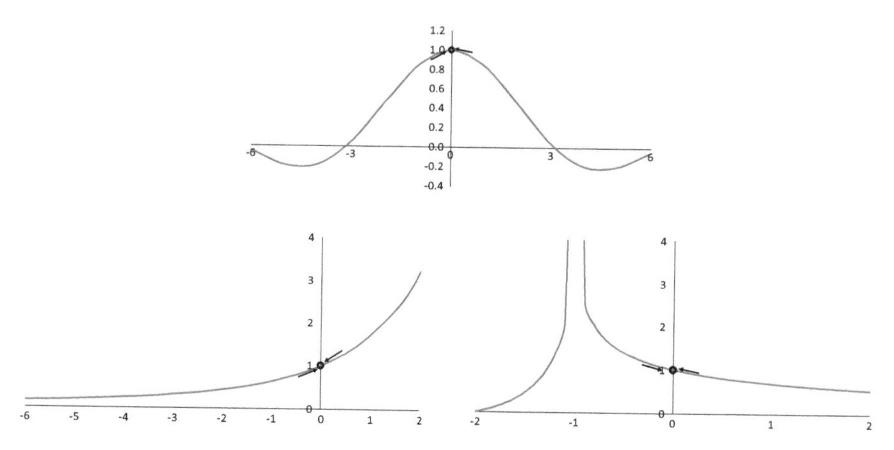

FIGURE 4.5 Discontinuous functions at point $x = 0$, but the limits exist.

Variable Substitution

A variable substitution is utilized many times, for example, making $h = \Delta x = x - a$, and directly expressing the deviation or perturbation of variable x; such substitution can help with the limit factorization, and the parameter h tends to 0 when variable x approaches a:

$$x = a + h \Leftrightarrow h = x - a$$

$$x \to a \Rightarrow h \to 0$$

In this way, by replacing variable x with the parameter h, and updating the limit operator too,

$$\lim_{x \to a} f(x) = f(a) \Rightarrow \lim_{h \to 0} f(a + h) = f(a)$$

The flowchart in Figure 4.6 outlines an alternative pathway to verify the function continuity at a given point, a, and several examples follow too.

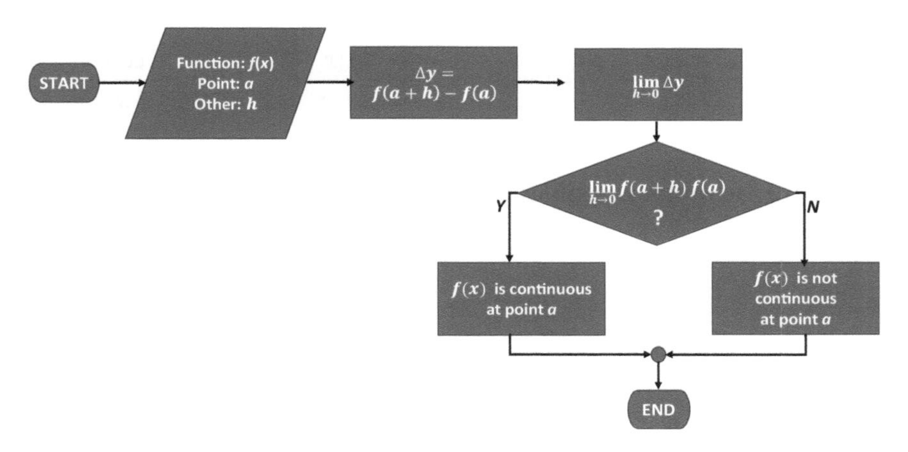

FIGURE 4.6 Flowchart outlining an alternative verification of function continuity at point a.

Example—Exponential

Let it be $y(x) = e^x$; then the relation $\lim_{x \to a} [y(x) - y(a)]$ can take both the forms,

$$\lim_{x \to a} (e^x - e^a) = \lim_{h \to 0} (e^{a+h} - e^a)$$

Manipulating the difference of exponentials in the second relation,

$$\Delta y(a+h) = e^{a+h} - e^{a} = e^{a}(e^{h}-1) = e^{a}\left(\frac{e^{h}-1}{h}\right)h$$

Taking limit when h approaches zero, and considering the remarkable limit inside parentheses,

$$\lim_{h \to 0} \Delta y(a+h) = e^{a} \lim_{h \to 0}\left(\frac{e^{h}-1}{h}\right)h = e^{a} \times 1 \times 0 = 0$$

In this way, the continuity of exponential function is verified for any point on its domain:

$$\lim_{x \to a}(e^{x} - e^{a}) = 0$$

$$\lim_{x \to a} e^{x} = e^{a}, \quad \forall\, a$$

Example—Natural Logarithm

At now, let it be $y(x) = \ln|x|$; then the relation $\lim_{x \to a}[y(x) - y(a)]$ takes the forms,

$$\lim_{x \to a}[\ln|x| - \ln|a|],\ a \neq 0$$

$$\lim_{h \to 0}[\ln|a+h| - \ln|a|]$$

Manipulating the difference of logarithms in the last relation,

$$\Delta y(a+h) = \ln|a+h| - \ln|a| = \ln\left|\frac{a+h}{a}\right| = \frac{\ln\left|1+\frac{h}{a}\right|}{(h/a)}(h/a)$$

Taking limit when h approaches zero, and considering the remarkable limit in the quotient,

$$\lim_{h \to 0} \Delta y(a+h) = \lim_{h \to 0} \frac{\ln\left|1+\frac{h}{a}\right|}{(h/a)}(h/a) = 1 \times 0 = 0$$

In this way, the continuity of logarithmic function is verified too for any point on its domain:

$$\lim_{x \to a} \ln|x| - \ln|a| = 0$$

$$\lim_{x \to a} \ln|x| = \ln|a|, \quad \forall\, a \neq 0$$

Example
Let it be $y(x) = 1/x$, for non-zero values of variable x; then the relation $\lim_{x \to a} [y(x) - y(a)]$ takes either the form,

$$\lim_{x \to a}\left(\frac{1}{x} - \frac{1}{a}\right), \text{ or also } \lim_{h \to 0}\left(\frac{1}{a+h} - \frac{1}{a}\right)$$

Manipulating the difference between the ratios in the last relation,

$$\Delta y(a+h) = \frac{1}{a+h} - \frac{1}{a} = \frac{a-(a+h)}{(a+h)a} = \frac{-h}{a^2 + ah}$$

Taking limit when h approaches zero, and considering the quotient of the two limits,

$$\lim_{h \to 0} \Delta y(a+h) = \lim_{h \to 0}\left(\frac{-h}{a^2 + ah}\right) = \frac{\lim_{h \to 0}(-h)}{\lim_{h \to 0}(a^2 + ah)} = \frac{0}{a^2 + 0} = 0$$

Thus, the continuity of function, $y(x) = 1/x$, is verified too:

$$\lim_{x \to a}\left(\frac{1}{x} - \frac{1}{a}\right) = 0$$

$$\lim_{x \to a}\left(\frac{1}{x}\right) = \frac{1}{a}, \quad a \neq 0, \quad x \neq 0$$

Example
And now, let it be $y(x) = 1/x^2$, for non-zero values of variable x; then the relation $\lim_{x \to a} [y(x) - y(a)]$ takes either the form,

$$\lim_{x \to a}\left(\frac{1}{x^2} - \frac{1}{a^2}\right), \text{ or also } \lim_{h \to 0}\left[\frac{1}{(a+h)^2} - \frac{1}{a^2}\right]$$

Manipulating the difference between the ratios in the second relation,

$$\Delta y(a+h) = \frac{1}{(a+h)^2} - \frac{1}{a^2} = \frac{a^2-(a+h)^2}{(a+h)^2a^2} = \frac{a^2-(a^2+2ah+h^2)}{a^2(a+h)^2}$$
$$= \frac{-2ah-h^2}{a^2(a+h)^2}$$

Taking limit when h approaches zero, and considering the quotient of the two limits,

$$\lim_{h\to 0} \Delta y(a+h) = \lim_{h\to 0}\left(\frac{-2ah-h^2}{a^2(a+h)^2}\right) = \frac{\lim\limits_{h\to 0}(-2ah-h^2)}{a^2\lim\limits_{h\to 0}(a+h)^2} = \frac{0}{a^2.\,a} = 0$$

Finally, the continuity of function $y(x) = 1/x^2$ is also verified:

$$\lim_{x\to a}\left(\frac{1}{x^2} - \frac{1}{a^2}\right) = 0$$

$$\lim_{x\to a}\left(\frac{1}{x^2}\right) = \frac{1}{a^2}, \quad a \neq 0, \quad x \neq 0$$

Limit Calculation at a Point for Continuous Functions

In practice, to find the limit of a continuous function when x approaches point a, then replace the variable x in the $f(x)$ expression by its value a. In fact, for the continuous function at point a,

$$\lim_{x\to a} f(x) = f(a)$$

The continuity of linear function, $y(x) = x$, is already shown in the prior section; in the reverse sense,

$$\lim_{x\to a}(x) = a \Rightarrow a = \lim_{x\to a}(x)$$

Then the limit calculation can directly substitute point a in the continuous function's expression:

$$\lim_{x \to a} f(x) = f\left[\lim_{x \to a} (x)\right]$$

In practice, the referred substitution of variable with the point at hand is directly performed, and typically some of the intermediate steps presented in the following examples are neglected.

Examples

- $\lim_{x \to 1} (x^4) = \left[\lim_{x \to 1} (x)\right]^4 = 1^4 = 1$
- $\lim_{x \to 0} (e^x - 1) = e^{\lim_{x \to 0}(x)} - 1 = e^0 - 1 = 1 - 1 = 0$
- $\lim_{x \to 0} \ln|1 + x| = \ln\left|1 + \lim_{x \to 0} (x)\right| = \ln|1 + 0| = \ln|1| = 0$
- $\lim_{x \to 1}\left(\frac{400}{x}\right) = \frac{400}{\lim_{x \to 1}(x)} = \frac{400}{1} = 400$
- $\lim_{x \to 1}\left(\frac{-800}{x^2}\right) = \frac{-800}{\left[\lim_{x \to 1}(x)\right]^2} = \frac{-800}{1^2} = -800$

4.3 CONTINUITY ON A RANGE

The continuity of a function in a range is focused on; first, the continuity attributes are considered in the interior points of the interval, and then the lateral limits in the border points are completing the study for the interval continuity. This topic is important for many developments, namely, for the theorems and results that are dedicated to closed intervals only; in fact, the applicability of such results is very restricted and needs specific analysis in the case of open intervals or semi-intervals.

Continuity on an Open Interval

A function $f(x)$ is continuous on an open interval of its domain, $]a, b[$, if and only if it is continuous at every point on that interval. In this case, the interval's border points are not under consideration, neither right continuous at point a, nor left continuous at point b, nor even continuous in the two border points. Several instances are presented in Figure 4.7, respectively, for the function continuity in the open interval $]0.8, 1.2[$, as well as in both the semi-open intervals $[0.8, 1.2[$ and $]0.8, 1.2]$.

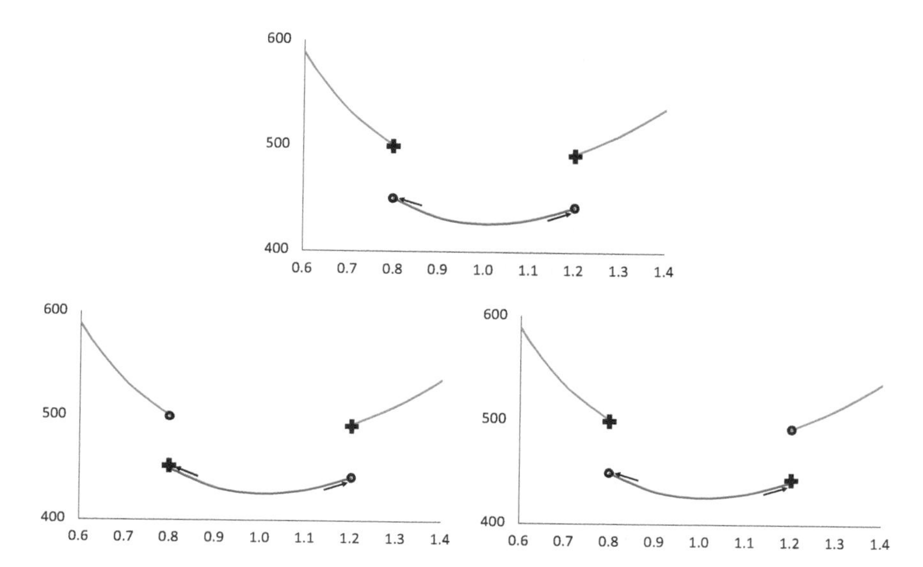

FIGURE 4.7 Function continuity on an open interval, and on the semi-open intervals.

Continuity on a Closed Interval

A function $f(x)$ is continuous on a closed interval of its domain, $[a, b]$, if and only if, it is continuous in the open interval, $]a, b[$, right continuous at point a and left continuous at point b. An instance is presented in Figure 4.8, for a continuous function in the closed interval $[0.8, 1.2]$.

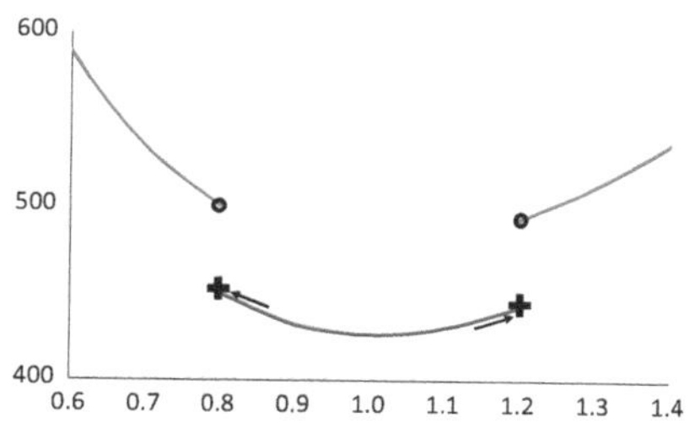

FIGURE 4.8 Continuous function on a closed interval.

4.4 PROPERTIES OF CONTINUOUS FUNCTIONS

In this section, the main properties concerning continuous functions are addressed, together with arithmetic operations with continuous functions. The continuity of composite function is referred, as well as the continuity of other functions of interest. These properties and results are analogous to those of function limits, namely, they are based on arithmetic operations and subsequent results of the function limit at a point, and then generalized to the entire function domain. The rigorous habits of mathematics and associated proofs are crucial; however, formal manipulations are neglected in here due to the computational approach on this text.

Operations with Continuous Functions—If $f(x)$ and $g(x)$ are continuous functions at a point, a, common to both domains, they are also continuous at that point, a:

- The sum function, $(f + g)(x)$;
- The product function, $(f.\ g)(x)$;
- The quotient function, $(f/g)(x)$;
- The power function, $f^n(x)$;
- The root function, $\sqrt[n]{f(x)}$.

Continuity of the Composite Function—If $g(x)$ is continuous at a point a on its domain, and $f(x)$ is continuous at $g(a)$, then the composite function fog also is continuous at that point a:

$$\lim_{x \to a} f[g(x)] = f\left[\lim_{x \to a} g(x)\right] = f[g(a)]$$

Continuity of Other Functions

- Every polynomial function is continuous on R—power functions are continuous functions, then their sums and products are continuous functions too.
- Every rational function, formed by the quotient between two polynomials, is continuous in its domain—since polynomials are continuous functions, then the ratio between two polynomials is continuous too, except for the roots of the denominator's polynomial.
- Every exponential function, of the form a^x, is continuous on R—the exponential e^x is a continuous function (section 4.2), and the bases manipulation, $a = e^{\ln(a)}$ and $e = a^{1/\ln(a)}$, allow the proper transformation between such continuous functions.

- Every logarithmic function, of the form *log* to x, is continuous in its R^+ domain—the natural logarithm $\ln|x|$ also is a continuous function (section 4.2), and the bases manipulation, $\log_a(e) = 1/\ln(a)$ and $\log_a(b) = 1/\log_b(a)$, allow the transformation between continuous functions too.

Note that both for exponential and logarithmic functions, base a is positive and different from 1.

4.4.1 Continuity of Cost Function

Let it be $y(x) = 100\left(\frac{\pi}{2}x^2 + \frac{3.2}{x}\right) = 50\pi\, x^2 + \frac{320}{x}$; and assuming point, $x = 1$, along with an infinitely small perturbation, h, at that point:

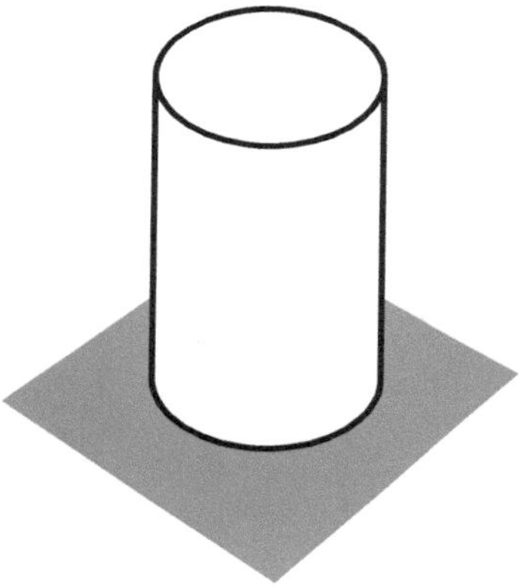

$$y(1) = 50\pi\,(1)^2 + \frac{320}{1} = 50\pi + 320$$

$$y(1 + h) = 50\pi\,(1 + h)^2 + \frac{320}{(1 + h)}$$

The relation $\lim_{\Delta x \to 0} \Delta y(x)$ can take both the forms,

$$\lim_{x \to 1}[y(x) - y(1)], \text{ or } \lim_{h \to 0}[y(1 + h) - y(1)]$$

Manipulating the difference in the second relation, and grouping similar terms,

$$\begin{aligned}
\Delta y(1) &= \left[50\pi\,(1 + h)^2 + \frac{320}{(1 + h)}\right] - (50\pi + 320) \\
&= [50\pi\,(1^2 + 2h + h^2) - 50\pi] + \frac{320 - 320(1 + h)}{(1 + h)} \\
&= 50\pi\,(1 + 2h + h^2 - 1) + \frac{320 - 320 - 320h}{(1 + h)}
\end{aligned}$$

By canceling the symmetric terms, and factoring parameter h,

$$= 50\pi(2h + h^2) - \frac{320h}{(1+h)}$$

$$= h\left[100\pi + 50\pi\ h - \frac{320}{(1+h)}\right]$$

Taking limit when h approaches zero, and considering the operations about limits,

$$\lim_{h\to 0} \Delta y(1) = \lim_{h\to 0} h\left[100\pi + 50\pi\ h - \frac{320}{(1+h)}\right]$$

$$= 0 \times \left[100\pi + 50\pi \times 0 - \frac{320}{(1+0)}\right]$$

$$= 0 \times [100\pi + 0 - 320] = 0$$

In this way, the continuity of cost function at point $x = 1$ is verified (Figure 4.9):

$$\lim_{x\to 1}\ [y(x) - y(1)] = 0$$

$$\lim_{x\to 1} y(x) = y(1)$$

Since the cost function only considers arithmetic operations of continuous functions, continuity holds in the entire cost function's domain; also

$$\lim_{x\to 1} y(x) = \lim_{x\to 1}\left(50\pi\, x^2 + \frac{320}{x}\right) = 50\pi\left(\lim_{x\to 1}(x)\right)^2 + \frac{320}{\lim_{x\to 1}(x)}$$

$$= 50\pi(1)^2 + \frac{320}{1} = 50\pi + 320 \cong 477.08$$

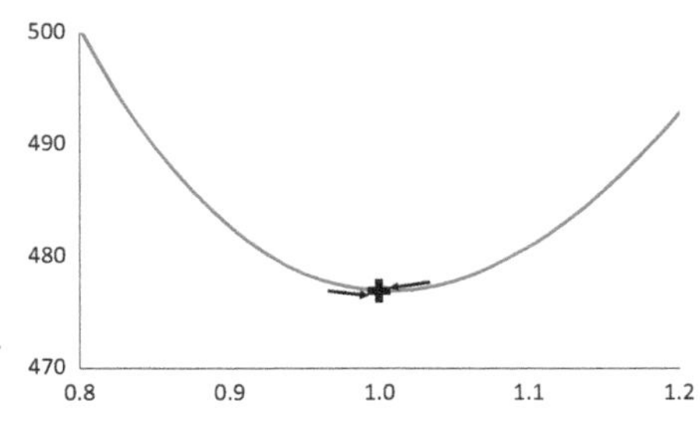

FIGURE 4.9 Continuity of cost function at point $x = 1$.

4.5 THEOREMS ABOUT CONTINUOUS FUNCTIONS

This section addresses important results about continuous functions, namely the Weierstrass theorem about extreme values theorem, and the Bolzano theorem that focuses on intermediate values. The simultaneous application of such results, as well as the well-known corollary for roots location are discussed too.

Extreme Values Theorem

If the function $f(x)$ is continuous on the closed interval $[a, b]$, then the function takes, at least once over this segment, its maximum value (M) and its minimum value (m).

Example

Let it be the function, $y(x) = 50\pi\, x^2 + \frac{400}{x}$, which is continuous in the closed interval [1.1, 1.3]. As presented in Figure 4.10, the function takes a minimum and a maximum in that interval, namely: $y(x)$ takes the minimum point at the left border, $x = 1.1$; the function is strictly increasing, thus it takes the maximum point on the right border, $x = 1.3$.

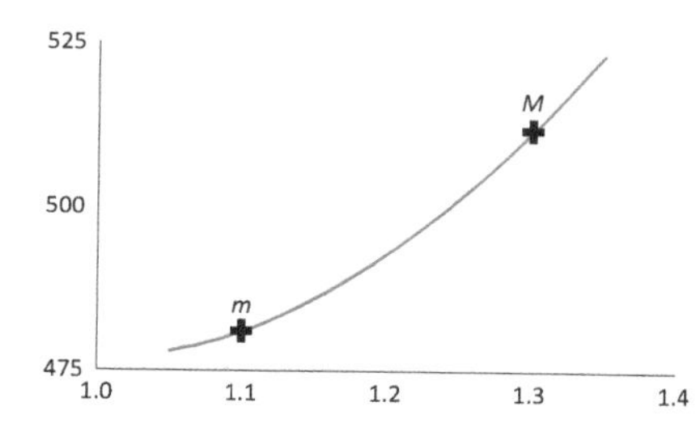

FIGURE 4.10 Weierstrass theorem in the closed interval [1.1, 1.3]: the extreme values exist, both maximum and minimum.

Note that if the interval is open at one of the border values, the Weierstrass theorem might not hold. Namely, in the case of open interval or semi-open intervals for strict monotonous functions, the extreme values theorem is missing the key border points. For instance, the strictly increasing function presented in Figure 4.11 will miss: (a) the maximum; (b) the minimum; and (c) both the maximum and the minimum.

In addition, the search of maximum and minimum points on a given interval needs also the recognition of continuity. Beyond the border points analysis, the Weierstrass theorem also requires continuity in the interior points, otherwise the existence of extreme points cannot be ensured, as follows from Figure 4.12 (infinite discontinuity at interior point, $x = 0$).

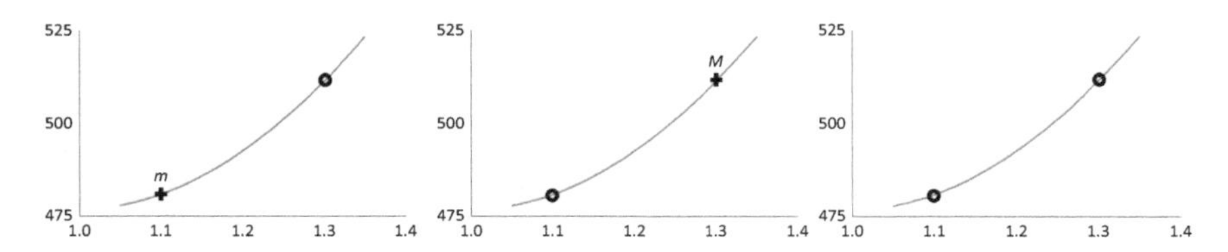

FIGURE 4.11 Non-application of Weierstrass theorem, for open or semi-open intervals. From left to right: (a) Only the minimum point exists, in the semi-interval [1.1, 1.3[; (b) Only the maximum point exists, in the semi-interval]1.1, 1.3]; (c) Neither the minimum nor the maximum exists, in the open interval]1.1, 1.3[.

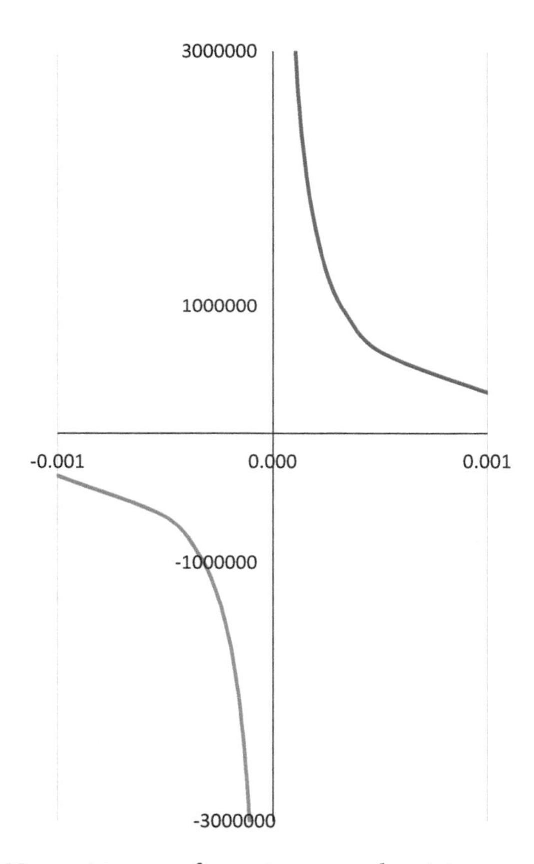

FIGURE 4.12 Non-existence of maximum and minimum points for discontinuous function at $x = 0$.

Intermediate Values Theorem

If function $f(x)$ is continuous on $[a, b]$, where u is a real number between the extreme and different values of $f(a)$ and $f(b)$, then there exists a number c in that interval, such that $f(c) = u$.

- The Bolzano's intermediate values theorem states the continuous function cannot vary from one value to another without going through all the intermediate values (Figure 4.13).

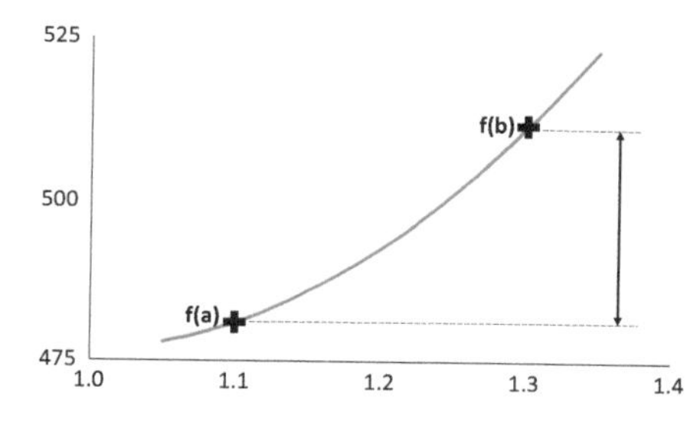

FIGURE 4.13 Bolzano theorem in the closed interval $[a, b]$: the function takes all the intermediate values between $f(a)$ and $f(b)$.

Corollary

If the function f is continuous on a closed interval $[a, b]$, and if it reaches its maximum, M, and minimum, m, then the function takes, at least once, any intermediate value between the maximum and minimum values.

- If the interval is closed, this corollary always holds; Figure 4.14 presents a continuous function in $[0.9, 1.3]$, where an intermediate

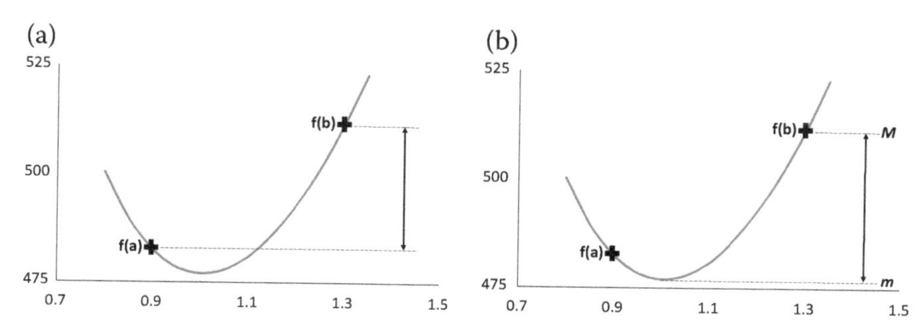

FIGURE 4.14 Simultaneous application of Bolzano and Weierstrass theorems: (a) The function takes all the intermediate values between $f(0.9)$ and $f(1.3)$; (b) The function takes all the intermediate values between the minimum and maximum in $[0.9, 1.3]$.

point is an extreme point, namely, the minimum; thus the variation range of function $f(x)$ is not only $[f(0.9), f(1.3)]$, but will be $[m, f(1.3)]$.

- Similarly, for the continuous function in the closed interval [1, 2], as presented in Figure 4.15, the simultaneous application of Weierstrass and Bolzano theorems will enlarge the function's variation range; in this instance, both the minimum and maximum occur in intermediate points; then, the variation range of function $f(x)$ is not strictly $[f(1), f(2)]$, but will consider the two existing extrema in the closed range, $[m, M]$.

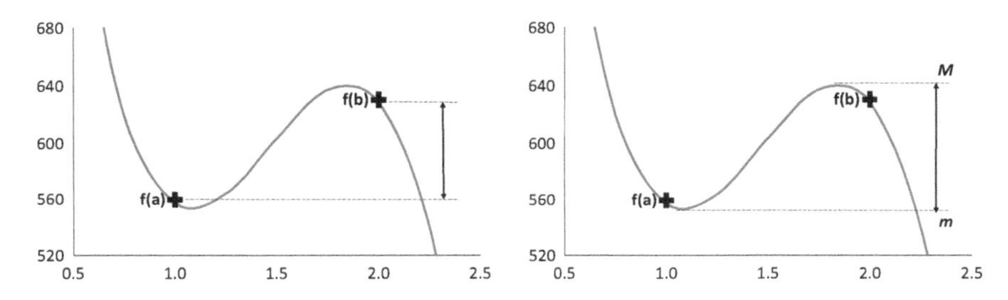

FIGURE 4.15 Simultaneous application of Bolzano and Weierstrass theorems.

Corollary—If $f(x)$ is continuous on $[a, b]$, and $f(a).f(b) < 0$, then the function becomes zero at least once in the open interval $]a, b[$:

$$f(c) = 0, \quad a < c < b$$

Example—Third-Order Polynomial

Let it be the third-order polynomial, $P_3(x) = x^3 - 6x + 2$.

This polynomial function admits three real roots, and the roots location can be initiated by applying the Bolzano corollary. In fact, by selecting the closed intervals where the polynomial changes sign either from positive to negative or from negative to positive, the function continuity enforces that the polynomial takes at least one root on such sub-intervals.

- As follows from Figure 4.16, the first approach for the roots location of polynomial $P_3(x)$ considers the observation of either the sign changes in the numerical table, or the function graphic when

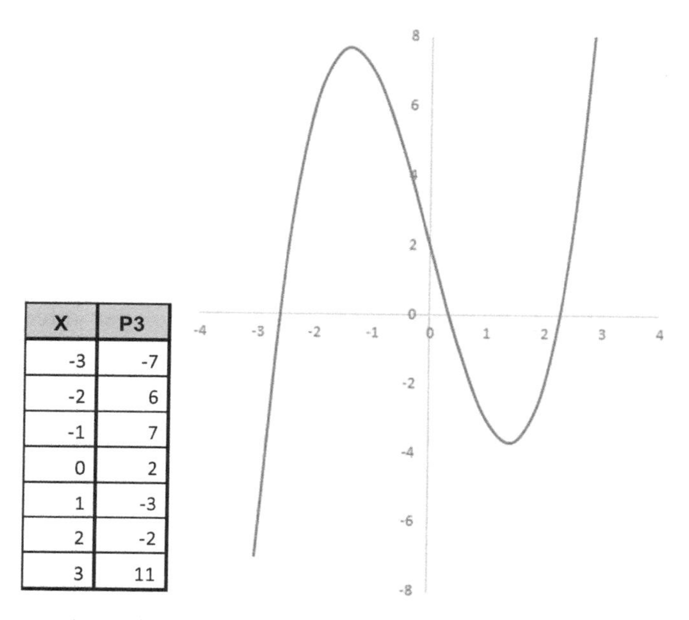

X	P3
-3	-7
-2	6
-1	7
0	2
1	-3
2	-2
3	11

FIGURE 4.16 Roots location for third-order polynomial, $P_3(x)$.

the line crosses the horizontal X-axis. In that way, variable x is altered in unit steps, $h = 1$, and the $P_3(x)$ evaluations follow:

$$P_3(0) = 0^3 - 6 \times 0 + 2 = 0 - 0 + 2 = 2$$

$$P_3(1) = 1^3 - 6 \times 1 + 2 = 1 - 6 + 2 = -3$$

then the polynomial $P_3(x)$ becomes zero at least once in the open interval $]0, 1[$:

$$P_3(c) = 0, \quad 0 < c < 1$$

- In addition, the location range for a second root considers,

$$P_3(-3) = (-3)^3 - 6 \times (-3) + 2 = -27 + 18 + 2 = -7$$

$$P_3(-2) = (-2)^3 - 6 \times (-2) + 2 = -8 + 12 + 2 = 6$$

and then $P_3(x)$ becomes zero at least once in the open interval $]-3, -2[$:

$$P_3(c) = 0, \quad -3 < c < -2$$

- Finally, the range for the third root considers,

$$P_3(2) = 2^3 - 6 \times 2 + 2 = 8 - 12 + 2 = -2$$

$$P_3(3) = 3^3 - 6 \times 3 + 2 = 27 - 18 + 2 = 11$$

and then $P_3(x)$ becomes zero at least once in the open interval]2, 3[:

$$P_3(c) = 0, \quad 2 < c < 3$$

In order to improve the accuracy for the third root, the polynomial $P_3(x)$ is evaluated in the open interval]2, 3[, using smaller steps, $h = 0.1$, for the alteration of variable, x; again, the observation of either the sign changes in the table, or the line crossing X-axis in Figure 4.17, provides a better estimate for the root location.

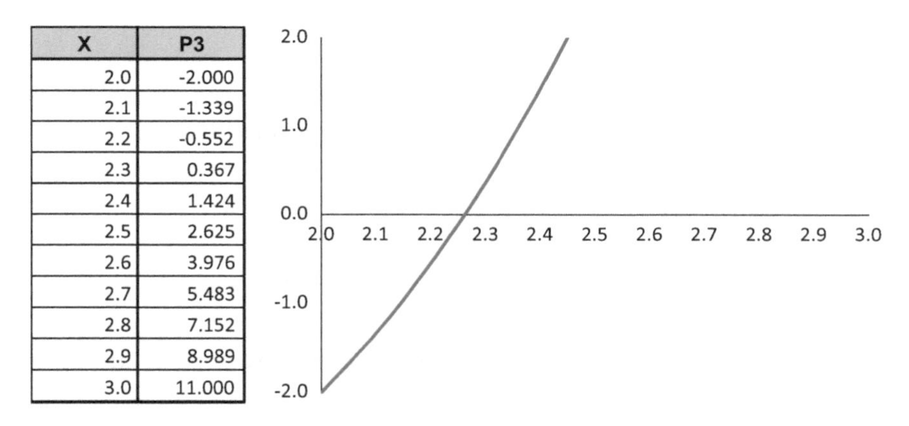

X	P3
2.0	-2.000
2.1	-1.339
2.2	-0.552
2.3	0.367
2.4	1.424
2.5	2.625
2.6	3.976
2.7	5.483
2.8	7.152
2.9	8.989
3.0	11.000

FIGURE 4.17 Accuracy of root location for third-order polynomial, $P_3(x)$: the first decimal digit.

In this way, the root accuracy can be improved with the first decimal digit, since

$$P_3(2.2).P_3(2.3) < 0$$

and then $P_3(x)$ becomes zero at least once in the open interval]2.2, 2.3[:

$$P_3(c) = 0, \quad 2.2 < c < 2.3$$

X	P3
2.20	-0.552
2.21	-0.466
2.22	-0.379
2.23	-0.290
2.24	-0.201
2.25	-0.109
2.26	-0.017
2.27	0.077
2.28	0.172
2.29	0.269
2.30	0.367

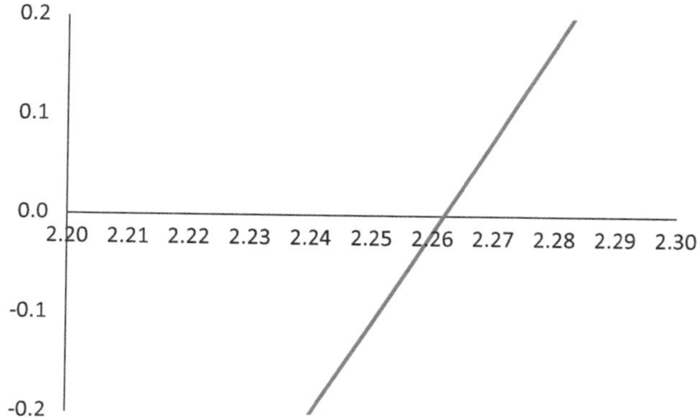

FIGURE 4.18 Accuracy of root location for third-order polynomial, $P_3(x)$: the second decimal digit.

This procedure corresponds to important applications, namely, in the search for function zeros with the required accuracy. In order to improve again the accuracy for this root, the polynomial $P_3(x)$ is now evaluated in $]2.2, 2.3[$ with smaller steps, $h = 0.01$, for the alteration of variable, x; once again, the observation of numerical table, or the X-axis cross in Figure 4.18 provides a centesimal accuracy for the root location.

In fact, the root accuracy can be improved with the second decimal digit, since

$$P_3(2.26).P_3(2.27) < 0$$

and then $P_3(x)$ becomes zero at least once in the open interval $]2.26, 2.27[$:

$$P_3(c) = 0, \quad 2.26 < c < 2.27$$

In this manner, locating the first decimal digit required 10 evaluations of polynomial $P_3(x)$, as in Figure 4.17; and the second decimal digit needed 10 evaluations of $P_3(x)$ too (Figure 4.18); then, three decimal digits will need a total of 30 evaluations, and six decimal digits will require 60 evaluations of $P_3(x)$ with such procedure. The location of roots with the required accuracy is crucial for a number of scientific and technological applications, and this topic deserves additional details and a more elaborated approach.

4.6 ROOTS OF NON-LINEAR EQUATIONS

This section addresses the very important issue of obtaining roots of equations: beyond the point x at which the function becomes zero, the function maxima and minima can be obtained from the derivative roots, and the inflection points can be associated with the roots for the second-order derivatives. In addition, in cases the inverse function is not available, the point x at the function takes a given value k, $y(x) = k$, can be obtained by simple manipulation too:

$$y(x) - k = 0$$

Therefore, the bisection method is introduced: ***i)*** In a closed interval $[a, b]$, it is verified that the continuous function $f(x)$ changes sign, and then there is at least one root in such interval; ***ii)*** To improve the estimate, the interval is divided into two equal parts by obtaining the midpoint, m, and it is verified in which of these halves the referred sign change occurs; and ***iii)*** The procedure will be repeated with this new sub-interval, whose size is half of the previous one, until the result is sufficiently accurate. In the following diagram, related with the third-order polynomial, the root search will continue in the left part of the midpoint, m:

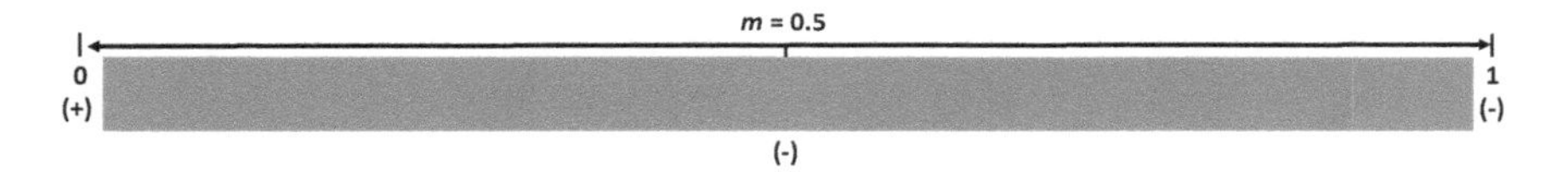

4.6.1 A Root of Third-Order Polynomial

Again, let it be the third-order polynomial, $P_3(x)$, already introduced in last section,

$$P_3(x) = x^3 - 6x + 2$$

- By selecting closed intervals, as in Figure 4.16, the observation of changing signs indicates root existence on such interval. In $[0, 1]$, the $P_3(x)$ evaluations indicate such occurrence,

$$P_3(0) = 0^3 - 6 \times 0 + 2 = 0 - 0 + 2 = 2$$

$$P_3(1) = 1^3 - 6 \times 1 + 2 = 1 - 6 + 2 = -3$$

and thus the midpoint in that unit interval is evaluated too, $m = (0 + 1)/2 = 0.5$:

$$P_3(0.5) = (0.5)^3 - 6 \times (0.5) + 2 = 0.125 - 3 + 2 = -0.875$$

Since $P_3(0.5)$ is negative, then the interval location is bisected, reduced to half,

$$P_3(0).P_3(0.5) < 0$$

and $P_3(x)$ becomes zero at least once in]0, 0.5[, with the interval spanning only 0.5.

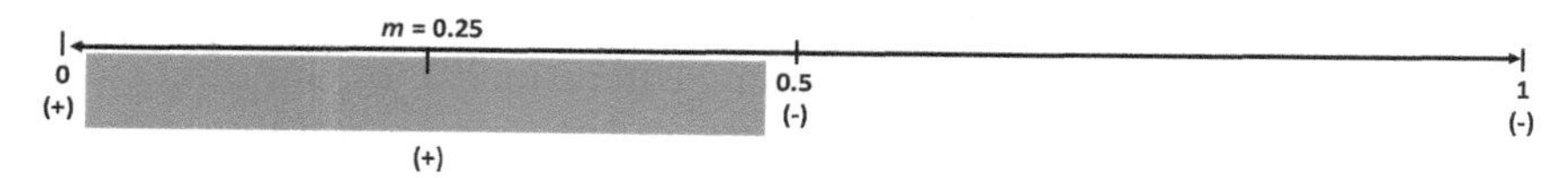

- The midpoint is evaluated again, at now in a smaller interval, $m = (0 + 0.5)/2 = 0.25$:

$$P_3(0.25) = (0.25)^3 - 6 \times (0.25) + 2 = 0.015625 - 1.5 + 2 = +0.515625$$

Since $P_3(0.25)$ is positive, the interval location is reduced to half again, since

$$P_3(0.25).P_3(0.5) < 0$$

and then $P_3(x)$ becomes zero in]0.25, 0.5[, with the interval spanning 0.25 at present.

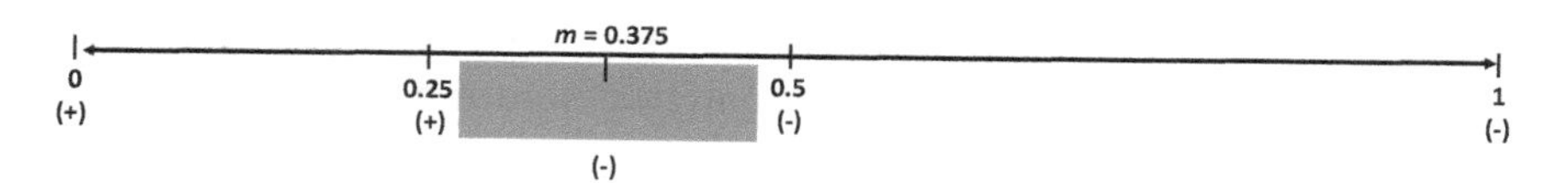

- Again, the medium point of this smaller interval is evaluated, $m = (0.25 + 0.5)/2 = 0.375$:

$$P_3(0.375) = (0.375)^3 - 6 \times (0.375) + 2 = 0.052734375 - 2.25 + 2$$
$$= -0.197265625$$

Since $P_3(0.3755)$ is negative, the interval location is reduced to half again,

$$P_3(0.25).P_3(0.375) < 0$$

and then $P_3(x)$ becomes zero in]0.25, 0.375[, with the interval spanning only 0.125. And rounding the first decimal, the root estimate is 0.3.

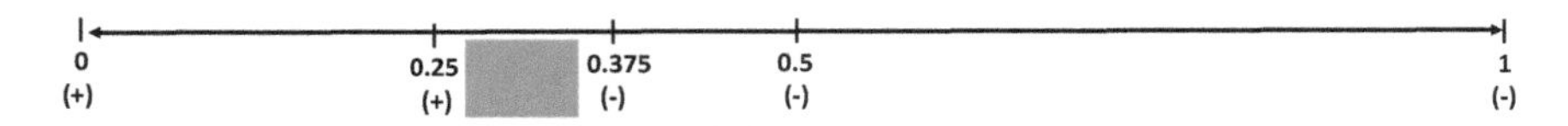

Through the bisection method that halves the root interval in each iteration, the location of each decimal digit requires about three function evaluations, as occurred for the first decimal of $P_3(x)$ root. With such procedure, about 18 to 21 evaluations of polynomial $P_3(x)$ are required to obtain six accurate decimal digits; after 21 iterations and rounding to the seventh decimal digit, the root estimate and the corresponding polynomial value follow:

$$P_3(0.3398768) = 0.0000004$$

4.6.2 A Barrel That Costs 500 (I)—The Bisection Method

The inverse rationale of defining the cost, and then obtaining the barrel diameter is revisited, due to the importance of an approach with quantified objectives or "target-values". In section 1.6, the diameter for the barrel costing less than 500 € is treated, by analyzing both the numerical table and function graphic; suitable diameter estimates are then obtained, namely, in the range [0.8, 1.2] the barrel cost is lower than 500.

However, the cost is now accurately defined, $C(d) = 500$, neither more nor less. The bisection method is proposed for solving the cost equation, since it does not present analytical solutions:

$$50\pi\ d^2 + \frac{320}{d} = 500$$

The numerical table and the corresponding graphic follow in Figure 4.19.

Once again by observation of table and graphic, the solutions either are in the interval [0.8, 0.9], or in [1.2, 1.3]; the function images involve the target-value 500, respectively in-between 500.53 and 482.79 or in-between 492.86 and 511.62, which can be found in an intermediate point of such intervals. In the other way, the function

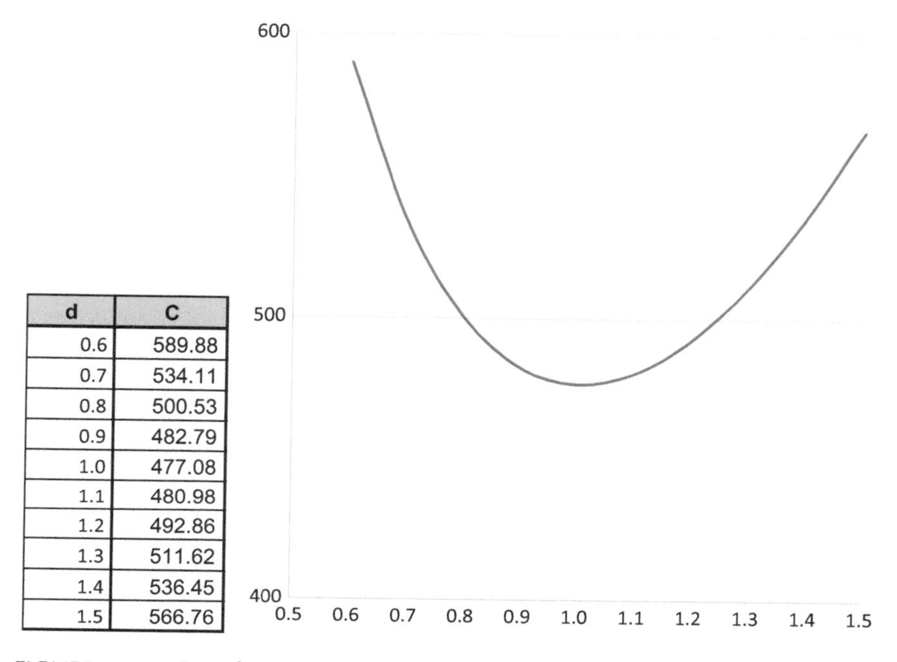

d	C
0.6	589.88
0.7	534.11
0.8	500.53
0.9	482.79
1.0	477.08
1.1	480.98
1.2	492.86
1.3	511.62
1.4	536.45
1.5	566.76

FIGURE 4.19 Cost function around the target-cost, 500 €.

"*Cost-minus-*500", $C_{500}(d)$, allows the application of bisection method, with the typical analysis of sign changes:

$$C_{500}(d) = 50\pi \ d^2 + \frac{320}{d} - 500 = 0$$

- As in Figure 4.20, the changing signs indicate roots existence in two intervals, in [0.8, 0.9] and [1.2, 1.3]. In the latter interval, for example, the function evaluations are,

$$C_{500}(1.2) = 50\pi(1.2)^2 + \frac{320}{(1.2)} - 500 = -7.14$$

$$C_{500}(1.3) = 50\pi(1.3)^2 + \frac{320}{(1.3)} - 500 = 11.62$$

and thus the midpoint in that interval will be evaluated too, $m = (1.2 + 1.3)/2 = 1.25$:

$$C_{500}(1.25) = 50\pi(1.25)^2 + \frac{320}{(1.25)} - 500 = 1.43$$

d	C-500
0.6	89.88
0.7	34.11
0.8	0.53
0.9	-17.21
1.0	-22.92
1.1	-19.02
1.2	-7.14
1.3	11.62
1.4	36.45
1.5	66.76

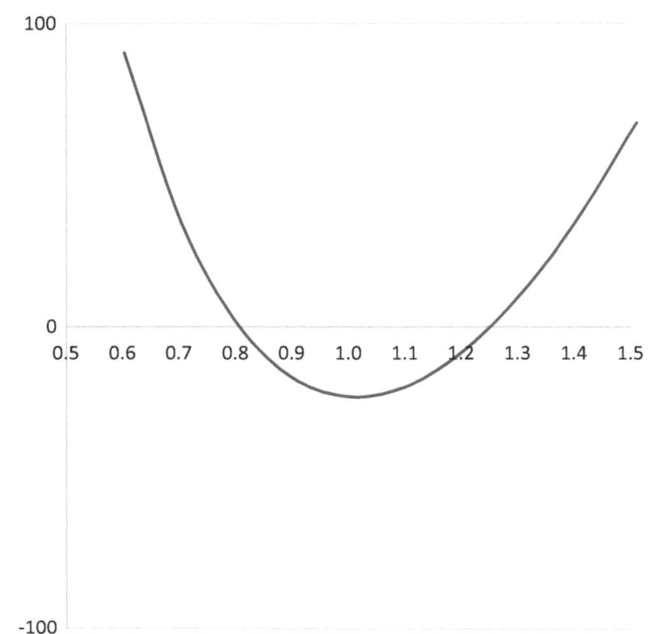

FIGURE 4.20 The function *Cost-minus-*500 around the zero.

Since $C_{500}(1.25)$ is positive, then the interval location is bisected, reduced to half,

$$C_{500}(1.2).\,C_{500}(1.25) < 0$$

and $C_{500}(x)$ becomes zero at least once in]1.2, 1.25[, with the interval spanning only 0.05.

- The new medium point is evaluated too, for the updated interval, $m = (1.2 + 1.25)/2 = 1.225$:

$$C_{500}(1.225) = 50\pi(1.225)^2 + \frac{320}{(1.225)} - 500 = -3.05$$

Since $C_{500}(1.225)$ is negative, the interval location is reduced to half again,

$$C_{500}(1.225).\,C_{500}(1.25) < 0$$

and then $C_{500}(x)$ becomes zero in]1.225, 1.25[, with the interval spanning only 0.025.

- Again, the medium point of the latter interval is evaluated, $m = (1.225{+}1.25)/2 = 1.2375$:

$$C_{500}(1.2375) = 50\pi(1.2375)^2 + \frac{320}{(1.2375)} - 500 = -0.86$$

Since $C_{500}(1.2375)$ is negative, the interval location is reduced to half again,

$$C_{500}(1.2375).\,C_{500}(1.25) < 0$$

and then $C_{500}(x)$ becomes zero in $]1.2375, 1.25[$, with the interval spanning only 0.0625. And rounding the second decimal, the root estimate is 1.24.

Following the bisection method, after 21 iterations and rounding to the seventh decimal digit, the root estimate and the corresponding function value follow:

$$C_{500}(1.2422497) = -0.000010$$

4.7 CONCLUDING REMARKS

Continuity and important notions are presented through a generalization approach that first treated the continuity of a function at a point, and then extended the continuity concept into a range. Recalling lateral limits, it is considered both the continuity on the left side and on the right side at a point; about the continuity in a range, the continuity attributes are first considered in the interior points, and then at the lateral limits of the interval border points.

- In addition, continuity assumes a very important role in the calculation of limits for elementary functions, by directly substituting the point at hand in the function argument to obtain the final result.

  The main properties of continuous functions are referred, together with the typical operations, the composite function, while the continuity of important functions is treated in convenient points (e.g., cost function at point 1).

- The existence of extreme values in a closed interval is discussed, as well as the intermediate values theorem in such closed range. The simultaneous application of both the Weierstrass and Bolzano theorems, and the well-known corollary for roots location are described too.
- The roots of non-linear equations are also addressed. Beyond the point x at which the function becomes zero, the function maxima and minima can be obtained from derivative roots, and inflection points can be associated with the roots of second-order derivative. In addition, the value x at the function takes a given value k, can be obtained by simple manipulation too.

The bisection method utilizes a closed interval that is cut in half, step after step, and the equation root is included in that interval due to the Bolzano corollary. However, it is also recognized as a slow method, since three iterations are usually needed to obtain one more decimal digit for the root estimate. Other methods are faster, and perform well under suitable conditions, such as the Newton-Raphson method (Chapter 6), but it requires the derivative calculation at each step.

Derivatives are thus very useful, and many results are associated with differentiable functions; namely, if the derivative function takes a finite value at a point, the function is also continuous at that point. In this manner, derivation tools are presented in next chapter. It first introduces the derivative function and also includes:

- The calculation of derivatives at a point, combining the practice of derivative definition with the control approach of verifying and confirming results.
- Beyond derivation properties, and derivatives of important transcendent functions, the inverse function's derivative is also featured.

CHAPTER 5

Derivative of a Function

This chapter presents important topics related to the function derivative. First, basic notions about the derivative function are introduced, in a way to better support the derivatives of elementary functions and the usual operations about derivatives. The derivation of various important functions is also presented (e.g., exponential and logarithmic, sine and cosine, quadratic power and square root), along with their geometric instances, together with the tangent to the function graphic at the point under analysis. The derivation of inverse function is treated and a comparison analysis is also developed for a specific case: inverting the cost function and after that deriving the inverse function; then comparing the instances treated via the derivation rules and via the analytical development. Finally, derivatives of different orders are obtained, recalling the nth-order derivative, $f^{(n)}(x)$, is obtained by deriving n times the original function $f(x)$.

5.1 INTRODUCTION

In general, the study of function derivatives is supported in the continuity attributes, either on the symbolic manipulations with different working parameters (e.g., the independent variable, the increment parameter), or in the recognition of binomial structures for the continuity proofs of power functions, where similar tasks occurred.

- Derivation and related properties are also suitable to verification procedures, namely, by using the associated definition to obtain the derivative at a point, and then verifying it through the operating rules and derivation expressions. For the instances at hand, important functions and the corresponding geometric representation are treated, including the respective tangent instance to the function graphic at the point under analysis.
- The reciprocal of function derivative is resulting in the derivative of inverse function; in that way, the cost function and simple instances are suitable for evaluating the container's diameter (or volume) that satisfies the target-cost.

DOI: 10.1201/9781003461876-5

- In a similar manner, derivatives of different orders are treated, both for the cost function and its Taylor polynomial: at the point under analysis, they present coincident images for the various derivatives; then the related values, rates of change, and graphic curvatures will be similar too.

The rest of this chapter is structured as follows: in the first place, the derivative of a function at a point is revisited, as well as the notions of derivative function and lateral derivatives; in the second step, the derivatives of elementary functions and the usual derivation rules are revisited; in a similar approach, the derivatives of transcendent functions are presented, including exponential and logarithmic functions, or trigonometric functions, while convenient examples are treated too; from a different point of view, important results about the derivative of inverse function are presented, and a comparison analysis is developed, namely, by both inverting and deriving the cost function; finally, derivatives of different orders are focused on, namely, to obtain the nth-order derivative, the function is derived n times.

5.2 DERIVATIVES AND GEOMETRIC INTERPRETATION

In this section, the derivative of a function at a point is revisited, along with the associated geometric interpretation, the notions of derivative function and lateral derivatives, in a way to better support the bridging of derivability and continuity.

The derivative of a function $f(x)$ at a point a, represented by $f'(a)$, if existing, is the limit of the incremental ratio:

$$f'(a) = \lim_{x \to a} \frac{f(x) - f(a)}{x - a} = \lim_{h \to 0} \frac{f(a + h) - f(a)}{h}$$

- When function $f(x)$ admits derivative at a point, a, the function is said to be differentiable.

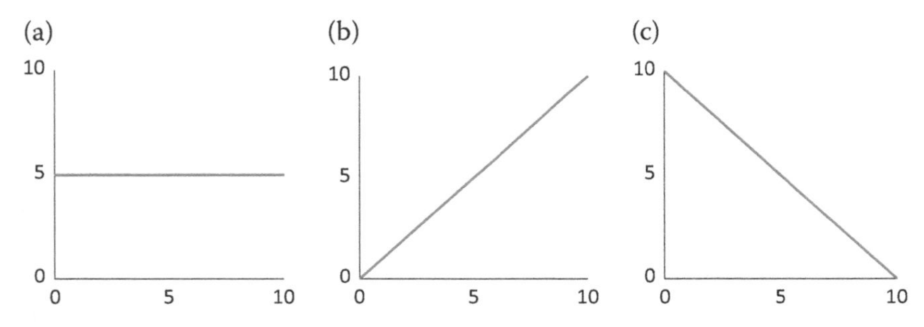

FIGURE 5.1 Constant and linear functions.

Examples—Monotony of Linear Functions

- For the constant function, $y(x) = 5$ (Figure 5.1a); the derivative at any point, a, is zero,

$$\lim_{x \to a} \frac{y(x) - y(a)}{x - a} = \lim_{x \to a} \frac{(5) - (5)}{x - a} = 0, \quad \forall\, a$$

And consider the linear increasing function, $y(x) = x$ (Figure 5.1b); the function's derivative at any point, a, follows:

$$\lim_{x \to a} \frac{y(x) - y(a)}{x - a} = \lim_{x \to a} \frac{(x) - (a)}{x - a} = 1, \quad \forall\, a$$

And now, for the linear decreasing function, $y(x) = 10 - x$ (Figure 5.1c) the function's derivative at any point, a, follows too:

$$\lim_{x \to a} \frac{y(x) - y(a)}{x - a} = \lim_{x \to a} \frac{(10 - x) - (10 - a)}{x - a} = \lim_{x \to a} \frac{-x + a}{x - a}$$

$$= -\lim_{x \to a} \frac{x - a}{x - a} = -1, \quad \forall\, a$$

Derivative Function

Let $f(x)$ be a function defined on a given set, and let D be the set of real numbers where $f(x)$ has a finite derivative; the derivative function is represented by $f'(x)$,

$$\begin{aligned} D &\to R \\ x &\to f'(x) \end{aligned}$$

Note the point, a, where the derivative is calculated can be any point in the function's domain. The derivative function at that point thus takes the value of its finite derivative, if it exists, and the flowchart in Figure 5.2 outlines a pathway to verify it.

Lateral Derivatives

At an interior point, a, recall the existence and equality of the two lateral limits (on the left side, by lower values; on the right side, by greater values) in order to satisfy the limit existence.

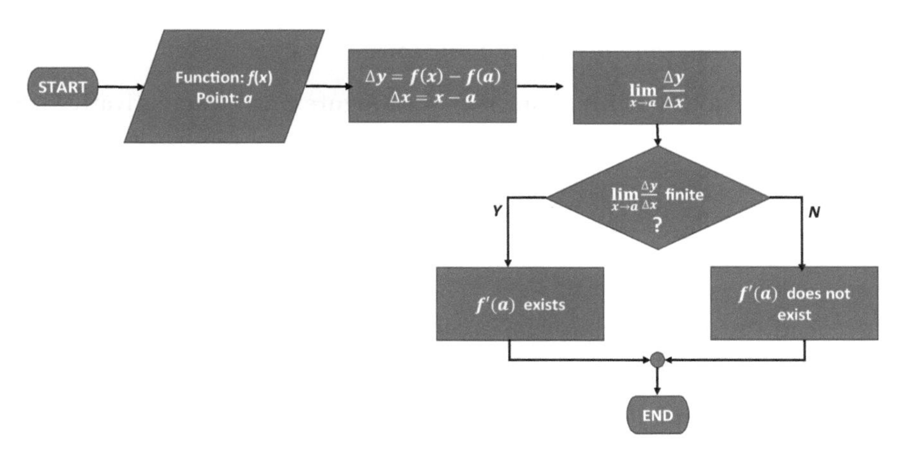

FIGURE 5.2 Flowchart outlining the verification of function derivative at point a.

The left derivative of a function $f(x)$ at point a, $f'(a^-)$, corresponds to the lateral limit, if it exists,

$$f'(a^-) = \lim_{x \to a^-} \frac{f(x) - f(a)}{x - a}$$

In the same way, the derivative on the right side, $f'(a^+)$, corresponds to the lateral limit,

$$f'(a^+) = \lim_{x \to a^+} \frac{f(x) - f(a)}{x - a}$$

The derivative at that interior point exists, if and only if, the lateral derivatives exist and take the same value:

$$\text{If } f'(a^-) = f'(a^+) \text{ then } f'(a) \text{ exists}$$

$$\text{and } f'(a) = f'(a^-) = f'(a^+)$$

Therefore, the function derivative at a point is one, and only one, similarly to the function limit.

Example—Absolute Value Function

- Let it be $y(x) = |2x|$, in Figure 5.3, and the positive and negative branches are separated:

$$y(x) = |2x| \Leftrightarrow \begin{cases} 2x & x \geq 0 \\ -2x & x < 0 \end{cases}$$

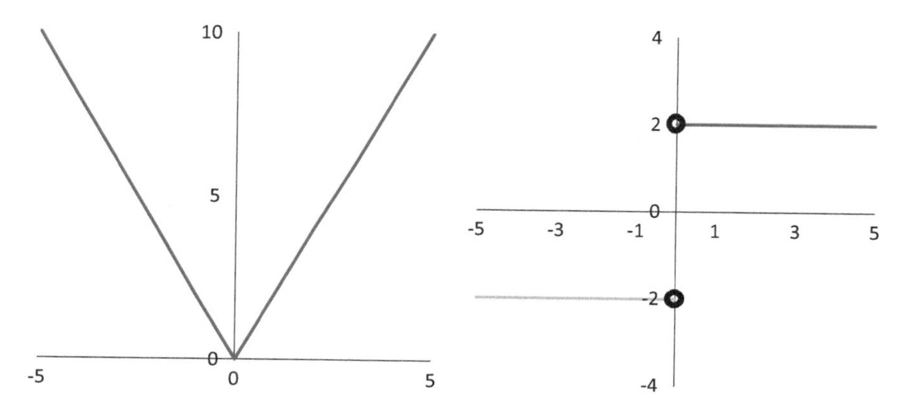

FIGURE 5.3 Absolute value function, $y = |2x|$, and its derivative, $y'(x)$.

The function's derivative at any positive point, a,

$$\lim_{x\to a}\frac{y(x)-y(a)}{x-a}=\lim_{x\to a}\frac{(2x)-(2a)}{x-a}=2\lim_{x\to a}\frac{x-a}{x-a}=2,\quad\forall\ a>0$$

And the function's derivative in every negative point, a, takes the symmetric value,

$$\lim_{x\to a}\frac{y(x)-y(a)}{x-a}=\lim_{x\to a}\frac{(-2x)-(-2a)}{x-a}=-2\lim_{x\to a}\frac{x-a}{x-a}=-2,\quad\forall\ a<0$$

However, the function does not admit derivative at $x = 0$ because the lateral derivatives at this point are different:

$$\lim_{x\to 0^+}\frac{y(x)-y(a)}{x-a}=\lim_{x\to 0^+}\frac{(2x)-(2a)}{x-a}=2\lim_{x\to 0^+}\frac{x-0}{x-0}=2$$

$$\lim_{x\to 0^-}\frac{y(x)-y(a)}{x-a}=\lim_{x\to 0^-}\frac{(-2x)-(-2a)}{x-a}=-2\lim_{x\to 0^-}\frac{x-0}{x-0}=-2$$

And the derivative function, $y'(x)$, also presents separate positive and negative branches (Figure 5.3), while it does not exist at that point:

$$y'(x) = \begin{cases} 2 & x > 0 \\ -2 & x < 0 \end{cases}$$

The Tangent Line Representation

The derivative at a point, a, indicates the slope (inclination) of the tangent line to the graphic at that point, $[a, f(a)]$. In addition, the derivative $f'(x)$ at, $x = a$, is equal to the tangent of the angle, α, between the positive part of X-axis and the referred tangent line.

$$f'(a) = \tan(\alpha)$$

Examples—Quadratic and Cubic Function

Let it be $y(x) = x^2$; the derivative relation is verified at point, a,

$$\lim_{x \to a} \frac{f(x) - f(a)}{x - a} = \lim_{x \to a} \frac{x^2 - a^2}{x - a} = \lim_{x \to a} \frac{(x - a)(x + a)}{x - a} = \lim_{x \to a} (x + a)$$
$$= 2a$$

And let it be now $y(x) = x^3$; the limit relation is verified too,

$$\lim_{x \to a} \frac{f(x) - f(a)}{x - a} = \lim_{x \to a} \frac{x^3 - a^3}{x - a} = \lim_{x \to a} \frac{(x - a)(x^2 + ax + a^2)}{x - a}$$
$$= \lim_{x \to a} (x^2 + ax + a^2) = 3a^2$$

The graphics in Figure 5.4 present tangent lines to the graphic of power functions at point 1. Namely, the straight line $y(x) = 2x - 1$ is tangent to the graphic of x^2, while the straight $y(x) = 3x-2$ is tangent to the graphic of x^3, both at point 1. In that sense:

- The first straight line, $y(x) = 2x - 1$, crosses the Y-axis at $x = -1$, taking $(0, -1)$, and also takes $(1, 1)$; the segment's inclination corresponds to the ratio, $[1-(-1)]/(1-0) = 2/1 = 2$, and it verifies the derivative at point 1, $y'(1) = 2$; the associated angle is, $\alpha = \arctan(2)$.

- The latter line, $y(x) = 3x - 2$, crosses the Y-axis at $x = -2$, taking $(0, -2)$, and also takes $(1, 1)$; the segment's inclination corresponds to the ratio, $[1-(-2)]/(1-0) = 3/1 = 3$, and it also equals the derivative at point 1, $y'(1) = 3$; then the associated angle is, $\alpha = \arctan(3)$.

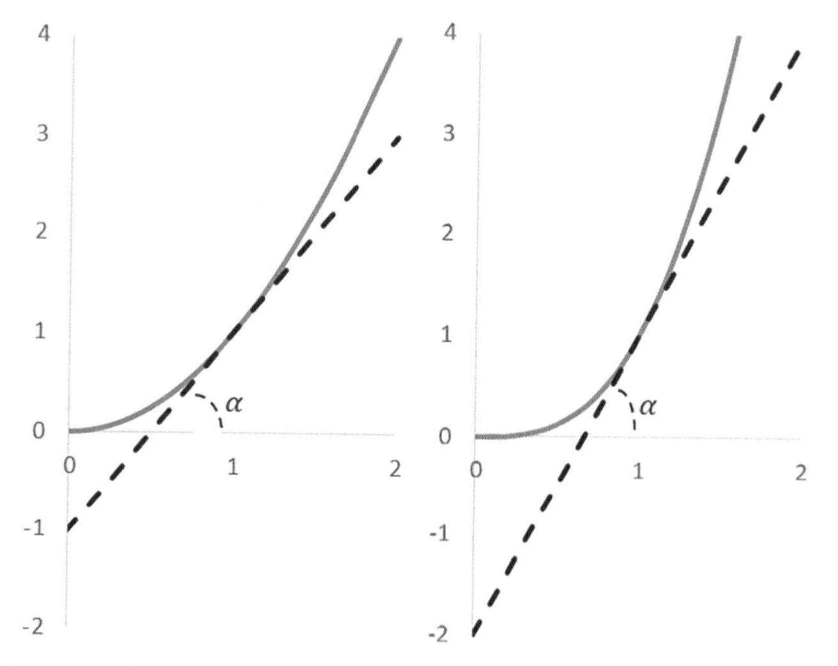

FIGURE 5.4 Tangent lines at point 1, respectively, for the quadratic and cubic functions.

Without loss of generalization, consider the tangent line evaluated at two consecutive integer points, a and $(a + 1)$, as in Figure 5.5. The straight line, $y(x) = y'(a)x + b$, where b is the ordinate at Y-axis, holds at both the points,

$$y(a) = y'(a).a + b$$

$$y(a + 1) = y'(a).(a + 1) + b$$

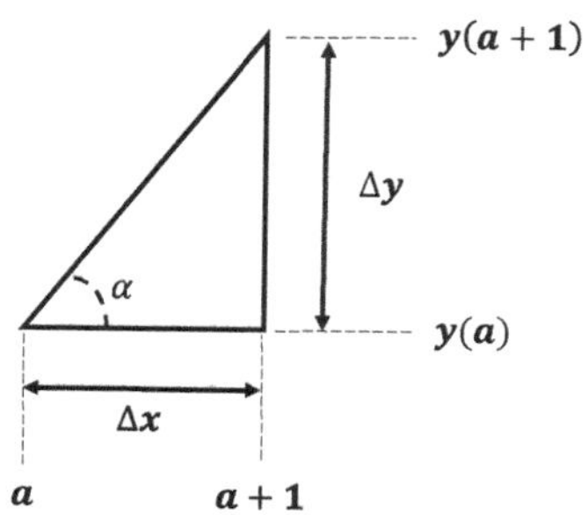

FIGURE 5.5 Geometric representation of tangent line points, $y(a)$ and $y(a+1)$.

The angle α is such that, $\tan(\alpha) = \frac{\Delta y}{\Delta x}$; by substitution of variable increments and canceling the symmetric terms, such relation transits also to the derivative at point, *a*:

$$\begin{aligned}\tan(\alpha) &= \frac{\Delta y}{\Delta x} = \frac{y(a+1) - y(a)}{(a+1) - a} = \frac{[y'(a).(a+1) + b] - [y'(a).a + b]}{1} \\ &= y'(a).a + y'(a).1 - y'(a).a = y'(a)\end{aligned}$$

Derivability and Continuity

- Every function with a finite derivative at a point is continuous at that point.

  In fact, assuming the existence of a finite derivative at point a, $f'(a)$,

$$\lim_{x \to a} \frac{f(x) - f(a)}{x - a} = f'(a)$$

  The quotient's limit corresponds to the quotient of limits in numerator and denominator,

$$\frac{\lim_{x \to a} [f(x) - f(a)]}{\lim_{x \to a} (x - a)} = f'(a)$$

  Manipulating the limit in denominator to the second member, and noting the limit in second member results zero, then

$$\lim_{x \to a} [f(x) - f(a)] = f'(a). \lim_{x \to a} (x - a)$$

$$\lim_{x \to a} f(x) - f(a) = 0$$

$$\lim_{x \to a} f(x) = f(a)$$

  The equality between the function limit at point, a, and the associated image, $f(a)$, represents continuity for the function under analysis.

- The converse is not true; namely, the absolute value function, $f(x) = |2x|$, is continuous at point 0, but the derivative at this point does not exist (the lateral derivatives are different).

5.3 DERIVATION RULES

At this point, the derivatives of arithmetic operations are focused on, and the usual derivation rules are revisited. Some properties and results are analogous to those of function limits, typically, they are based both on operations for the function limit at a given point, a, and then generalized to the derivative domain.
Despite the importance of symbolic developments and mathematical proofs, complex manipulations are neglected in this topic and general instances follow for derivation rules.

Linearity—The linearity attributes of function limits also apply to function derivatives, namely, when treating the derivative of functions sum (it results the sum of derivatives) or the derivative of a constant, k, multiplying a function (it results constant times the function derivative).

$$\begin{aligned}(k.f)' &= k.f' \\ (f+g)' &= f' + g'\end{aligned}$$

- In a general way the linearity attributes of function limits apply again:

  $[(k+1)x]' = (kx+x)' = (kx)' + (x)' = k+1$, by mathematical induction;
  $(k\,x + c)' = k\,(x)' + 0 = k$, since the derivation of constant is zero.

Other derivation rules follow, in conjunction with instances of interest, focusing on functions multiplication and suitable instances to integer powers, either negative (e.g., derivation of reciprocal function) or fractional powers (the non-zero parameter, q, also holds for rational functions, e.g., square root), or even derivative of functions division (if possible, when the function in denominator is not zero).

$$(f.g)' = f'.g + f.g' \qquad (f/g)' = \frac{f'.g - f.g'}{g^2}$$

$$(f^q)' = q.f^{q-1}.f' \qquad (1/g)' = -\frac{g'}{g^2}$$

$$(\sqrt{f})' = \frac{f'}{2\sqrt{f}}$$

In that sense, some examples follow.

- For integer power functions:

  $(x^2)' = (x.x)' = (x)'x + x(x)' = 1.x + x.1 = 2x$, the multiplication rule applies.

  $(x^3)' = (x^2.x)' = (x^2)'x + x^2(x)' = (2x).x + x^2.1 = 3x^2$, again with the same rule.

  By mathematical induction, assuming $(x^n)' = n\ x^{n-1}$, then

  $$\begin{aligned}(x^{n+1})' &= (x^n.x)' = (x^n)'x + x^n(x)' = (n\ x^{n-1}).x + x^n.1\\ &= n\ x^n + x^n.1 = (n+1)x^n\end{aligned}$$

  And just recalling that linearity applies and the derivative of constant is zero:

  $$(k\ x^2 + c)' = k(2x) + 0 = 2kx$$

  $$(x^3 + c)' = 3x^2 + 0 = 3x^2$$

- For negative (integer) power functions:

  $$\left(\frac{1}{x}\right)' = [x^{-1}]' = (-1).(x)^{-1-1} = -x^{-2} = -\frac{1}{x^2}$$

  $$\left(\frac{1}{x^2}\right)' = (x^{-2})' = -2(x)^{-2-1} = -2\ x^{-3} = -\frac{2}{x^3}$$

  Linearity still holds, for instance,

  $$\begin{aligned}\left(\frac{-4}{x-5}\right)' &= -4[(x-5)^{-1}]' = -4(-1)(x-5)^{-1-1} = +4(x-5)^{-2}\\ &= \frac{4}{(x-5)^2}\end{aligned}$$

- And for non-integer (positive) power functions:

  $$(\sqrt{x})' = \left(x^{(1/2)}\right)' = \frac{1}{2}x^{(1/2-1)} = \frac{1}{2}x^{(-1/2)} = \frac{1}{2\sqrt{x}}$$

  $$(\sqrt[3]{x})' = \left(x^{(1/3)}\right)' = \frac{1}{3}x^{(1/3-1)} = \frac{1}{3}x^{(-2/3)} = \frac{1}{3\sqrt[3]{x^2}}$$

Linearity holds, and added constants do not alter the derivative calculation:

$$(\sqrt{x+9}+3)' = \left[(x+9)^{(1/2)}\right]' + 0 = \frac{1}{2}(x+9)^{(1/2-1)}(1+0)$$
$$= \frac{1}{2}(x+9)^{(-1/2)} = \frac{1}{2\sqrt{x+9}}$$

- About deriving a sixth order polynomial, linearity is extending to all its terms,

$$P_6(x) = 2400x - 3000\,x^2 + \frac{(50\pi + 8000)}{3}x^3 - 1500\,x^4 + 480\,x^5 - \frac{200}{3}x^6 + c$$

each term is separately treated, and a fifth order polynomial, $P_5(x)$, is thus obtained:

$$\begin{aligned} P_6'(x) &= 2400 - 3000(2x) + \frac{(50\pi + 8000)}{3}(3x^2) - 1500(4x^3) \\ &\quad + 480(5x^4) - \frac{200}{3}(6x^5) + 0 \\ &= 2400 - 6000x + (50\pi + 8000)x^2 - 6000x^3 + 2400x^4 \\ &\quad - 400x^5 \end{aligned}$$

5.4 DERIVATION OF IMPORTANT FUNCTIONS

In this section, the derivatives of important functions are presented, including the derivatives for the exponential function and the logarithmic function, the sine and cosine functions, or the tangent and cotangent functions. Both the simple and the composite forms are addressed for the referred functions, together with the derivative of inverse function, while convenient instances are treated too. The flowchart in Figure 5.6 outlines the proposed pathway.

5.4.1 Exponential Functions

The exponential function with base, e, has finite derivative in its entire domain, and this derivative is equal to the function itself, that is:

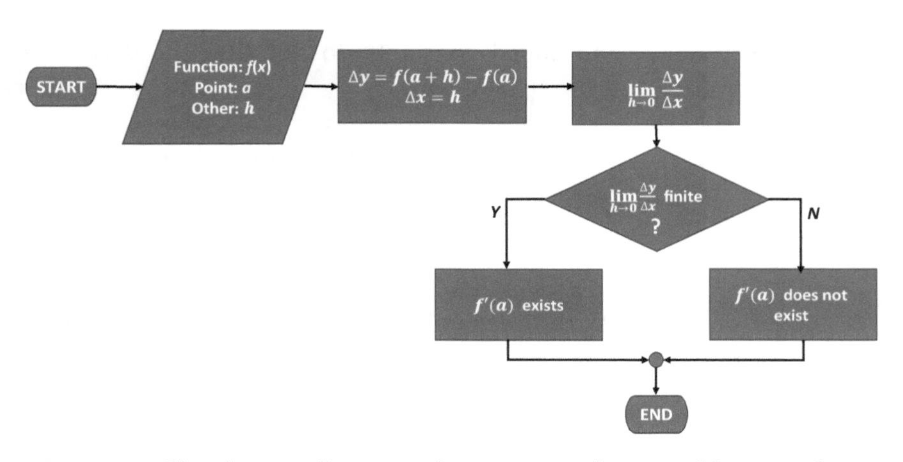

FIGURE 5.6 Flowchart outlining an alternative verification of function derivative at point a.

$$(e^x)' = e^x, \quad \forall\, x$$

Note the only function whose derivative coincides with the function itself is the exponential function with natural base, e^x.

Derivation of Natural Base Exponential, e

Let it be $y(x) = e^x$; then the derivative at point, $x = 2$ (Figure 5.7), can present both the forms,

$$\lim_{x \to 2}\left(\frac{e^x - e^2}{x - 2}\right) = \lim_{h \to 0}\left(\frac{e^{2+h} - e^2}{h}\right)$$

Manipulating the difference of exponentials in the second relation, taking limit when h approaches zero, and considering the remarkable limit inside parentheses,

$$\lim_{h \to 0}\left(\frac{e^{2+h} - e^2}{h}\right) = \lim_{h \to 0}\frac{e^2(e^h - 1)}{h} = e^2 \lim_{h \to 0}\left(\frac{e^h - 1}{h}\right) = e^2.1 = e^2$$

Beyond point 2, the operations and manipulations developed to derivate the exponential function are possible for every point, x; in this way,

$$(e^x)' = e^x, \quad \forall\, x$$

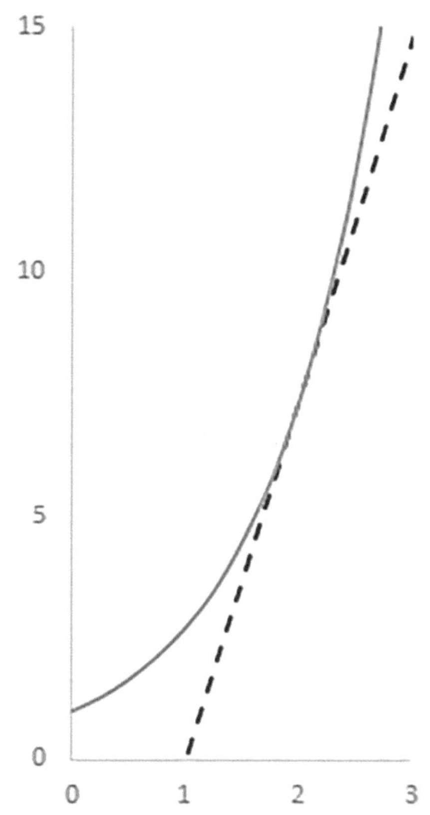

FIGURE 5.7 Exponential function and tangent at point 2.

Derivation of Hyperbolic Functions

- Consider the hyperbolic sine function $f(x) = \sinh(x)$. By derivation,

$$f'(x) = \left(\frac{e^x - e^{-x}}{2}\right)' = \frac{e^x - (-e^{-x})}{2} = \frac{e^x + e^{-x}}{2} = \cosh(x)$$

- And consider the hyperbolic cosine function, $f(x) = \cosh(x)$. Then,

$$f'(x) = \left(\frac{e^x + e^{-x}}{2}\right)' = \frac{e^x + (-e^{-x})}{2} = \frac{e^x - e^{-x}}{2} = \sinh(x)$$

5.4.2 Logarithmic Functions

The function defined by the natural logarithm (base e) has derivative in the real positive domain, which observes the following equality:

$$(\ln x)' = 1/x, \quad \forall\, x \in R^+$$

Derivation of Natural Logarithm

Now, let it be $y(x) = \ln|x|$; then the derivative at $x = 2$ (Figure 5.8) presents the two forms,

$$\lim_{x\to 2} \frac{\ln|x| - \ln|2|}{x-2}, \ x \neq 2, \text{ or also } \lim_{h\to 0} \frac{\ln|2+h| - \ln|2|}{h}$$

Manipulating the difference of exponentials in the second relation,

$$\lim_{h\to 0} \frac{\ln|2+h| - \ln|2|}{h} = \lim_{h\to 0} \frac{\ln\left|\frac{2+h}{2}\right|}{h} = \lim_{h\to 0} \frac{\ln\left|1+\frac{h}{2}\right|}{h}$$

Taking limit when h approaches zero, and considering the remarkable limit in the quotient,

$$\lim_{h\to 0} \frac{\ln\left|1+\frac{h}{2}\right|}{(h/2)}(1/2) = 1 \times (1/2) = 1/2$$

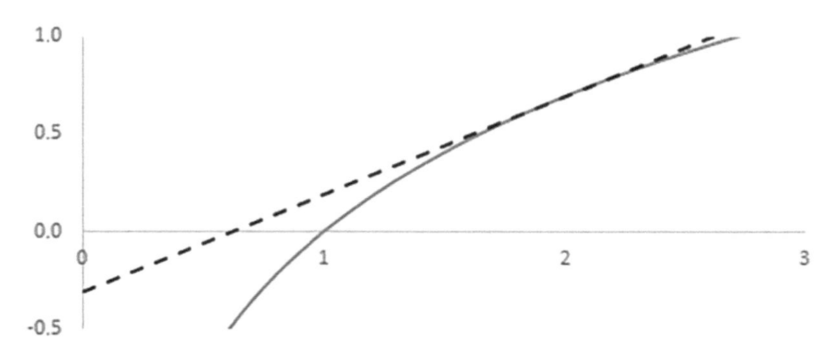

FIGURE 5.8 Logarithm function and tangent at point 2.

Beyond point 2, note the operations performed in the logarithmic's derivation are all possible in the real positive domain; thus,

$$\ln'|x| = 1/x, \quad \forall\, x \neq 0$$

Derivation of Logarithm of Positive Base, $a \neq 1$

And the derivative of logarithmic with base, $a \neq 1$, positive and different than 1:

$$(\log_a x)' = \frac{1}{x.\ln a}, \qquad \forall\, x \in R^+$$

- For example, let it be, $y(x) = \log_{10} x$; note $\log_{10}(e^{\ln x}) = \ln x.\log_{10}(e) = \ln x/\ln 10$, then the derivative follows:

$$(\log_{10} x)' = \frac{\ln' x}{\ln 10} = \frac{1}{x.\ln 10}$$

  And for instance, $y(x) = \log_2 x$; then,

$$(\log_2 x)' = \frac{\ln' x}{\ln 2} = \frac{1}{x.\ln 2}$$

Derivative of Composite Function

If $u = g(x)$ admits a derivative at the point x and $f'(u)$ exists for the corresponding point u; then, if the composite function $f[g(x)]$ is defined, the derivative of this function is determined using the chain derivation rule:

$$(fog)'(x) = f'[g(x)].g'(x)$$

Or with $u = g(x)$,

$$[f(u)]' = f'(u).u'$$

Setting $u = g(x)$, and applying this composite function rule to exponential functions, the following relations apply:

$$\begin{aligned}(e^u)' &= e^u.u' \\ (a^x)' &= a^x.\ln a \\ (a^u)' &= a^u.u'.\ln a\end{aligned}$$

- For example, let it be, $y(x) = e^{-x^2}$; the derivative

$$\left(e^{-x^2}\right)' = (e^u)'_{u=-x^2} = (e^u.u')_{u'=-2x} = -2x \;\; e^{-x^2}$$

And let it be, $y(x) = 2^x = (e^{\ln 2})^x \; = e^{(\ln 2)x}$; in this way,

$$\begin{aligned}(2^x)' &= [e^{(\ln 2)x}]' = (e^u)'_{u=(\ln 2)x} = (e^u.u')_{u'=\ln 2}\\ &= e^{(\ln 2)x}.\ln 2 = 2^x.\ln 2\end{aligned}$$

And for instance, $y(x) = 10^{x^3}$; then, the derivative for the base-10 exponential,

$$\left(10^{x^3}\right)' = (a^u)'_{u=x^3} = (a^u.u')_{u'=3x^2} = 3x^2 \;\; 10^{x^3}\ln 10$$

Similarly, following the composite derivation rule to logarithmic functions, the equalities apply:

$$\begin{aligned}(\ln|u|)' &= \frac{u'}{u}\\ (\log_a u)' &= \frac{u'}{u.\ \ln a}\end{aligned}$$

- Let it be, $y(x) = \ln|x^2+1|$; the derivative of the composite function:

$$\ln'|x^2+1| = (\ln|u|)'_{u=x^2+1} = \left(\frac{u'}{u}\right)_{u'=2x} = \frac{2x}{x^2+1}$$

And consider, $y(x) = \log_{10}|x-3|$; in the same way,

$$\log'_{10}|x-3| = (\log_{10}u)'_{u=x-3} = \left(\frac{u'}{u.\ln a}\right)_{u'=1} = \frac{1}{(x-3).\ln 10}$$

And now, let it be, $y(x) = \ln\left|\frac{1+x}{1-x}\right| = \ln|1+x| - \ln|1-x|$; deriving the two terms in the subtraction will be preferable to deriving the quotient of square roots, then,

$$\begin{aligned}\ln'|1+x| - \ln'|1-x| &= (\ln|u|)'_{u=1+x} - (\ln|u|)'_{u=1-x}\\ &= \left(\frac{u'}{u}\right)_{u'=1} - \left(\frac{u'}{u}\right)_{u'=-1} = \frac{1}{1+x} - \frac{-1}{1-x}\\ &= \frac{1-x}{1^2-x^2} + \frac{1+x}{1^2-x^2} = \frac{2}{1-x^2}\end{aligned}$$

5.4.3 Trigonometric Functions

Due to their importance on modeling a number of contexts, either on real world situations, social life, or economics, trigonometric functions and associated tools are widely used, and this way their derivatives are also key for new developments.

Derivation of Sine and Cosine Functions

The sine function is differentiable over the entire domain and its derivative is cosine function:

$$\sin'(x) = \cos(x)$$

And deriving cosine function results in the symmetric of the sine function:

$$\cos'(x) = -\sin(x)$$

Sine Function

Now, let it be $y(x) = \sin(x)$; then the derivative at $x = \pi/4$ (Figure 5.9) follows,

$$\lim_{h \to 0} \frac{\sin(\pi/4 + h) - \sin(\pi/4)}{h}$$

Manipulating the difference of sines in numerator, as in the trigonometric relation,

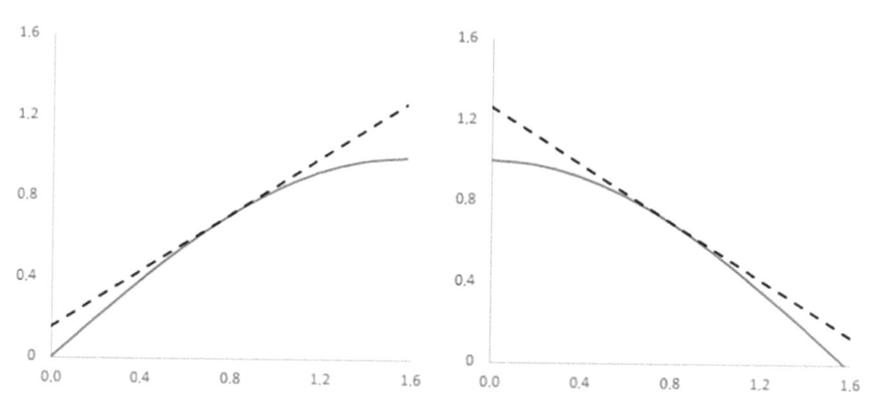

FIGURE 5.9 Tangent lines at point $\pi/4$, for both the sine (at left) and cosine (at right) functions.

$$\sin(a) - \sin(b) = 2\sin\left(\frac{a-b}{2}\right)\cos\left(\frac{a+b}{2}\right)$$

where $a = \pi/4 + h$ and $b = \pi/4$, then

$$\lim_{h\to 0}\frac{2\sin\left(\frac{\pi/4+h-\pi/4}{2}\right)\cos\left(\frac{\pi/4+h+\pi/4}{2}\right)}{h} = \lim_{h\to 0}\frac{2\sin\left(\frac{h}{2}\right)\cos\left(\pi/4+\frac{h}{2}\right)}{h}$$

Taking limit when h approaches zero, and considering the remarkable limit in the first factor,

$$\lim_{h\to 0}\frac{\sin(h/2)}{(h/2)}.\lim_{h\to 0}\cos\left(\pi/4+\frac{h}{2}\right) = 1.\cos(\pi/4+0) = \cos(\pi/4)$$

Therefore, the derivation of sine function is verified at point, $\pi/4$; note the presented operations are possible in all the points of the real domain, that is, $\forall\ x \in R$.

Cosine Function

Let it be $y(x) = \cos(x)$; the cosine derivative, again at $x = \pi/4$ (Figure 5.9), takes the form,

$$\lim_{h\to 0}\frac{\cos(\pi/4+h) - \cos(\pi/4)}{h}$$

The following trigonometric relation allows to treat the difference of cosines,

$$\cos(a) - \cos(b) = -2\sin\left(\frac{a+b}{2}\right)\sin\left(\frac{a-b}{2}\right)$$

where $a = \pi/4 + h$ and $b = \pi/4$, then

$$\lim_{h\to 0}\frac{-2\sin\left(\frac{\pi/4+h+\pi/4}{2}\right)\sin\left(\frac{\pi/4+h-\pi/4}{2}\right)}{h}$$

$$= \lim_{h\to 0}\frac{-2\sin\left(\pi/4+\frac{h}{2}\right)\sin\left(\frac{h}{2}\right)}{h}$$

When h approaches zero, and taking the remarkable limit in the second factor,

$$-\lim_{h\to 0}\sin\left(\pi/4+\frac{h}{2}\right).\lim_{h\to 0}\frac{\sin(h/2)}{(h/2)}=-\sin(\pi/4+0).1=-\sin(\pi/4)$$

The derivative of cosine function is obtained at point, $\pi/4$; again the presented operations are suitable for all the points in the real domain, that is, $\forall\, x \in R$.

- Applying the composite function rule to the sine and cosine functions:

$$\begin{aligned}\sin'(u) &= \cos(u).u'\\ \cos'(u) &= -\sin(u).u'\end{aligned}$$

For example, let it be, $y(x) = \sin(x^3+1)$; the derivative of the composite function,

$$\sin'(x^3+1) = (\sin u)'_{u=x^3+1} = [\cos(u).u']_{u'=3x^2} = 3x^2\cos(x^3+1)$$

And now consider, $y(x) = \ln|\sin(x)|$; following the composite derivation rule,

$$\ln'|\sin(x)| = (\ln|u|)'_{u=\sin(x)} = \left(\frac{u'}{u}\right)_{u'=\cos(x)} = \frac{\cos(x)}{\sin(x)} = \text{cotan}(x)$$

In the same way, consider $y(x) = \ln|\cos(x)|$; by derivation,

$$\ln'|\cos(x)| = (\ln|u|)'_{u=\cos(x)} = \left(\frac{u'}{u}\right)_{u'=-\sin(x)} = \frac{-\sin(x)}{\cos(x)} = -\tan(x)$$

Derivation of Tangent and Cotangent Functions

The tangent function, $\tan(x) = \sin(x)/\cos(x)$, is differentiable over its entire domain and

$$\tan'(x) = 1/\cos^2(x) = 1+\tan^2(x), \quad \text{with } \cos(x)\neq 0.$$

In fact, through the quotient's derivative,

$$\tan'(x) = [\sin(x)/\cos(x)]' = \left[\frac{f'.g - f.g'}{g^2}\right]_{\substack{f=\sin(x)\\ g=\cos(x)}}$$

$$= \frac{\sin'(x).\cos(x) - \sin(x).\cos'(x)}{\cos^2(x)}$$

$$= \frac{\cos(x).\ \cos(x) + \sin(x).\ \sin(x)}{\cos^2(x)} = \frac{1}{\cos^2(x)};$$

or separating the terms in numerator,

$$= 1 + \frac{\sin^2(x)}{\cos^2(x)} = 1 + \tan^2(x)$$

In a similar approach, the cotangent function, $\text{cotan}(x) = \cos(x)/\sin(x)$, is differentiable over its entire domain and its derivative is also obtained by treating the associated quotient,

$$\text{cotan}'(x) = -1/\sin^2(x) = -1 - \text{cotan}^2(x), \quad \text{with } \sin(x) \neq 0$$

- By the composite function rule, the associated derivatives follow:

$$\begin{aligned} \tan'(u) &= u'/\cos^2(u) \\ \text{cotan}'(u) &= -\,u'/\sin^2(u) \end{aligned}$$

5.5 DERIVATIVE OF INVERSE FUNCTION

In this section, important results about the derivative of inverse function are treated, including the inverse trigonometric functions and convenient examples; in particular, the derivation of the inverse cost function.

If function $y = f(x)$ admits an inverse function, $x = f^{-1}(y)$, and if the derivative function $f'(x)$ at a given point is different from zero, then $[f^{-1}(y)]' = 1/f'(x)$

In the suitable domain, let it be both, $y = f(x)$, and the inverse function, $x = f^{-1}(y)$. In the latter relation, by derivation of variable x, and noting the composite function, $f^{-1}(f(x)$,

$$(x)' = [f^{-1}(y)]' \Leftrightarrow 1 = [f^{-1}(f(x)]'$$

From the derivative of composite function in the second member,

$$1 = [f^{-1}(y)]' \, . \, f'(x)$$

And the reciprocal of function's derivative corresponds to the derivative of inverse function:

$$1/f'(x) = [f^{-1}(y)]'$$

Example—Cubic Root

- Let it be, $y(x) = \sqrt[3]{x}$, and the inverse function in the suitable domain, $x(y) = y^3$.
  From the relation for the inverse function derivative,

  $$[y(x)]' = 1/x'(y)$$

  Noting that,

  $$x'(y) = (y^3)' = 3y^2$$

  By substitution of the original relation, $y(x)$, then it results,

  $$(\sqrt[3]{x})' = \frac{1}{3y^2} = \frac{1}{3(\sqrt[3]{x})^2} = \frac{1}{3\sqrt[3]{x^2}}$$

  Such result can be directly verified, namely, recalling the power derivation and the fractional power, 1/3.

Example—Natural Logarithm

- Let it be, $y(x) = \ln|x|$, $x > 0$, and the inverse function in the suitable domain, $x(y) = e^y$.
  For the inverse function derivative, and noting that, $x'(y) = (e^y)' = e^y$,

  $$[y(x)]' = 1/x'(y) = 1/e^y$$

Substituting the original relation, $y(x)$, it follows,

$$\ln'|x| = 1/e^{y} = 1/e^{\ln|x|} = 1/x$$

And such result directly verifies the derivative for the natural logarithm (Section 5.3).

5.5.1 Inverse Trigonometric Functions

To derivate the inverse trigonometric functions, arctan(x) and arccotan(x), the same approach follows too. With an additional step that recalls the trigonometric fundamental relation to better reverse the variables at hand, $y(x)$ and $x(y)$, then arcsin(x) and arccos(x) are also treated.

Derivation of Arctangent

- Let it be, $y(x) = \arctan(x), \quad x \in R$, and the inverse function in the suitable domain,

$$x(y) = \tan(y), \quad y \in \left]-\frac{\pi}{2}, \frac{\pi}{2}\right[$$

From the inverse function derivative, and also noting that, $x'(y) = 1 + \tan^2(y)$,

$$\arctan'(x) = \frac{1}{\tan'(y)} = \frac{1}{1 + \tan^2(y)}$$

Substituting the original relation, $y(x)$, then it results,

$$\arctan'(x) = \frac{1}{1 + \tan^2(\arctan(x))} = \frac{1}{1 + x^2}$$

With a very similar procedure, the inverse trigonometric function of arccotan(x) can be derived too, and the corresponding derivative is:

$$\text{arccotan}'(x) = -1/(1 + x^2)$$

Derivation of Arcsine

- Let it be, $y(x) = \arcsin(x), \quad x \in [-1, 1]$, and the inverse function in the suitable domain,

$$x(y) = \sin(y), \quad y \in \left[-\frac{\pi}{2}, \frac{\pi}{2}\right]$$

For the inverse function derivative, and noting that, $x'(y) = \cos(y)$,

$$\arcsin'(x) = \frac{1}{\sin'(y)} = \frac{1}{\cos(y)}$$

In order to adequately express the cosine in denominator as function of variable, x, the trigonometric fundamental equality indicates a non-negative cosine in $[-\pi/2, \pi/2]$:

$$\sin^2(y) + \cos^2(y) = 1 \Leftrightarrow \cos^2(y) = 1 - \sin^2(y) \Leftrightarrow$$
$$\cos(y) = +\sqrt{1 - \sin^2(y)}$$

By substitution of the original relation, $y(x)$,

$$\arcsin'(x) = \frac{1}{\sqrt{1 - \sin^2(y)}} = \frac{1}{\sqrt{1 - \sin^2(\arcsin(x))}} = \frac{1}{\sqrt{1 - x^2}}$$

With a very similar procedure, the derivative for the inverse trigonometric arccos(x) can be obtained too, and then the corresponding derivative is:

$$\text{arcos}'(x) = -1/\sqrt{1 - x^2}$$

For derivation of inverse trigonometric functions, in case of composite functions, consider:

$$\arctan'(u) = u'/(1 + u^2) \qquad \text{arccotan}'(u) = -u'/(1 + u^2)$$
$$\arcsin'(u) = u'/\sqrt{1 - u^2} \qquad \text{arcos}'(u) = -u'/\sqrt{1 - u^2}$$

Example

- Let it be, $y(x) = x.\arctan(x) - \frac{1}{2}\ln|1 + x^2|$; the derivative considers the multiplication in the first term, and separately, the logarithm in the second term:

$$\begin{aligned} y'(x) &= x'.\arctan(x) + x.\arctan'(x) - \frac{1}{2}\ln'|1 + x^2| \\ &= 1.\arctan(x) + x.\frac{1}{1 + x^2} - \frac{1}{2}.\frac{2x}{x^2 + 1} \\ &= \arctan(x) + \frac{x}{1 + x^2} - \frac{x}{x^2 + 1} = \arctan(x) \end{aligned}$$

5.5.2 The Inverse Cost Function

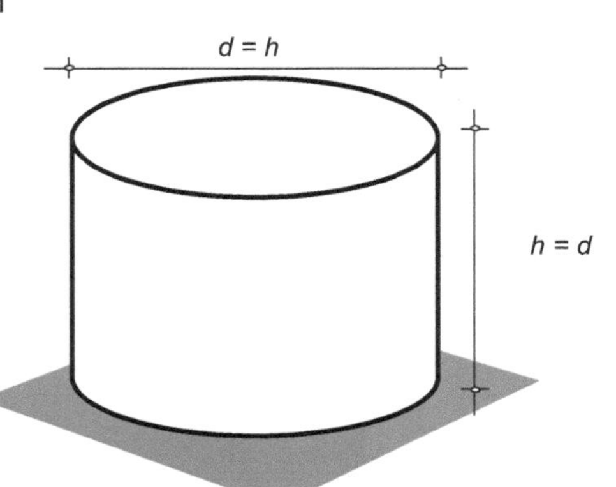

For the barrel's minimum cost, when assuming equal height and diameter, $h = d$, then the associated volume, $V(d)$, corresponds to:

$$V = h\ A_{Bas} = h\left(\frac{\pi}{4}d^2\right) = \frac{\pi}{4}d^3$$

- The expression for the cost with such optimal sizing, $C(d)$, becomes

$$\begin{aligned} C(d) &= 100\left(\frac{\pi}{2}d^2 + \frac{4\ V}{d}\right) = 50\ \pi d^2 + \frac{400}{d}\left(\frac{\pi}{4}d^3\right) \\ &= 50\ \pi\ d^2 + 100\pi\ d^2 = 150\ \pi\ d^2 \end{aligned}$$

  And this expression allows us to obtain the diameter, d, corresponding to a given cost, C:

$$d(C) = +\sqrt{\frac{C}{150\pi}} = \frac{1}{\sqrt{150\pi}}C^{1/2}$$

- In addition, the optimal cost can be associated with the volume too, $C(V)$,

$$C(V) = 150\pi\left(\sqrt[3]{\frac{4\ V}{\pi}}\right)^2 = 150\sqrt[3]{16\pi\ V^2}$$

  The inverse relation to calculate the volume for a given cost, $V(C)$, also follows:

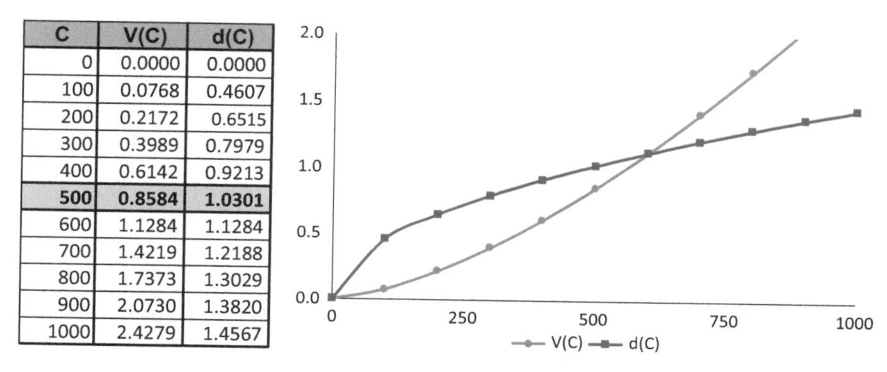

C	V(C)	d(C)
0	0.0000	0.0000
100	0.0768	0.4607
200	0.2172	0.6515
300	0.3989	0.7979
400	0.6142	0.9213
500	**0.8584**	**1.0301**
600	1.1284	1.1284
700	1.4219	1.2188
800	1.7373	1.3029
900	2.0730	1.3820
1000	2.4279	1.4567

FIGURE 5.10 Evolution of optimal volume (m^3) and diameter (m) with the cost (€), assuming equal height and diameter, $h = d$.

$$V(C) = +\sqrt{\frac{C^3}{16\pi .150^3}} = \frac{1}{4\sqrt{150^3\pi}}C^{3/2}$$

With such relations and the target cost of 500 €, the optimal volume is 858.4 liters, and the associated size for both diameter and height ($h = d$) is 1.030 meters, as in Figure 5.10.

Example—Cost Function on Diameter, C(d), and Inverse Function, d(C)

- Let it be, $d(C) = +\sqrt{\frac{C}{150\pi}} = \frac{1}{\sqrt{150\pi}}C^{1/2}$, and the inverse function in the positive domain, $C(d) = 150\,\pi\, d^2$.

  The symmetric evolution of the two variables, $d(C)$ and $C(d)$, is presented in Figure 5.11, and the relation for the inverse function derivative satisfies the relation:

$$d'(C) = \frac{1}{C'(d)}$$

Note that,

$$C'(d) = 150\pi\,(d^2)' = 150\pi\,(2d)$$

By substituting the original relation, $d(C)$, results in,

$$d'(C) = \frac{1}{150\pi\,(2d)} = \frac{1}{150\pi\,(2)\sqrt{\frac{C}{150\pi}}} = \frac{1}{2\sqrt{150\pi\ C}}$$

Such result can be directly confirmed, namely, from the square root derivative:

x	C(d)	d(C)
0.0	0.00000	0.000
0.1	0.00471	0.461
0.2	0.01885	0.651
0.3	0.04241	0.798
0.4	0.07540	0.921
0.5	0.11781	1.030
0.6	0.16965	1.128
0.7	0.23091	1.219
0.8	0.30159	1.303
0.9	0.38170	1.382
1.0	0.47124	1.457

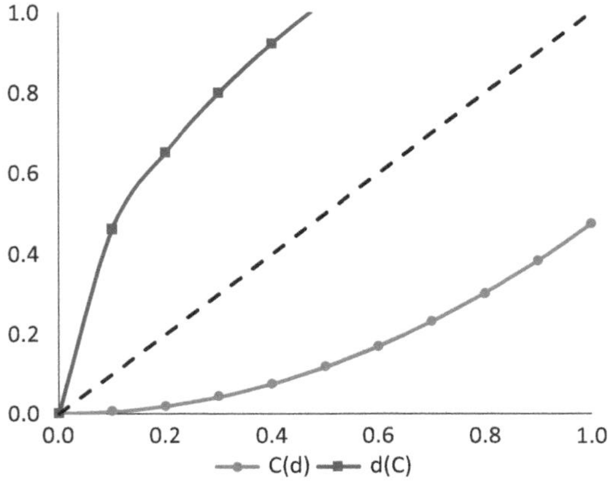

FIGURE 5.11 Symmetric evolution of optimal cost (10^3 €) and diameter (m), assuming equal height and diameter, $h = d$.

$$d'(C) = \frac{1}{\sqrt{150\pi}}(C^{1/2})' = \frac{1}{\sqrt{150\pi}} \cdot \frac{1}{2\sqrt{C}} = \frac{1}{2\sqrt{150\pi\ C}}$$

Example—Cost Function on Volume, C(V), and Inverse Function, V(C)

- Let it be, $V(C) = \left(\sqrt{\frac{C}{150\sqrt[3]{16\pi}}}\right)^3 = \frac{1}{\sqrt{16\pi.150^3}}C^{3/2} = \frac{1}{\sqrt{154\pi.10^6}}C^{3/2}$, and the inverse function in the positive domain, $C(V) = 150\sqrt[3]{16\pi}\ V^{2/3}$.

  In Figure 5.12, the symmetric evolution of the two variables, $V(C)$ and $C(V)$, is presented and the relation for the inverse

x	C(V)	V(C)
0.0	0.00000	0.000
0.1	0.11927	0.077
0.2	0.18932	0.217
0.3	0.24808	0.399
0.4	0.30053	0.614
0.5	0.34873	0.858
0.6	0.39381	1.128
0.7	0.43643	1.422
0.8	0.47706	1.737
0.9	0.51603	2.073
1.0	0.55358	2.428

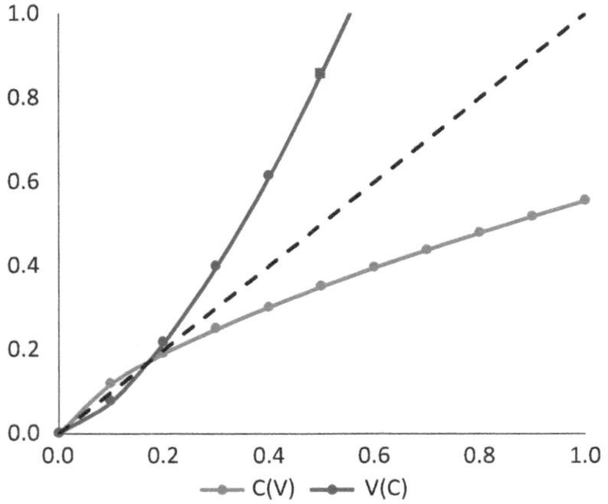

FIGURE 5.12 Symmetric evolution of optimal cost (10^3 €) and volume (m^3), assuming equal height and diameter, $h = d$.

function derivative follows too, $V'(C) = 1/C'(V)$.

Note that,

$$C'(V) = 150\sqrt[3]{16\pi}\,(V^{2/3})' = 150\sqrt[3]{16\pi}\left(\frac{2}{3}V^{-1/3}\right) = \frac{2 \times 150\sqrt[3]{16\pi}}{3\sqrt[3]{V}}$$

Again, by substitution of the original relation, $V(C)$, it follows:

$$V'(C) = \frac{1}{\frac{2 \times 150\sqrt[3]{16\pi}}{3\sqrt[3]{V}}} = \frac{3\sqrt[3]{V}}{2 \times 150\sqrt[3]{16\pi}} = \frac{3\sqrt{\frac{C}{150\sqrt[3]{16\pi}}}}{2 \times 150\sqrt[3]{16\pi}} = \frac{3}{2}\sqrt{\frac{C}{54\pi.10^6}}$$

Such result can be verified too, namely, from the fractional power derivative:

$$V'(C) = \frac{1}{\sqrt{154\pi.10^6}}(C^{3/2})' = \frac{1}{\sqrt{154\pi.10^6}}.\frac{3\sqrt{C}}{2} = \frac{3}{2}\sqrt{\frac{C}{54\ \pi.10^6}}$$

5.6 DERIVATIVES OF DIFFERENT ORDERS

By derivation of the derivative function $f'(x)$ in the suitable domain, the second-order derivative $f''(x)$ is obtained; then, to obtain the third derivative function, $f'''(x)$, the second derivative function is derived too; and so on, to obtain the nth-order derivative, the function is derived n times.

- The second-order derivative $f''(x)$ of a function $f(x)$ is determined by differentiating the derivative function $f'(x)$, alias, the first-derivative function:

$$[f(x)]'' = [f'(x)]' = f''(x)$$

- The derivative of order n, $f^{(n)}(x)$, of a function f is obtained by differentiating the derivative function of the previous order, $f^{(n-1)}(x)$,

$$f^{(n)}(x) = [f^{(n-1)}(x)]'$$

Example—The Hyperbolic Sine Function

Let it be $f(x) = \sinh(x)$.

The associated derivative corresponds to the hyperbolic cosine function,

$$f'(x) = \sinh'(x) = \cosh(x)$$

And the second derivation results again in the hyperbolic sine function,

$$f''(x) = \cosh'(x) = \sinh(x)$$

The successive derivations follow,

$$f'''(x) = \sinh'(x) = \cosh(x)$$

$$f^{(IV)}(x) = \cosh'(x) = \sinh(x)$$

A cycle of two derivatives occurs, and the general formula for nth-order derivative oscillates between:

$$f^{(n)}(x) = \begin{cases} \cosh(x), & \text{if } n \text{ is odd} \\ \sinh(x), & \text{if } n \text{ is even} \end{cases}$$

Example—The Sine Function

Let it be now, $f(x) = \sin(x)$.

The associated derivative corresponds to the cosine function,

$$f'(x) = \sin'(x) = \cos(x) = \sin\left(x + \frac{\pi}{2}\right)$$

The second derivation results in the symmetric sine function,

$$f''(x) = \cos'(x) = -\sin(x) = \sin(x + \pi)$$

And successive derivations follow, among other trigonometric manipulations,

$$f'''(x) = -\sin'(x) = -\cos(x) = \sin\left(x + \frac{3\pi}{2}\right)$$

$$f^{(IV)}(x) = -\cos'(x) = +\sin(x) = \sin(x + 2\pi)$$

A new cycle of four derivatives thus starts with the fifth-order's derivation:

$$f^{(V)}(x) = \sin(x) = -\cos'(x) = \sin\left(x + \frac{5\pi}{2}\right) = \sin\left(x + \frac{\pi}{2}\right)$$

And the nth-order derivative for the sine function presents the general formula:

$$f^{(n)}(x) = \sin\left(x + \frac{n\pi}{2}\right)$$

5.6.1 Derivatives of Higher Order to the Cost Function

Recall again the cost function, $y(x) = 50\pi\, x^2 + \frac{400}{x}$.

- For the first-order derivative, $y'(x)$, the first term is a quadratic term; the derivation of the negative power in the second term leads to:

$$\begin{aligned} y'(x) = \left(50\pi\, x^2 + \frac{400}{x}\right)' &= 50\pi(2x) + 400.(-1)x^{-1-1} \\ &= 100\pi\, x - 400x^{-2} = 100\pi x - \frac{400}{x^2} \end{aligned}$$

- For the second-order derivative, $y''(x)$, by deriving the first-order derivative,

$$\begin{aligned} y''(x) = (100\pi\, x - 400x^{-2})' &= 100\pi(1) - 400.(-2)x^{-2-1} \\ &= 100\pi + 800x^{-3} = 100\pi + \frac{800}{x^3} \end{aligned}$$

- Similarly, the third-derivative is obtained too:

$$\begin{aligned} y'''(x) &= 0 + 2.400.(-3)x^{-3-1} \\ &= -(2.3).400x^{-4} \\ &= \frac{-2400}{x^4} \end{aligned}$$

- Following with the derivation procedures, the second member is treated as a negative power and the derivatives of higher order can be obtained straightforward:

$$y^{(IV)}(x) = -400.\,3!(-4)\,x^{-4-1} = +400.\,4!\,x^{-5} = \frac{400 \times 4!}{x^5}$$

$$y^{(V)}(x) = 400.\,4!(-5)\,x^{-5-1} = -400.\,5!\,x^{-6} = \frac{-400 \times 5!}{x^6}$$

And the derivative of nth order,

$$y^{(n)}(x) = (-1)^n\, 400\; n!\; x^{-n-1} = \frac{(-1)^n\, 400\; n!}{x^{n+1}}$$

Example—Fifth Order Polynomial, $P_5(x)$

Let the fifth order polynomial be $P_5(x)$, corresponding to a Taylor series (Chapter 10):

$$P_5(x) = 2400 - 6000x + (50\pi + 8000)x^2 - 6000x^3 + 2400x^4 - 400x^5$$

The associated derivatives are obtained by successively deriving the polynomials, term after term, since the derivative of a power function is also a power function of lower order: $(x^n)' = n\, x^{n-1}$. In that manner:

$$\begin{aligned}
P_5'(x) &= -\,6000 + (100\pi + 16000)x - 18000x^2 + 9600x^3 - 2000x^4 \\
P_5''(x) &= (100\pi + 16000) - 36000x + 28800x^2 - 8000x^3 \\
P_5'''(x) &= -\,36000 + 57600x - 24000x^2 \\
P_5^{(IV)}(x) &= 57600 - 48000x \\
P_5^{(V)}(x) &= -\,48000 \\
P_5^{(VI)}(x) &= 0
\end{aligned}$$

At point $x = 1$, this Taylor polynomial takes the same value as the cost function, $y(x)$, as well as their derivatives at that point; for instance, the fifth-order derivatives for both functions at that point result −48000:

$$P_5^{(V)}(1) = -48000$$

$$y^{(5)}(1) = \frac{(-1)^5 400\; 5!}{(1)^{5+1}} = -400 \times 120 = -48000$$

5.7 CONCLUDING REMARKS

Functions derivation and the related results are key for many developments and applications, including the mathematical conditions for minima or maxima (e.g., cost or return), or also geometric and numerical enhancements with large impacts in different scientific fields, technologies, engineering, not counting the economic and management sciences.

- The derivative of a function at a point is revisited, along with the associated geometric interpretation, with convenient instances for a number of important functions. The study of lateral derivatives is complementing the interconnection between derivation, continuity, and lateral limits; in addition, instances for derivation rules (sum and subtraction, multiplication and division, power and roots) are treated, not to mention the derivation of elementary functions.
- In that way, the derivatives of exponential and logarithmic functions, the sine and cosine functions, or the tangent and cotangent functions are detailed. The composite form is also addressed for the referred functions, while convenient examples are treated.
- In addition, important results about the derivative of inverse function are presented, including the inverse trigonometric functions; a simple comparison analysis is also developed for the cylinder case, namely, by inverting and then deriving (both in relation to diameter and volume) the inverse cost function.
- Finally, derivatives of different orders are presented; by derivation of the derivative function $f'(x)$ in the suitable domain, the second-order derivative $f''(x)$ is obtained; then, to obtain the third derivative function, $f'''(x)$, the second derivative function is derived too, and so on, to obtain the nth–order derivative, the function is derived n times.

These results about differentiable functions are very useful to functions sketching, which is presented in the next chapter. The location of first derivative zeros is very important to identify extreme points, and important results and theorems (Rolle, Lagrange, Cauchy) about differentiable functions will follow too. Among others of interest, such results are also utilized for numerical estimates and error bounding, in the treatment of indeterminate limits, in developments of integral calculus, or even in creative approaches to more difficult functions from the moment they are differentiable. In addition, beyond the function's inflection points that are related with the second derivative zeros, the evaluation of asymptotes usually requires the survey of indeterminacies. In general, the study of functions variation is combining roots location, function monotony, identification of maxima/minima, in conjuction with graph concavity and intersection points.

CHAPTER 6

Sketching Functions and Important Theorems

To better address the sketching of functions, the role of first derivative to the function monotony, and to identify roots and extreme points is discussed in first place; thus, important results about differentiable functions are presented through a reduction approach, including Rolle's theorem about the derivative roots, and Lagrange's theorem about the derivative mean value. After that, the second-order derivative and the function concavity is completing the study of maxima and minima. Asymptotes are also treated, either vertical, oblique, or horizontal asymptotes; Cauchy's theorem that relates to the increments of two functions is presented, and L'Hôpital's rule is applied to cope with the asymptotes indeterminate limits. In addition, sketching procedures for the cost function are described, and a table summarizes the overall function's variation, including the extension to the negative domain. Other applications of interest are detailed too, such as the Newton-Raphson method to size up a barrel of 500 €, and the error evolution along the method iterations.

6.1 INTRODUCTION

In general, the study of functions considers the intersection with axis (e.g., roots location), function monotony (increasing and decreasing), identification of extreme points (maxima and minima), as well as the graphic concavity (downward and upward) and inflection points. Thus, the evaluation of first derivative (if positive, the function increases, and vice versa) and the alteration of its sign (from positive to negative, and vice versa), along with the second derivative (if negative, the curvature is downward, and vice versa) and the respective sign alterations, are all important elements to functions sketching. In addition, beyond the extrema and inflection points that are usually related with the derivatives zeros, the evaluation of asymptotes usually requires the survey of indeterminacies.

And several theorems about differentiable functions are very useful to functions sketching. On a generalization path, the Rolle's theorem about the

DOI: 10.1201/9781003461876-6

derivative roots is treated in the first place; in a second step, the Lagrange theorem that focuses on the derivative's mean value is addressed too; after that, the Cauchy theorem that treats the relation between increments of two functions is revisited, and L'Hôpital's rule is addressed too.

- However, a reduction approach is also suitable, by reversing the referred path; namely, by defining $g(x) = x$ in the increments relation, and then obtaining the slope at intermediate point, c, in the closed range $[a, b]$; also, the horizontal slope occurs when function $f(x)$ takes two times the same value, and so $f'(c) = 0$.
- The application of L'Hôpital's rule to the treatment of indeterminate limits is very effective, including the treatment of asymptotes for the cost function. Nevertheless, the potential pitfall of circular thinking may occur, namely, in the evaluation of remarkable limits (the utilization of derivatives shall take account if the result was already utilized to obtain the derivative itself!) or when calculating a function derivative through its definition (usually an indeterminacy occurs, and L'Hôpital's rule proposes to utilize the derivative, when it is the derivative itself to be calculated!).

The variation of cost function $y(x)$ is studied along this chapter, as follows: first, important theorems are revisited; then, the role of first derivative to study functions variation and to identify extreme points, both maxima and minima, is discussed; after that, function asymptotes are presented, due to their importance for function sketching; then sketching procedures for the cost function are described, and the typical attributes of functions variation are summarized. Finally, other applications are presented, both the Newton-Raphson method to cope with the *Barrel That Costs-500* and the error evolution in this kind of methods.

Recall that defining the volume, $V = 1\ \text{m}^3$, the cost function depends only on diameter d,

$$C(1, d) = C(d) = 100\left(\frac{\pi}{2}d^2 + \frac{4}{d}\right)$$

By substitution of the variables pair, $y = C$ and $x = d$, then the instance under analysis follows:

$$y(x) = 100\left(\frac{\pi}{2}x^2 + \frac{4}{x}\right) = 50\pi\, x^2 + \frac{400}{x}$$

6.2 IMPORTANT THEOREMS ON DIFFERENTIABLE FUNCTIONS

This section presents important results about differentiable functions, including Rolle's theorem about derivative roots and the joint application with Bolzano theorem, as well as Lagrange's theorem that focuses on the derivative mean value or the function' finite increments. In addition, Cauchy's theorem that treats the relation between the increments of two functions and the application of L'Hôpital's rule to cope with limit indeterminacies are described.

6.2.1 The Derivative Roots Theorem

If function f(x) is continuous and differentiable at any interior point of interval [a, b], and if it is zero at the border points, f(a) = f(b) = 0, then there exists at least one intermediate point c, a < c < b, where the derivative f'(x) becomes zero, f'(c) = 0.

That is, between two roots of a continuous and differentiable function, $f(x)$, there is at least one root of the derivative function, $f'(x)$, as follows in Figure 6.1a.

$$\sin(0) = \sin(\pi) = 0 \Rightarrow \exists \ c \in \,]0, \pi[: \sin'(c) = 0$$

In this instance, the cosine (sine derivative) becomes zero at the intermediate point, $\pi/2$, and it coincides with the sine's local maximum too.

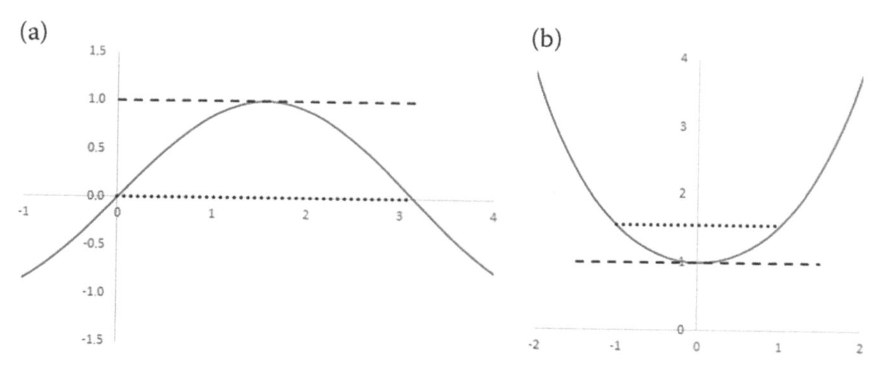

FIGURE 6.1 Roots of derivative function and zero value for slope lines. (a) Trigonometric sine, $\sin(x)$, in $[0, \pi]$ and (b) hyperbolic cosine, $\cosh(x)$, in $[-1, 1]$.

However, the equality of images in two different points is sufficient condition for the occurrence of a derivative zero, even if the function is not zero, as in Figure 6.1b. In fact, the hyperbolic cosine is even function, and it takes the same value at points 1 and −1:

$$\cosh(1) = \frac{e^{1} + e^{-1}}{2} = \frac{e^{-1} + e^{1}}{2} = \cosh(-1) \Rightarrow \exists \ c \in]-1, \ 1[: \cosh'(c) = 0$$

Now, the hyperbolic sine (derivative of hyperbolic cosine) becomes zero at the intermediate point, $x = 0$, and it coincides again with an extreme point, the function minimum in this case.

Between two consecutive zeros of the derivative, f′(x), there cannot be more than one root of the function under analysis, f(x).

By contradiction, consider the hypothesis that two function roots would occur between two consecutive zeros of the derivative, $f'(x)$; in such case, the prior result should apply, and the derivative would become zero in some intermediate point, c; thus, the hypothesis of two consecutive zeros for the derivative is not verified; if suitable, one and only one function root can occur between the two consecutive zeros for the derivative.

In order to confirm the existence of a function root between the two consecutive zeros of the derivative, the combined application with the Bolzano corollary is proposed: by evaluating the function sign in the consecutive zeros for the derivative, in case the sign alters, then at least one function root occurs; if the function sign still is the same, then no root exists in the range at hand.

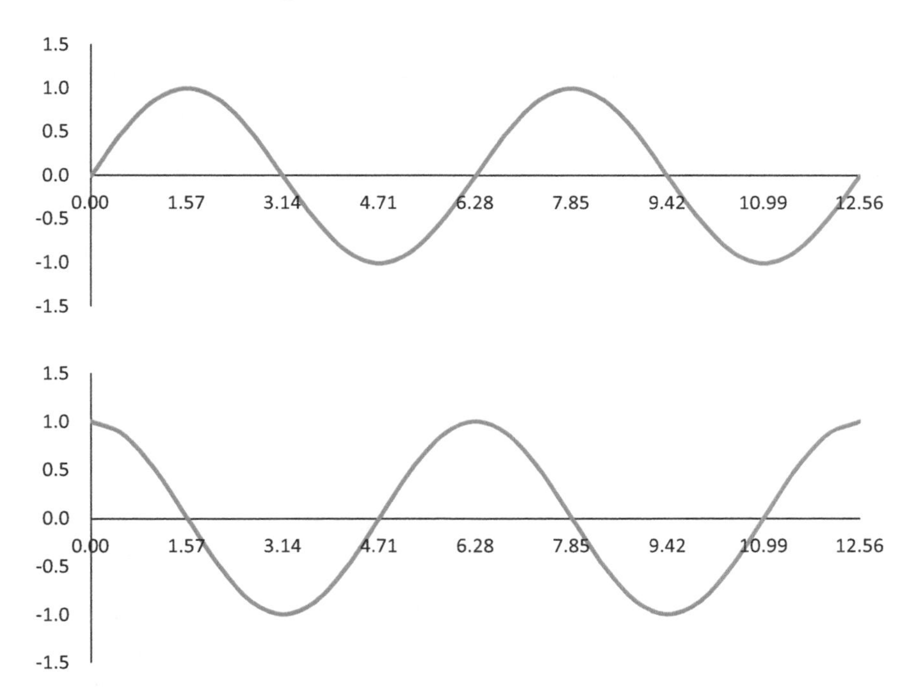

FIGURE 6.2 The roots of derivative function, $\sin'(x) = \cos(x)$. From top to down: $\sin(x)$; $\cos(x)$.

A close observation is proposed for this combined procedure. The sine function and its derivative, cosine, are both represented in Figure 6.2; for instance, the derivative function (cosine) becomes zero at consecutive points, $\pi/2$ and $3\pi/2$; the sine function is positive at the first point, $\sin(\pi/2) = 1$, and it then alters to negative, $\sin(3\pi/2) = -1$. By the intermediate values theorem, the function admits at least one root at some intermediate point in the interval $]\pi/2, 3\pi/2[$; in fact, the root occurs at point π, since $\sin(\pi) = 0$.

6.2.2 The Derivative Mean Value Theorem

If function f(x) is continuous and differentiable at any interior point of interval [a, b], then exists at least one intermediate point c, a < c < b, such that,

$$\frac{f(b) - f(a)}{b - a} = f'(c)$$

Or better, there is an intermediate point, c, whose tangent has the same slope as the secant drawn between the border points, as in Figure 6.3.

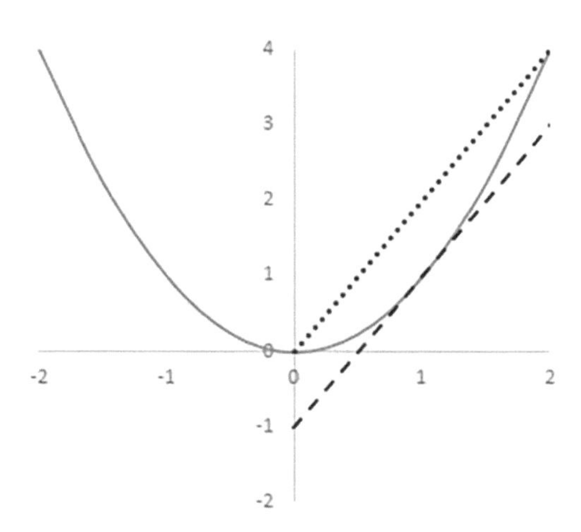

FIGURE 6.3 The derivative mean value, $f'(c)$.

Also, it relates the increment of function $f(x)$ with the increment of variable x:

$$f(b) - f(a) = f'(c) \,.\, (b - a)$$

The derivative mean value allows the evaluation of function increments, and it is very effective for numerical analysis and computational developments; namely, it allows to evaluate and bound the error committed in a number of approximations (e.g., roots of equations, series).

Examples

- Let $Y(x) = 50\pi\frac{x^3}{3} + 400 \quad \ln|x| - \frac{50\pi}{3}$, in the range [1, 3] and noting $Y(1) = 0$; from the mean value theorem in that range, there is an intermediate point, c, whose tangent has the same inclination as the secant drawn between points 1 and 3 (Figure 6.4).

$$\frac{Y(3) - Y(1)}{3 - 1} = \frac{\left[50\pi\frac{3^3}{3} + 400 \quad \ln|3| - \frac{50\pi}{3}\right] - 0}{2} = \frac{1800.802}{2} = 900.401$$

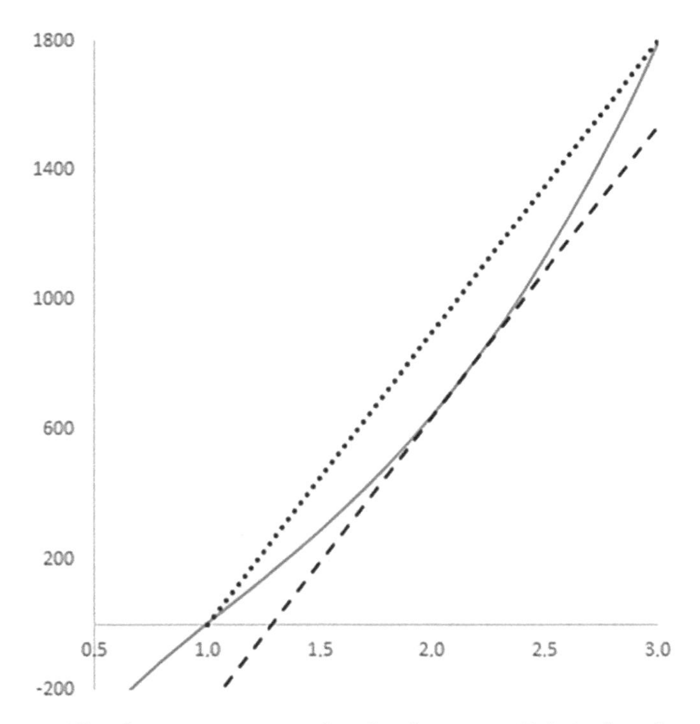

FIGURE 6.4 The derivative mean value for function, $Y(x)$, in [1, 3].

The tangent inclination is obtained from the first-order derivative, $Y'(x)$,

$$Y'(x) = \left[50\pi\frac{x^3}{3} + 400\ln|x| - \frac{50\pi}{3}\right]' = 50\pi\left(\frac{3x^2}{3}\right) + 400\left(\frac{1}{x}\right) - 0$$
$$= 50\pi\ x^2 + \frac{400}{x}$$

Then there is at least one intermediate point where the derivative results 900.401, and such point can be obtained by numerical methods (e.g., bisection, Newton-Raphson):

$$\exists\ c \in\ ]1, 3[: Y'(c) = 900.401, \text{ and } c = 2.129919908.$$

Note the function, $Y(x)$, after derivation results in the cost function; in this way, $Y(x)$ is and anti-derivative function and is crucial to area estimates, namely, the mean cost in the range [1, 3] can be evaluated through the area between the cost graphic and X-axis.

- Recalling a prior instance for the hyperbolic cosine in [−1, 1], the secant drawn between the border points is horizontal because the function is even (Figure 6.1b); the derivative mean value thus becomes zero, and there exists at least one intermediate point where the derivative root occurs:

$$\cosh(1) - \cosh(-1) = \cosh'(c)\ .\ [1 - (-1)] = 0$$
$$\Rightarrow \exists\ \ c \in\ ]-1, 1[: \cosh'(c) = 0$$

6.2.3 The Relation between the Increments of Two Functions

If functions f(x) and g(x) are continuous and differentiable at any interior point of [a, b], and let g(x) be such that g'(x) does not vanish at any point of [a, b], then there is at least one intermediate point c, a < c < b, such that,

$$\frac{f(b) - f(a)}{g(b) - g(a)} = \frac{f'(c)}{g'(c)}$$

This important result allows estimates about the relation between increments of two functions, $f(x)$ and $g(x)$, and thus support the treatment of indeterminate limits (also, *ind.*), with developments aligned with L'Hôpital's rule.

In addition, a reduction approach involving these important theorems is summarized:

- Let the relation between two functions $f(x)$ and $g(x)$, both continuous and differentiable in the closed interval $[a, b]$, and $g'(x)$ does not vanish at any point in that segment; then there is at least a point c, $a < c < b$, such that the function increments,

$$\frac{f(b) - f(a)}{g(b) - g(a)} = \frac{f'(c)}{g'(c)}$$

- Now, let $g(x) = x$; by substituting it in the prior relation, the increment in function, $f(x)$, can be estimated via the derivative at an intermediate point, $f'(c)$,

$$\frac{f(b) - f(a)}{b - a} = \frac{f'(c)}{1} \text{ or}$$

$$f(b) - f(a) = f'(c) \,.\, (b - a)$$

  Such value, $f'(c)$, where $a < c < b$, is said the derivative mean value, and it is very often utilized to bound the errors in numerical calculations (since the exact error is not known!).

- And if $f(b) = f(a)$, then the function increment in the interval $[a, b]$ is zero, and so it is the associated mean value; then the derivative becomes zero in that segment, $f'(c) = 0$.

  In addition, if $f(b) = f(a) = 0$, then $f'(c) = 0$ too, and a root of derivative thus occurs in-between these two roots of function, $f(x)$, since $a < c < b$.

6.2.4 L'Hôpital's Rule

Let be two functions $f(x)$ and $g(x)$ that satisfy the conditions of Cauchy's theorem over the range $[a, b]$, and that both cancel at the point $x = a$,

$f(a) = g(a) = 0$; if the limit of derivatives quotient $f'(x)/g'(x)$ exists as x approaches a, then there exists the limit of quotient $f(x)/g(x)$ at point a, and

$$\lim_{x \to a} \frac{f(x)}{g(x)} = \lim_{x \to a} \frac{f'(x)}{g'(x)}$$

A very common application is the survey of indeterminacies in infinitely small quotient limits; thus, it only requires the existence of derivatives at point, a, that are not simultaneously zero or infinite.

Examples and Pitfalls

- Two simple instances follow and show the typical procedure:

$$\lim_{x \to 0} \frac{x^3 - x^2}{x^2 - x} = \left(\frac{0}{0}\right), \ ind.$$

$$= \lim_{x \to 0} \frac{(x^3 - x^2)'}{(x^2 - x)'} = \lim_{x \to 0} \frac{3x^2 - 2x}{2x - 1} = \frac{0 - 0}{0 - 1} = \frac{0}{-1} = 0$$

And also

$$\lim_{x \to 0} \frac{e^x - e^{-x}}{2x} = \left(\frac{0}{0}\right), \ ind.$$

$$= \lim_{x \to 0} \frac{(e^x - e^{-x})'}{(2x)'} = \lim_{x \to 0} \frac{e^x + e^{-x}}{2} = \frac{1 + 1}{2} = 1$$

A common pitfall in using L'Hôpital's rule is the occurrence of some kind of circular thinking, namely, by calculating derivatives by definition or even remarkable limits.

- Consider the **quadratic function,** $y(x) = x^2$; the derivative is well known, $y'(x) = 2x$; when treating the derivative definition, the typical indeterminacy occurs:

$$\lim_{x \to a} \frac{f(x) - f(a)}{x - a} = \lim_{x \to a} \frac{x^2 - a^2}{x - a} = \left(\frac{0}{0}\right), \ ind.$$

However, the numerator derivation would require the usage of the derivative itself, and it just is being proven. Circular thinking occurs, because it is being assumed and utilized the final result within the demonstration steps.

- Consider the **remarkable limits**, in below; their result is also well known, 1; when treating these limits, the usual indeterminacy occurs:

$$\lim_{x\to 0}\frac{\sin(x)}{x} = \left(\frac{0}{0}\right),\ ind.$$
$$\lim_{x\to 0}\frac{e^x-1}{x} = \left(\frac{0}{0}\right),\ ind.$$
$$\lim_{x\to 0}\frac{\ln|x+1|}{x} = \left(\frac{0}{0}\right),\ ind.$$

Once again, the numerator derivation would require the usage of the corresponding derivatives: sine, exponential, and natural logarithm. However, circular thinking would occur, because it is assumed the related derivatives are correct and robust, but the final result for these limits, 1, was already utilized during the derivatives proof (Section 5.4).

Other Applications of L'Hôpital's Rule

- If $f'(a) = g'(a) = 0$, and if the derivative functions $f'(x)$ and $g'(x)$ satisfy the conditions required by L'Hôpital's rule, the rule can be applied again to the quotient, $f'(x)/g'(x)$:

$$\lim_{x\to a}\frac{f(x)}{g(x)} = \lim_{x\to a}\frac{f'(x)}{g'(x)} = \lim_{x\to a}\frac{f''(x)}{g''(x)}$$

- The rule is also applicable when a is an improper (infinite) point:

$$\lim_{x\to\pm\infty}\frac{f(x)}{g(x)} = \lim_{x\to\pm\infty}\frac{f'(x)}{g'(x)}$$

- The rule is also applicable to quotients between infinitely large instances,

$$\lim_{x\to a}\frac{f(x)}{g(x)} = \left(\frac{\infty}{\infty}\right) \Rightarrow \lim_{x\to a}\frac{f'(x)}{g'(x)}$$

6.3 MAXIMA AND MINIMA

The first derivative is key to study functions variation and to identify extreme points, both maxima and minima. First, the sufficient condition for relative extrema is focusing on the first-order derivative signs, and then the necessary condition is completing the analysis with the concavity analysis.

The First Derivative and the Function Variation

Being $f(x)$ a differentiable function on a closed interval $[a, b]$ of its domain, where it is verified for all points on that range:

- **If $f'(x) > 0$, the function is strictly increasing.**
  The derivative mean value in the range indicates, for all intermediate points, $x_a < c < x_b$,

$$\frac{f(x_b) - f(x_a)}{x_b - x_a} = f'(c) > 0, \quad \forall\, c \in \,]x_a, x_b[$$

  Since both the fraction denominator, $x_b - x_a > 0$, and the derivative in second member are positive, the numerator is also positive; then

$$f(x_b) - f(x_a) > 0 \Rightarrow f(x_b) > f(x_a)$$

- **If $f'(x) < 0$, the function is strictly decreasing.**
  In the opposite sense, if the second member's derivative is negative while the fraction denominator still is positive, $x_b - x_a > 0$, then the numerator is negative; thus,

$$f(x_b) - f(x_a) < 0 \Rightarrow f(x_b) < f(x_a)$$

- **If $f'(x) = 0$, the function is constant.**
  Similarly, when the derivative in second member is zero for all the intermediate points, $x_a < c < x_b$, within the interval $]a, b[$, the fraction numerator also results zero, and

$$f(x_b) - f(x_a) = 0 \Rightarrow f(x_b) = f(x_a)$$

Sufficient Condition for Relative Extrema—If function $f(x)$ is continuous and its first-order derivative $f'(x)$ changes sign at a point $x = a$ in its domain, then the function presents a relative (local) extremum at that point.

- **$f(a)$ is a maximum, if the derivative $f'(x)$ changes from positive to negative.**
  For all points on the left side of point a, $x_a < c < a$, the derivative mean value is positive; the fraction numerator is also positive and the function is strictly increasing:

$$\frac{f(a) - f(x_a)}{a - x_a} = f'(c) > 0, \quad \forall\, c \in \,]x_a, a[$$

$$f(a) - f(x_a) > 0 \Rightarrow f(a) > f(x_a)$$

  And for the points on the right side, $a < c < x_b$, the derivative mean value is negative; then the fraction numerator is negative too, and the function is strictly decreasing:

$$\frac{f(x_b) - f(a)}{x_b - a} = f'(c) < 0 \,, \quad \forall\, c \in \,]a, \; x_b[$$

$$f(x_b) - f(a) < 0 \Rightarrow f(x_b) < f(a)$$

  In this way, the conditions for existence of local maximum are satisfied at point a, both from the left side and the right side:

$$f(a) > f(x_a), \;\forall \;\; x_a < a$$
$$f(a) > f(x_b), \;\forall \;\; x_b > a$$

- **$f(a)$ is a minimum, if the derivative $f'(x)$ changes from negative to positive.**
  In the opposite sense, for all points on the left side of point a, $x_a < c < a$, the derivative mean value is negative and the function is strictly decreasing; thereafter, on the right side, $a < c < x_b$, the derivative mean value is positive, and the function is strictly increasing:

$$f(a) - f(x_a) < 0 \Rightarrow f(a) < f(x_a)$$

$$f(x_b) - f(a) > 0 \Rightarrow f(x_b) > f(a)$$

Therefore, the conditions of local minimum are satisfied both from the left side and the right side of point a:

$$f(a) < f(x_a), \ \forall \ \ x_a < a$$
$$f(a) < f(x_b), \ \forall \ \ x_b > a$$

These are the sufficient conditions for the existence of relative extrema, either local maximum or local minimum: the first derivative is changing signs. In addition, if the derivative function is defined at such extreme point a, it is necessary condition it becomes zero at that point, $f'(a) = 0$.

Necessary Condition for Relative Extrema—If the differentiable function $f(x)$ takes a relative extreme at point a, maximum or minimum, then its first derivative is zero at that point, $f'(a) = 0$.

- By the intermediate values theorem to the continuous derivative function, the changing signs indicates the derivative function results zero in such intermediate point; the equality of lateral derivatives is necessary for the derivative existence at point a, and then

$$f'(a^-) = 0^+ \wedge f'(a^+) = 0^- \Rightarrow f'(a) = 0, \text{ if point } a \text{ is local maximum;}$$

$$f'(a^-) = 0^- \wedge f'(a^+) = 0^+ \Rightarrow f'(a) = 0, \text{ if point } a \text{ is local minimum.}$$

Any other value than zero, either positive or negative, would make false these equalities.

Example—Minimum Point

- By recalling the cost function, and obtaining the first derivative,

$$y'(x) = 100\left(\frac{\pi}{2}x^2 + \frac{4}{x}\right)' = 100\left[\frac{\pi}{2}(2x) + 4\left(\frac{-1}{x^2}\right)\right] = 100\left(\pi \ x - \frac{4}{x^2}\right)$$

Then, the derivative zero can be obtained too,

$$y'(x) = 0 \Leftrightarrow 100\left(\pi\, x - \frac{4}{x^2}\right) = 0 \Leftrightarrow \pi\, x = \frac{4}{x^2} \Leftrightarrow x^3 = \frac{4}{\pi} \Leftrightarrow x = \sqrt[3]{\frac{4}{\pi}}$$

From a qualitative approach to the factor in parentheses: when positive variable x increases, the first term increases while the second term is subtracting a decreasing ratio; thus, the derivative is always increasing with the variable x, it is positive on the right side of its zero. And vice versa, the derivative was negative on the left side of its zero; in this way, the derivative sign alters from negative to positive, and the function monotony alters from decreasing to increasing: the point, $x = \sqrt[3]{4/\pi}$, is a local minimum.

Example—Maximum Point

- Let it be

$$y_1(x) = 100\left(\pi\, x - \frac{4}{x^2}\right)$$

Obtaining the first derivative, $y_1'(x)$,

$$y_1'(x) = 100\left(\pi\, x - \frac{4}{x^2}\right)' = 100\left[\pi - 4(-2x^{-3})\right] = 100\left(\pi + \frac{8}{x^3}\right)$$

and then calculating its zero,

$$y_1'(x) = 0 \Leftrightarrow 100\left(\pi + \frac{8}{x^3}\right) = 0 \Leftrightarrow x^3 = -\frac{8}{\pi} \Leftrightarrow \pi = -\frac{8}{x^3} \Leftrightarrow x = -\sqrt[3]{\frac{8}{\pi}}$$

From a qualitative approach to the factor in parentheses: when variable x increases, the second term corresponds to a decreasing ratio, for both the positive and negative domain; in the negative domain, when variable x is increasing toward zero, the ratio absolute value increases but with negative sign. Thus, the derivative $y_1'(x)$ is always decreasing in both domains, it is thus negative on the right side of its zero, and vice versa, the derivative $y_1'(x)$ was positive on the left side of its zero. Since the derivative sign alters from positive to negative, the function monotony alters from increasing to decreasing, and point, $x = -\sqrt[3]{8/\pi}$, is a local maximum of $y_1(x)$.

Example—No Extreme Points

- Now, let it be

$$y_2(x) = 100\left(\pi + \frac{8}{x^3}\right)$$

Obtaining the first derivative, $y_2'(x)$,

$$y_2'(x) = 100\left(\pi + \frac{8}{x^3}\right)' = 100\ [0 + 8(-3x^{-4})] = \frac{-2400}{x^4}$$

but calculating its zero is not possible, a false equality arises:

$$y_2'(x) = 0 \Leftrightarrow \frac{-2400}{x^4} = 0, \textit{ False}$$

The derivative function, $y_2'(x)$, is always negative in its domain; thus, the associated function $y_2(x)$ is always decreasing and local maxima and minima do not occur.

The functions $y_1(x)$ and $y_2(x)$ indeed correspond, respectively, to the first derivative and second derivative of cost function, $y(x)$; the denomination is adjusted to improve readability, otherwise these derivations and their relations with the cost function derivatives would be tiring.

The Second Derivative and the Graphic Concavity

If $f(x)$ is a differentiable function on an interval of its domain, where it is verified for all points of that interval:

- **If $f''(x) > 0$, the concavity is facing upward; then if $f'(a) = 0$, a local minimum occurs at that point a.**
  Similarly, applying the derivative mean value to the first derivative function, for all intermediate points in the range, $x_a < c < x_b$,

$$\frac{f'(x_b) - f'(x_a)}{x_b - x_a} = f''(c) > 0, \quad \forall\, c \in \,]x_a, x_b[$$

The numerator is thus positive, and the derivative function is strictly increasing;

$$f'(x_b) - f'(x_a) > 0 \Rightarrow f'(x_b) > f'(x_a)$$

At point a, $f'(a) = 0$; thereafter, on the right side the first derivative is positive, $f'(x) > 0$, while the derivative takes negative values on the left side, $f'(x) < 0$. Consequently, function $f(x)$ is decreasing before reaching point a, and $f(a) < f(x_a)$, $\forall \ x_a < a$; right after point a, function $f(x)$ is increasing and $f(a) < f(x_b)$, $\forall \ x_b > a$.

- **If $f''(x) < 0$, the concavity faces downward; then if $f'(a) = 0$, a local maximum occurs at that point a.**
  In the opposite sense, the first derivative function is strictly decreasing;

$$f'(x_b) - f'(x_a) < 0 \Rightarrow f'(x_b) < f'(x_a)$$

Again, let $f'(a) = 0$; thereafter, on the right side the first derivative is negative, $f'(x) < 0$, while the derivative takes positive values on the left side, $f'(x) > 0$. In this way, function $f(x)$ is increasing before reaching point a, and $f(a) > f(x_a)$, $\forall \ x_a < a$; right after point a, function $f(x)$ is decreasing and $f(a) > f(x_b)$, $\forall \ x_b > a$.

- **For inflection points, the second derivative, if any, cancels out: $f''(x) = 0$.**
  However, for inflection points, where the concavity or curvature changes direction, the second derivative becomes zero too, $f''(x) = 0$; in this case, the first derivative does not change sign and the function keeps on with its increase (or decrease).

Example—Inflection Point

- By recalling the cost function, and the second derivative,

$$y''(x) = 100\left(\pi + \frac{8}{x^3}\right)$$

Then, the zero of second derivative is obtained too,

$$y''(x) = 0 \Leftrightarrow 100\left(\pi + \frac{8}{x^3}\right) = 0 \Leftrightarrow x = -\sqrt[3]{\frac{8}{\pi}}$$

Recalling prior analysis, the corresponding function $y_2(x)$ is always decreasing in both domains, and local maxima and minima do not occur. Therefore, the second derivative is negative on the right side of its zero, while it was positive on the left side. Since the sign of second derivative alters from positive to negative, the graphic concavity alters from upward to downward and, $x = -\sqrt[3]{8/\pi}$, is an inflection point.

6.4 ASYMPTOTES

Very often, the points on the function graphic get closer and closer to a straight line, the asymptote. Indeed, the distance between the $f(x)$ graphic and the asymptote approaches zero when the independent variable tends to infinite, that is, the two referred lines are asymptotically closer. Due to their importance for function sketching, asymptotes are presented in this section.

Beyond vertical asymptotes that are found at points of infinite discontinuity, also oblique asymptotes with finite inclination are focused on; in addition, horizontal asymptotes can be seen as a special case of oblique asymptotes that occurs when the associated inclination is zero.

- ***Oblique Asymptote***—The equation line, $y = A.x + B$, is an oblique asymptote to the $f(x)$ graphic, if one of the following pairs of equality is verified:

$$A = \lim_{x \to +\infty} \frac{f(x)}{x}; \quad B = \lim_{x \to +\infty} [f(x) - A.x]$$
$$A = \lim_{x \to -\infty} \frac{f(x)}{x}; \quad B = \lim_{x \to -\infty} [f(x) - A.x]$$

For $A = 0$, there is a horizontal asymptote, $y = B$, and the following relations apply.

- ***Horizontal Asymptote***—The horizontal line of equation, $y = B$, is a horizontal asymptote to the $f(x)$ graphic if any of the following conditions are met:

$$\lim_{x \to +\infty} f(x) = B$$
$$\lim_{x \to -\infty} f(x) = B$$

Vertical Asymptote—The vertical line of equation $x = a$ is a vertical asymptote to the $f(x)$ graphic in case any of the following conditions are satisfied:

$$\lim_{x \to a} f(x) = \pm\infty$$
$$\lim_{x \to a^+} f(x) = \pm\infty$$
$$\lim_{x \to a^-} f(x) = \pm\infty$$

Vertical Asymptote for the Cost Function

In a way to illustrate the different asymptotes (vertical, oblique, and horizontal), consider the cost function once more:

$$f(x) = 100\left(\frac{\pi}{2}x^2 + \frac{4}{x}\right) = 50\pi\, x^2 + \frac{400}{x}$$

The vertical asymptote exists, both on the right side of $x = 0$,

$$\lim_{x \to 0^+} f(x) = \lim_{x \to 0^+} 100\left(\frac{\pi}{2}x^2 + \frac{4}{x}\right) = 0 + \infty = +\infty$$

and on the left side of the vertical Y-axis:

$$\lim_{x \to 0^-} f(x) = 0 - \infty = -\infty$$

These limits represent a vertical asymptote for the function graphic, both on the right side and left side of the vertical Y-axis, $x = 0$, as presented in Figure 6.5.

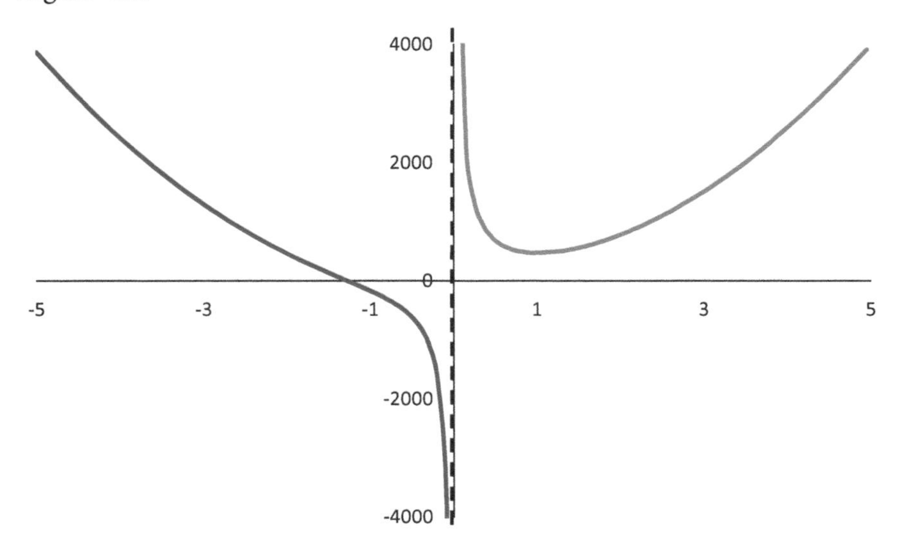

FIGURE 6.5 Vertical asymptote at $x = 0$ for cost function, $f(x)$.

Oblique Asymptote

Now, consider the first-order derivative for the cost function:

$$f'(x) = 100\left(\pi\ x - \frac{4}{x^2}\right) = 100\pi\ x - \frac{400}{x^2}$$

The slope A for the oblique asymptote follows, both for the positive and the negative domain.

- For the positive domain,

$$\lim_{x\to+\infty} \frac{f'(x)}{x} = \lim_{x\to+\infty} \frac{\left(100\pi\ x - \frac{400}{x^2}\right)}{x} = \lim_{x\to+\infty}\left(100\pi - \frac{400}{x^3}\right)$$

$$= 100\pi - 0 = 100\pi$$

- And also, for the negative domain,

$$A = \lim_{x\to-\infty} \frac{f'(x)}{x} = 100\pi + 0 = 100\pi$$

The estimate of coordinate B where the oblique asymptote crosses the vertical Y-axis follows too:

- For the positive domain,

$$B = \lim_{x\to+\infty} [f'(x) - A.x] - \lim_{x\to+\infty}\left[\left(100\pi\ x - \frac{400}{x^2}\right) - 100\pi\ x\right]$$

$$= \lim_{x\to+\infty}\left(-\frac{400}{x^2}\right) = 0^-$$

- And also, for the negative domain,

$$B = \lim_{x\to-\infty} [f'(x) - A.x] = \lim_{x\to-\infty}\left(-\frac{400}{x^2}\right) = 0^-$$

The equation line, $y(x) = 100\pi\ x + 0 = 100\pi\ x$, is thus an oblique asymptote for the $f'(x)$ graphic, as shown in Figure 6.6.

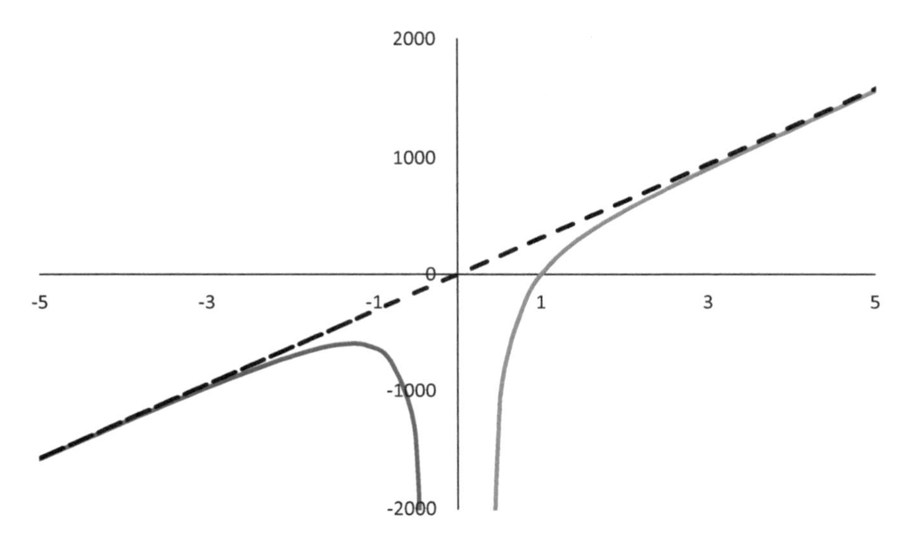

FIGURE 6.6 Oblique asymptote, $y(x) = 100\pi\ x$, for the function's first derivative, $f'(x)$.

Horizontal Asymptote

And now, consider the second-order derivative for the cost function:

$$f''(x) = 100\left(\pi + \frac{8}{x^3}\right) = 100\pi + \frac{800}{x^3}$$

The slope A for the asymptote results zero, both for the positive and the negative domain.

- For the positive domain,

$$\lim_{x\to+\infty} \frac{f''(x)}{x} = \lim_{x\to+\infty} \frac{\left(100\pi + \frac{800}{x^3}\right)}{x} = \lim_{x\to+\infty} \left(\frac{100\pi}{x} + \frac{800}{x^4}\right)$$
$$= 0 + 0 = 0$$

- And also, for the negative domain,

$$A = \lim_{x\to-\infty} \frac{f''(x)}{x} = 0 + 0 = 0$$

Since the inclination is zero, $A = 0$, the asymptote corresponds to a horizontal line, to a horizontal asymptote with the form:

$$y(x) = 0\ x + B \Rightarrow y(x) = B$$

The coordinate B where the horizontal asymptote crosses the vertical Y-axis follows in below.

- For the positive domain,

$$B = \lim_{x \to +\infty} [f'(x) - A.x] = \lim_{x \to +\infty} \left[\left(100\pi + \frac{800}{x^3} \right) - 0\ x \right]$$

$$= \lim_{x \to +\infty} \left(100\pi + \frac{800}{x^3} \right) = 100\pi^+$$

- And also for the negative domain,

$$B = \lim_{x \to -\infty} [f'(x) - A.x] = \lim_{x \to -\infty} \left(100\pi + \frac{800}{x^3} \right) = 100\pi^-$$

The equation for the straight line thus indicates a horizontal asymptote for the graphic of $f''(x)$, both for the positive and the negative domain, as in Figure 6.7.

$$y(x) = 0\ x + 100\pi \Rightarrow y(x) = 100\pi$$

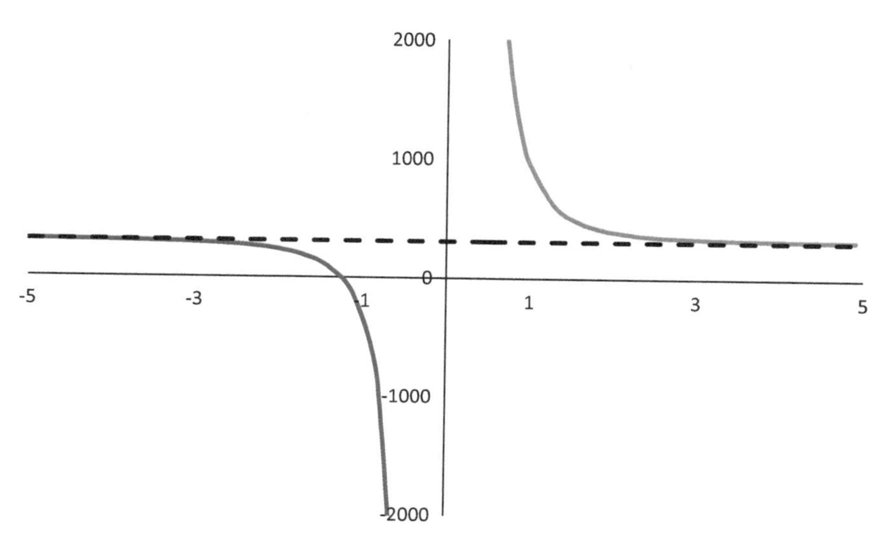

FIGURE 6.7 Horizontal asymptote, $y(x) = 100\pi$, for the function's second derivative, $f''(x)$.

6.5 SKETCHING THE EXTENDED COST FUNCTION

Typically, the sketching of functions assumes a number of attributes are studied, namely, domain, and image-domain, axis intersections and points of discontinuity; monotony intervals (increasing, decreasing), maximum and minimum points (local and general), concavity and inflection points, and asymptotes (vertical; oblique or horizontal), while it is also usual to draw up a table summarizing all the function variation. Note also that knowing if the function is even or odd allows to bound the study interval, and reduce the related works.

The extended function is studied both in the positive and negative domains (Figure 6.8),

$$y(x) = 100\left(\frac{\pi}{2}x^2 + \frac{4}{x}\right) = 50\pi\ x^2 + \frac{400}{x}$$

although it corresponds to the cost function when restricted to positive values of variable x.

Domain

The domain considers the operations are all performed, except at zero: $D = R_{-\{0\}}$.

Definition Range and Points of Discontinuity

The function graphic does not intersect the vertical Y-axis, since the point $x = 0$ does not belong to the function domain; continuity conditions hold in the entire domain, a linear combination of continuous functions is performed, x^2 and $(1/x)$, and points of discontinuity do not occur; in addition, there is no graphic symmetry, the function is neither even nor odd. The image domain (counterdomain), both by qualitative and graphic analysis: $CD =]-\infty,\ +\infty[$.

Intervals of Monotony and Extreme Points

The first-order derivative indicates the monotony ranges,

$$y'(x) = 100\left(\pi\ x - \frac{4}{x^2}\right)$$

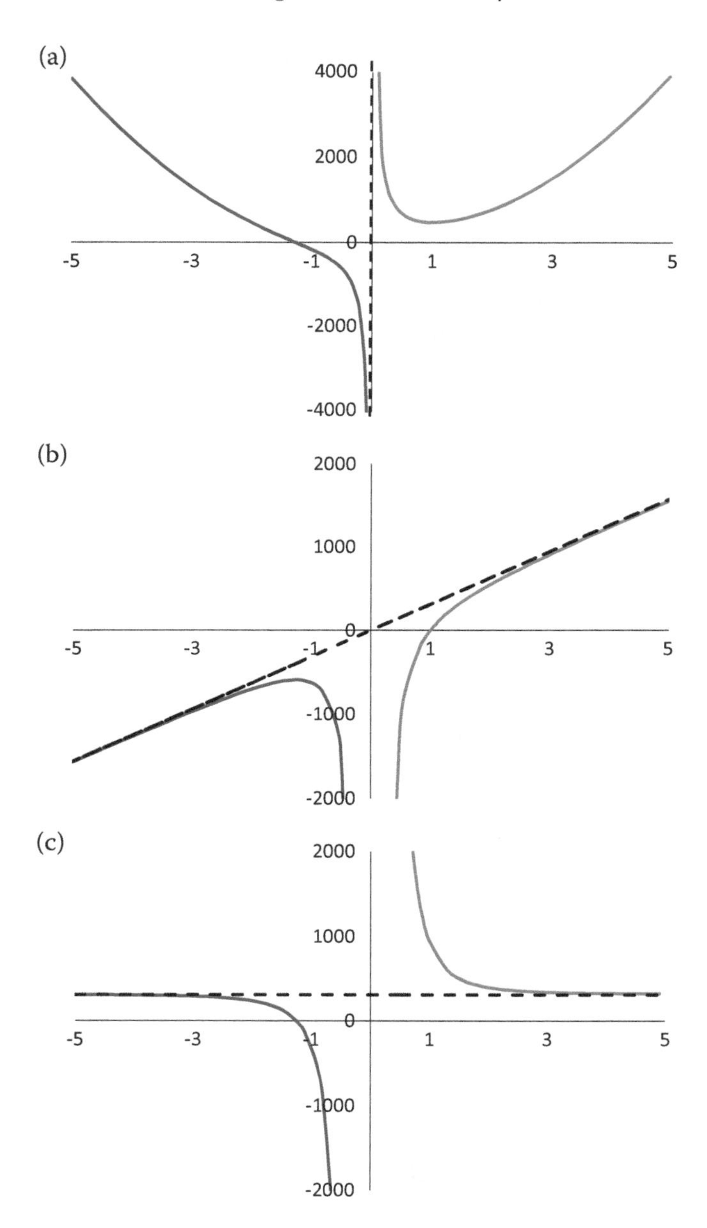

FIGURE 6.8 From top to down: (a) Extended cost function, $y(x)$; (b) the first derivative, $y'(x)$; and (c) the second derivative, $y''(x)$.

and the extreme points can be identified by the corresponding zero and sign alterations:

$$y'(x) = 0 \Leftrightarrow x = \sqrt[3]{\frac{4}{\pi}}$$

From a qualitative analysis, the first derivative is always negative when $x < 0$, and always increasing with the positive variable, $x > 0$; the derivative is positive on the right side of its zero and negative on the left side; $x = \sqrt[3]{4/\pi}$, is thus a local minimum.

Concavity and Points of Inflection

To focus on the concavity and inflection points, the second-order derivative is addressed:

$$y''(x) = 100\left(\pi + \frac{8}{x^3}\right)$$

If $x > 0$, the second-order derivative is strictly positive, $y''(x) > 0$, and the function curvature is upward. If $x < 0$, the second-order derivative $y''(x)$ becomes zero at $x = -\sqrt[3]{8/\pi}$, and it changes from positive to negative at that point. Then, concavity inflects from upward to downward at $x = -\sqrt[3]{8/\pi}$: this is a function's inflection point.

Asymptotes

A vertical asymptote to the $y(x)$ graphic occurs at $x = 0$, from upper values (right side),

$$\lim_{x \to 0^+} y(x) = \lim_{x \to 0^+} 100\left(\frac{\pi}{2}x^2 + \frac{4}{x}\right) = 0 + \infty = +\infty$$

and also from lower values (left side):

$$\lim_{x \to 0^-} y(x) = 0 - \infty = -\infty$$

- However, the graphic of $y(x)$ is not allowing oblique asymptotes, nor horizontal asymptotes, since the inclination A cannot be defined,

$$A = \lim_{x\to+\infty} \frac{y(x)}{x} = \lim_{x\to+\infty} \frac{100\left(\frac{\pi}{2}x^2 + \frac{4}{x}\right)}{x} = \lim_{x\to+\infty} 100\left(\frac{\pi}{2}x + \frac{4}{x^2}\right)$$
$$= +\infty + 0 = +\infty$$

And also for the negative infinite:

$$A = \lim_{x\to-\infty} \frac{y(x)}{x} = -\infty + 0 = -\infty$$

It is usual to draw up a summary table for the function variation, including the relevant attributes or items, and also for both the derivatives of first order and second order.

x	(...)	$-\sqrt[3]{\frac{8}{\pi}}$	(...)	0	(...)	$\sqrt[3]{\frac{4}{\pi}}$	(...)
$y'(x)$	Negative	Negative	Negative	***Not defined***	Negative	0	Positive
$y''(x)$	Positive	0	Negative	***Not defined***	Positive	Positive	Positive
$y(x)$	Decreasing, upward	0, Inflection point	Decreasing, downward	***Not defined***	Decreasing, upward	Minimum point	Increasing, upward

In Figure 6.8, by aggregating from top to down the plots of the extended cost function, $y(x)$, the first derivative, $y'(x)$, and the second derivative, $y''(x)$, then a global view for the function's variation is provided.

6.6 OTHER IMPORTANT APPLICATIONS

In this section, applications of specific interest are described, namely the Newton-Raphson method to define the container's diameter for a given cost (500 €, in the instance at hand), and the derivative mean value is utilized to bound the error estimates and related evolution.

6.6.1 A Barrel That Costs 500 (II)—The Newton-Raphson Method

For a continuously differentiable function on a given interval, $[x_{n-1}, x_n]$, the Lagrange theorem suggests the secant defined from those two points can be replaced by the value of the tangent to the curve $f(x)$, at an intermediate point, c. Therefore, it is reasonable to replace the intermediate and unknown point, c, by a point close to x_n and x_{n-1}, where the derivative $f'(x)$ can be calculated, obtaining the Newton-Raphson formula. Then, the tangent to the curve of $f(x)$ at point, x_n, is extrapolated to the X-axis, obtaining a new estimate of the root at x_{n+1}:

$$x_{n+1} = x_n - \frac{f(x_n)}{f'(x_n)}$$

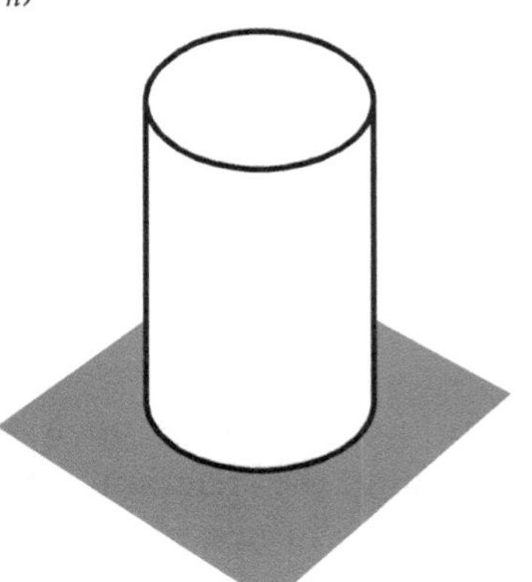

In Section 4.6, the diameter for the barrel (V = 800 liters) costing 500 € was obtained through the bisection method, based in the intermediate values property for continuous function in the closed range [1.2, 1.3]. Now, the Newton-Raphson method is proposed to deal again with the *Cost-minus*-500 function:

$$50\pi\ d^2 + \frac{320}{d} = 500 \Leftrightarrow C_{500}(d) = 50\pi\ d^2 + \frac{320}{d} - 500 = 0$$

In this way, let it be:

$$f(x) = 50\pi\ x^2 + \frac{320}{x} - 500, \text{ and } f'(x) = 100\pi\ x - \frac{320}{x^2}$$

Then, the next estimate by the Newton-Raphson method is obtained from the relation: $x_{n+1} = x_n - \frac{f(x_n)}{f'(x_n)}$

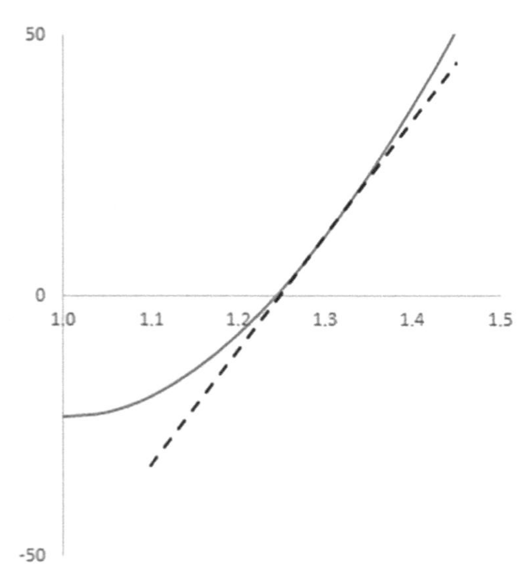

FIGURE 6.9 The Newton-Raphson method and the root of function *Cost-minus*-500: first estimate.

- As Figure 6.9, with initial guess, $x_0 = 1.3$, the function and derivative evaluations are,

$$f(1.3) = 50\pi\ (1.3)^2 + \frac{320}{(1.3)} - 500 = 11.61842538$$

$$f'(1.3) = 100\pi\ (1.3) - \frac{320}{(1.3)^2} = 219.0579$$

The first estimate is obtained from the Newton-Raphson relation:

$$x_1 = 1.3 - \frac{f(1.3)}{f'(1.3)} = 1.3 - \frac{11.61842538}{219.0579} = 1.24696186$$

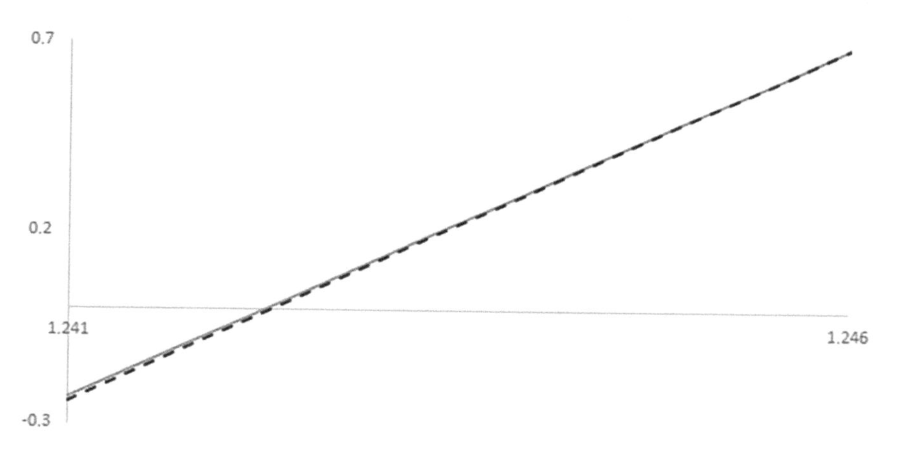

FIGURE 6.10 The Newton-Raphson method and the root of function *Cost-minus*-500: second estimate.

- With this estimate, $x_1 = 1.24696186$, the function and derivative take, respectively,

$$f(1.24696186) = 50\pi\ (1.24696186)^2 + \frac{320}{(1.24696186)} - 500$$
$$= 0.86902814$$

$$f'(1.24696186) = 100\pi\ (1.24696186) - \frac{320}{(1.24696186)^2} = 185.9454$$

The second estimate is obtained, the tangent nears well the function line (Figure 6.10),

$$x_2 = 1.24696186 - \frac{f(1.24696186)}{f'(1.24696186)} = 1.24696186 - \frac{0.86902814}{185.\ 9454}$$

$$= 1.24228829$$

As in the following table, after four iterations and rounding to the 8th-decimal digit, the root estimate and the associated function value follow: $C_{500}(1.24224976) = 0.00000000$.

n	x	f(x)	f′(x)
0	1.30000000	11.61842538	219.0579
1	1.24696186	0.86902814	185.9454
2	1.24228829	0.00704938	182.9258
3	1.24224976	0.00000048	182.9009
4	1.24224976	0.00000000	182.9009

6.6.2 The Error Evolution

- Note the Newton-Raphson method is a fixed-point method with iterative function, $g(x)$,

  $$x_{n+1} = g(x_n), \text{ where } g(x_n) = x_n - \frac{f(x_n)}{f'(x_n)},$$

  and the related numerical values where data input and output are coinciding follow:

n	x	g(x)
0	1.30000000	1.24696186
1	1.24696186	1.24228829
2	1.24228829	1.24224976
3	1.24224976	1.24224976
4	1.24224976	1.24224976

The function under analysis, $f(x)$, becomes zero when is evaluated at the root, $f(R) = 0$; thereafter the iterative function, $g(x)$, takes the same value as in prior iteration because the advance in the root search corresponds to the ratio, $-f(R)/f'(R)$, and it is also zero:

$$g(R) = R - \frac{f(R)}{f'(R)} = R - 0 = R$$

- Note the error at iteration n, E_n, corresponds to the difference between the function root, R, and the current estimate, x_n,

$$E_n = R - x_n$$

For a fixed-point method with iterative function, $g(x)$, then both the following relations hold, $x_n = g(x_{n-1})$, and $R = g(R)$. By substitution in the error expression,

$$E_n = g(R) - g(x_{n-1})$$

By the derivative mean value, there is an intermediate point c, $x_{n-1} < c < R$, such that,

$$g(R) - g(x_{n-1}) = g'(c) \,.\, (R - x_{n-1})$$

And making use of such mean value for the function derivative in the range at hand,

$$E_n = g(R) - g(x_{n-1}) = g'(c) \,.\, (R - x_{n-1}) = g'(c)E_{n-1}$$

The error will diminish from one iteration to another, in case the absolute value derivative is lower than one, $|g'(c)| \leq k < 1$; in that manner,

$$|E_n| \leq k\, |E_{n-1}|$$

By induction, a geometric sequence with ratio lower than one, $k < 1$, is obtained,

$$|E_n| \leq k\, |E_{n-1}| \leq k^2\, |E_{n-2}| \leq k^3\, |E_{n-3}| \leq \ldots \leq k^n\, |E_0|$$

And taking limit, when the number of iterations, n, increases without upper bound,

$$\lim_{n\to\infty}|E_n| = \lim_{n\to\infty} k^n \; |E_0| = 0$$

The error tends to zero, and the derivative mean value is very often utilized to bound the errors in computational methods; indeed, in different methods and numerical topics (e.g., differential equations, numerical integration, series), it is very common to bound the error expressed in the Lagrange form, assuming the occurrence of an intermediate point c, $a < c < b$, where the Lagrange result for the mean derivative holds.

6.7 CONCLUDING REMARKS

This chapter is directed at the study of functions variation, including important theorems for sketching functions, as well as graphical and numerical insights, and the treatment of convenient instances that allow multiple views about verification procedures and patterns recognition.

- The Rolle theorem about derivative roots is presented, together with the joint application with the Bolzano theorem; after that, the Lagrange theorem that focuses on the derivative's average value is addressed; the Cauchy theorem that treats the relation between increments of two functions is focused on too, and L'Hôpital's rule is then presented. A reduction approach is also described to complement the study of these important results.
- The importance of first derivative on function sketching and extreme points is described; the sufficient condition for relative extrema is mainly focusing on sign alterations of first derivative, the second-order derivative completes the study with the necessary condition for optima and the curvature analysis.
- Asymptotes are presented, due to their importance in function sketching. Beyond vertical asymptotes, oblique asymptotes with finite inclination are also treated; in addition, horizontal asymptotes can be seen as a specific case of oblique asymptotes that occur when the associated inclination is zero. L'Hôpital's rule is utilized in asymptotes calculation, namely, for the cost function and associated derivatives.
- In that way, sketching procedures for the cost function are summarized, including domain and counterdomain, axis

intersections and points of discontinuity; monotony intervals, maximum and minimum points, concavity and inflection points, asymptotes, and finally a table and graphics summarize all the function variations.

- Other important applications are described, namely, the Newton-Raphson method to calculate the diameter for the barrel costing 500 €, and the Lagrange form to bound the error evolution.

Note the anti-derivative function, $Y(x)$, is crucial to area estimates, namely, the mean cost in the range [1, 3] can be reached by evaluating the area below the cost line. From the other side, integral sums aim at area estimates too; for a non-negative function, $y(x)$, they evaluate the area below the function's line and above the horizontal X-axis in a closed interval $[a, b]$. In the next chapter, two different approaches are applied to the integral sums:

- One by introducing the Riemann integral, and the other using anti-derivative functions.
- Beyond the convergence of lower and upper sums associated with the limit for the integral sum, several instances of areas quadrature are treated, and results are confirmed through graphical analogies.

CHAPTER 7

First Steps on Integral Sums

Integral sums are widely used to solve pertinent problems, and also motivate scores of applications and computational methods in physics, statistics, and other important domains. Two approaches to the integral sum are focused on, the lower and the upper sum; they are associated with the limit convergence to the integral value, respectively, from both the inferior and superior sums sequences. The areas quadrature is described, and a graphical approach is presented too: for one side, illustrating simple examples of area squaring: (i) a rectangle, (ii) a triangle, and (iii) a circle; for the other side, by providing a verification tool that complements these first steps on integral sums. It is important to remark here the anti-derivative approach, since the area differential estimates the area variation due to an infinitesimal alteration, and the total variation of the function's area can be calculated through the integral sum of all those area differentials.

7.1 INTRODUCTION

The solution of a classic problem is focused on in this chapter, namely, the calculation of the area defined by the graph of a continuous and non-negative function and by the horizontal axis.

In the graphical approach, a number of real world problems can be represented by the region below the function's line, which can be represented through sums of area elements with different sizes and shapes (e.g., circles, rectangles, squares, trapezoids).

Thus, integral sums include the study of area estimates, namely, by evaluating both the lower sum and the upper sum as inferior and superior estimates, respectively; then, by studying the sums convergence when the number of area elements increases without bound; finally, the existence and uniqueness of limit for both sums indicate the integral value.

DOI: 10.1201/9781003461876-7

In this chapter, methods for the integral sums of a constant function, a linear function, and a linear function in polar coordinates, are treated. They are used, respectively, in the resolution of simple problems for squaring the area of a rectangle, a triangle, and a circle. In this way, the related results allow graphical confirmation, while a differential approach is developed.

That is, the area alteration due to the addition of a single area element (e.g., a very thin rectangle) can be evaluated by the associated differential of area. Or better, the area's differential can be estimated by the rectangle's height—the function value $y(x)$—multiplied by the differential width, dx—the infinitesimal change in variable x. When the total variation in the area is targeted, then the integral sum of all the area differentials is a pertinent approach, and the function's anti-derivative, the function $Y(x)$ that originates $y(x)$ when it is derived, performs an important role to obtain the solution.

- It should be noted that the possibility of squaring areas is greatly increased by integral sums when compared to previous studies in differential calculus or geometry. At now, only areas limited by polygons or the area of a circle are worked on, but then it will be possible to calculate a multiplicity of areas of figures, as long as it is possible to define them from a continuous function.
- Integral calculus is a powerful method for solving problems in contexts as diverse as calculating areas, volumes, and curve lengths, estimating accumulated probabilities and risk in statistics modelling, determining the pressure exerted by liquids in a wall or inside a recipient, or even obtaining centers of gravity for irregular figures.
- Suitable developments of interest are introduced too, including the exploitation of definite integral based on Riemann's formulation, by discussing it as an approximate calculation of integral sums and directed to computational methods.
- Obtaining the mean value of a continuous function in a given interval, among other statistical measures and operational tools, is also suitable by application of integrals topics.

In fact, integral sums can be used to measure the mean cost for a barrel when its diameter is changing over a closed interval, for example, between one and two meters. The mean cost can be reached by evaluating the area below the cost function's line, calculating both the lower and upper sums, and studying the sums convergence when the number of area subintervals tends to infinite.

7.2 INTEGRAL SUM AND GEOMETRIC INTERPRETATION

The integral sum is addressed in this section, together with the related geometric interpretation. In the graphic representation of a non-negative function, the integral value is associated with the area below the function's line, whether such value can be estimated from different approaches. Namely, the integral's estimate by excess using an upper sum, the estimate by default through a lower sum, and also studying the convergence of both estimates to the same and exact value.

Let a function, $f(x)$, be defined and bounded on a closed interval $[a, b]$, as in Figure 7.1.

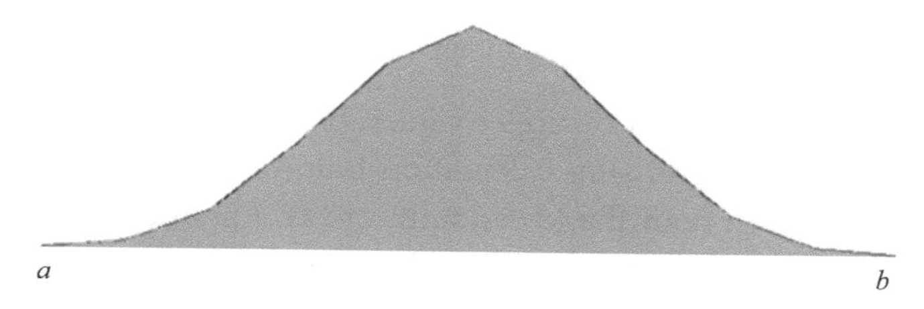

FIGURE 7.1 Area of the curvilinear trapezoid to be measured.

The area estimates are developed using the area of simple geometric figures, for example, rectangles; then the sum of such partial areas is refined more and more by increasing the number of thinner and thinner rectangles. The convergence of such estimates for the same value ensures the associated limit exists and it is unique; therefore, the shadowed area in Figure 7.1 can be calculated through the integral sum:

$$\int_a^b f(x)\,dx$$

In fact, the integral, $\int_a^b f(x)\,dx$, is numerically equal to the area of the curvilinear trapezoid in Figure 7.1 formed by the curve $y = f(x)$, by the vertical lines, $x = a$ and $x = b$, and by the horizontal axis.

Thus, the function, $f(x)$, is integrated in the closed interval $[a, b]$ if, and only if, for every positive convergence radius, r no matter how small, it is possible to obtain a partition P of the interval $[a, b]$, such that,

$$\forall\ r > 0,\ \exists\ P:\quad U[f(x), P] - L[f(x), P] < r$$

where $L[f(x), P]$ represents the lower sum, an estimate to the area by lower values, or by default; and $U[f(x), P]$ represents the upper sum, the

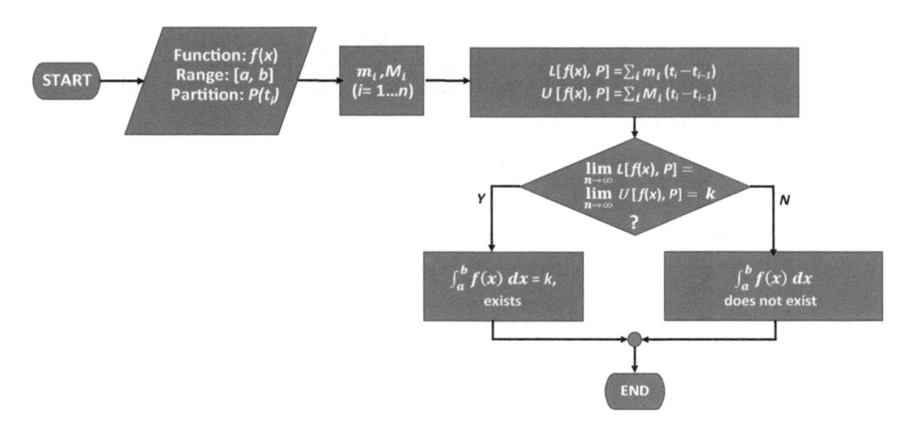

FIGURE 7.2 Flowchart outlining the integral sum of function in the range $[a, b]$.

related estimate to the area by upper values, or by excess. In this way, the flowchart for the integral sum calculation follows in Figure 7.2.

The pair of sums, both the lower sum, $L[f(x), P]$, and the upper sum, $U[f(x), P]$, are focused on in more detail in the following two sub-sections.

A. The Lower Sum, $L[f(x), P]$

- A first approximation by lower values to the area value follows; assuming only eight subintervals, then the error is large as illustrated in Figure 7.3.

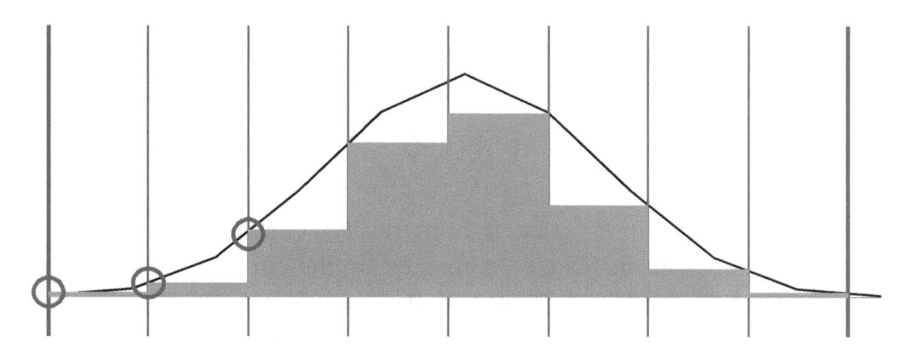

FIGURE 7.3 A first partition of the area under study—eight subintervals.

- A new approximation by lower values with 16 subintervals follows, and the error is being reduced; through new approximations with 32 subintervals, with 64 subintervals, and the error is reduced more and more; the area estimates are getting better and better, always nearing the exact value by lower values (Figure 7.4).

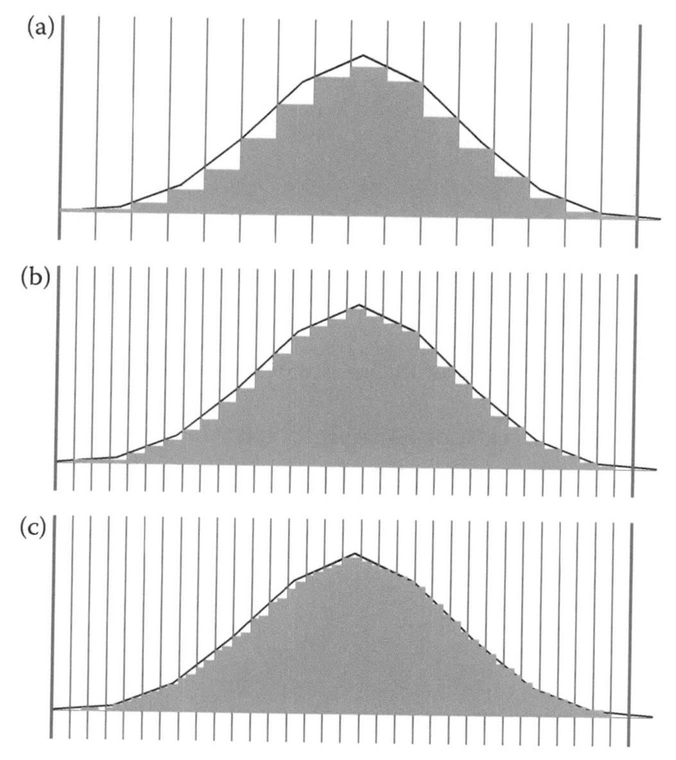

FIGURE 7.4 Other partitions of the area under study: (a) 16 subintervals; (b) 32 subintervals; and (c) 64 subintervals.

With new approximations of 128, 256, or more subintervals, the lower sum estimates are getting even better, as the error is being reduced again. Note that these approximations are always by the lower side of the function's line, that is, the approximations by the lower sum, $L[f(x), P]$, cannot be greater than the area under analysis.

$$L\,[f\,(x),\,P] \leq f\,(x)\,dx$$

B. The Upper Sum, $U[f(x), P]$

- A first approximation by upper values follows; assuming only eight subintervals, obviously the error is large (Figure 7.5).
- A new approximation by upper values is developed with 16 subintervals, and the error is being reduced; through new approximations

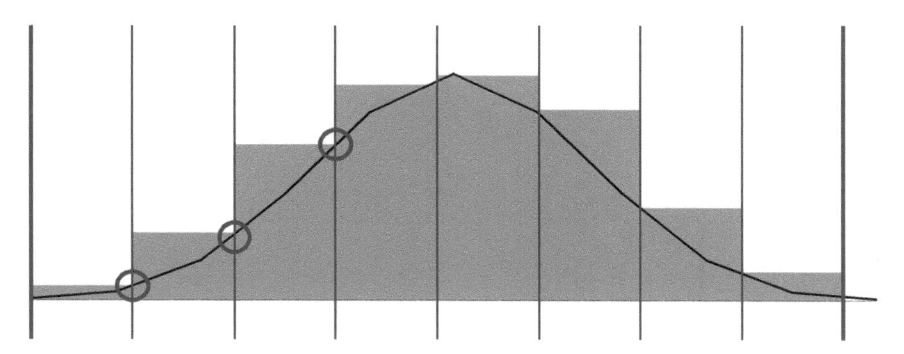

FIGURE 7.5 A first partition of the area under study—eight subintervals.

with 32 subintervals, or 64 subintervals, the error is reduced more and more; the area estimates are getting better and better, the approximations are always by excess, that is, by upper values (Figure 7.6).

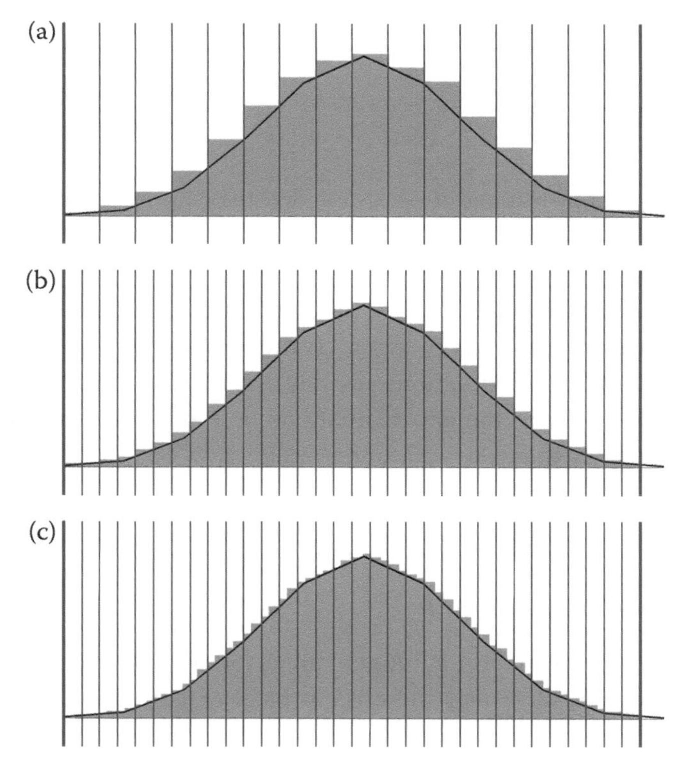

FIGURE 7.6 Other partitions of the area under study: (a) 16 subintervals; (b) 32 subintervals; and (c) 64 subintervals.

With new approximations of 128, 256, and more subintervals, the estimate is getting better as the error is being reduced. Note that these approximations are always by the upper side of the function's line, that is, the approximations by the upper sum $U[f(x), P]$ cannot be lower than the area under analysis.

$$U\,[f\,(x),\,P] \geq f\,(x)\,dx$$

Assuming the number of subintervals can be very high and increasing without any bound, that is, taking limit when the number of subintervals n tends to infinite, then the summation of upper areas $U[f(x), P]$ and the summation of lower areas $L[f(x), P]$ tend to the same value. That limit-value corresponds to the area between the function's line and the horizontal axis, and this integral sum is represented through a stylized S, the integral symbol $\int$:

$$\begin{aligned}\lim_{n\to\infty} U\,[f\,(x),\,P] &= \lim_{n\to\infty} L\,[f\,(x),\,P] \\ &= \int_a^b f\,(x)\,dx\end{aligned}$$

7.3 CALCULATION OF AREAS

Squaring the area, or area quadrature, is a classic problem that aims at measuring the area of a region, namely, estimating the associated value through the square with the same area. Such approaches are being naturally updated; currently calculus tools are used to evaluate the differential areas associated with three simple examples: (i) a rectangle, (ii) a triangle, and (iii) a circle. Thereafter, though simple integration or anti-derivation, the related area values are obtained.

Note that the primary purpose in here is to show both the effectiveness of the calculus approach, and the usefulness of well-known results for verification or confirmation of the approach in use. In this way, the graphical approach is complemented with the analytical development, while obtaining new knowledge or results is not the target here.

Squaring areas (Quadrature)

From geometry basics, it is possible to calculate the area of regular figures. However, to obtain the area of irregular regions or figures, it

is necessary to resort to more elaborate calculation tools. Through integral calculus, use the area's differential to develop expressions for the areas of:

a. **A rectangle of height 1 and length x.**

b. **A triangle with equal legs (isosceles) of length x.**

c. **A circle of radius r.**

7.3.1 Area of Rectangle

The relation for the area of the rectangle, A_r, of height 1 and width x is well known from geometry:

$$A_r(x) = Height \times Width = 1.x$$

The constant function $f(x) = 1$ is addressed, in conjunction with the differential approach to the area below the function's line (Figure 7.7).

The differential of area function, dA_r, allows us to estimate the increase in the rectangle's area. In fact, the differential dA_r is estimating the alteration in the rectangle's area, A_r, when the corresponding width, x, changes. Therefore, the area's differential for the rectangle of height, 1, and width differential, dx, is expressed by: $dA_r = 1\ dx$.

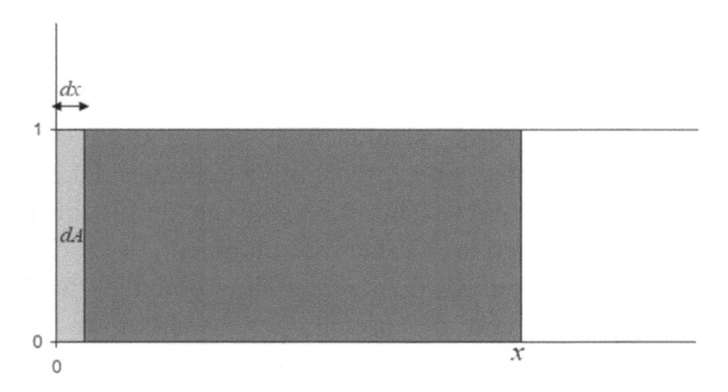

FIGURE 7.7 The rectangle's area and the related differential, $dA_r = 1\ dx$.

The first subinterval corresponding to the area differential is represented in Figure 7.7, assuming the initial width is $x = 0$ and the associated area is also zero, $A_r = 0$. This is a numerical approach only, the notion of a rectangle of width $x = 0$ would correspond to a void concept.

A second subinterval can be added to the first one, as represented in Figure 7.8a; that is, the second rectangle presents also height 1 and differential width, dx, and the related differential of area, dA, indicates the area increasing. Then, a third subinterval corresponding to a third rectangle of height 1 and differential width, dx, is also added to the area under analysis, while the area increase is considered by the associated differential of area, dA, as presented in Figure 7.8b. Successively, more and more subintervals of differential width, dx, are fully added to the area until the right border of the interval is reached (Figure 7.8c); in this way, the differential areas for all the rectangles are integrally summed.

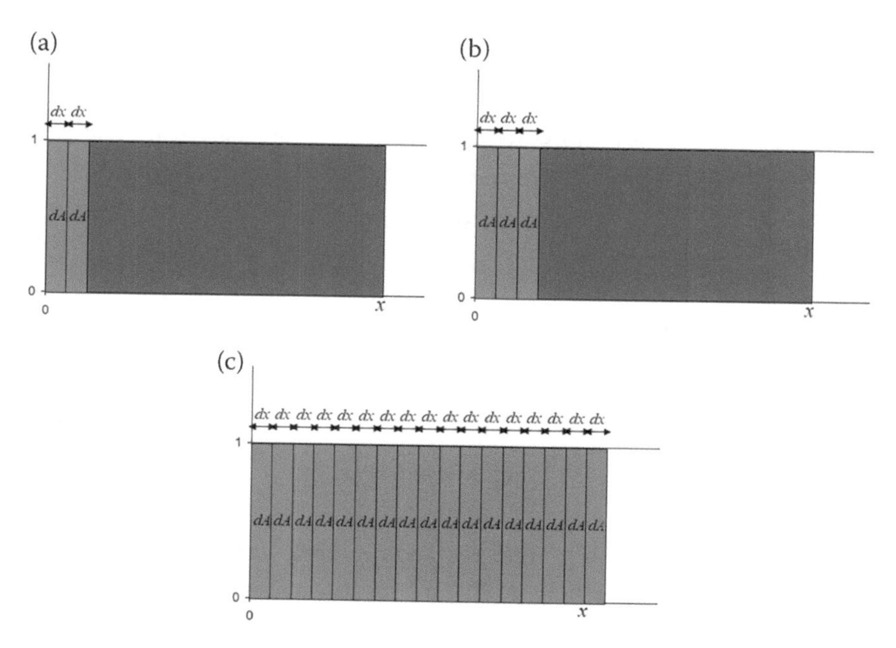

FIGURE 7.8 Adding more subintervals onto the rectangle's area: (a) Two subintervals; (b) three subintervals; and (c) a large number of subintervals.

The relation for the integral sum of all the area differentials, dA_r, with height, 1, and differential length, dx, follows:

$$\int dA_r = \int 1\ dx$$

Note the rectangle's area is obtained from the integral sum of all differential areas for the infinitesimal rectangles.

- That is, A_r is the result of the **left hand side** in the prior relation; and the initial rectangle is obtained when all the infinitesimal rectangles are concatenated all together.
- About the **right hand side**, the rectangle's area is obtained through the function $A_r(x)$ that becomes 1 when such function is derived (*Why*?); since the derivative of constant is zero, it corresponds to the linear function, x, because the result from its derivation is 1:

$$A_r(x) = x + c$$

In fact, the prior result is rapidly confirmed:

$$A_r'(x) = (x + c)' = 1 + 0 = 1$$

In addition, the integration constant, c, can be obtained too by studying the initial point:

$$x = 0 \Rightarrow A_r = 0$$

Then,

$$A_r(0) = 0 \Leftrightarrow 0 = 0 + c \Leftrightarrow 0 = c$$

Finally, the area relation follows:

$$A_r(x) = x + 0 = x$$

And it is widely known, the rectangle's area can be obtained by multiplying the related height, 1, and width, x, as confirmed in this simple example:

$$A_r(x) = 1.x = x$$

7.3.2 Area of Triangle

The relation for the area of the isosceles triangle, A_t, of equal legs of length, x, is also well known:

$$A_t(x) = \frac{Leg \times Leg}{2} = \frac{x.x}{2}$$

Now, the linear function $f(x) = x$ is treated, together with the differential approach to the area below the function's line.

The differential of area function, dA_t, allows us to estimate the increase in the triangle's area. In fact, the differential dA_t is estimating the alteration in the triangle's area, A_t, when the corresponding width, x, changes. Therefore, the area's differential for the triangle of height, x, and differential length, dx, is: $dA_t = x\ dx$.

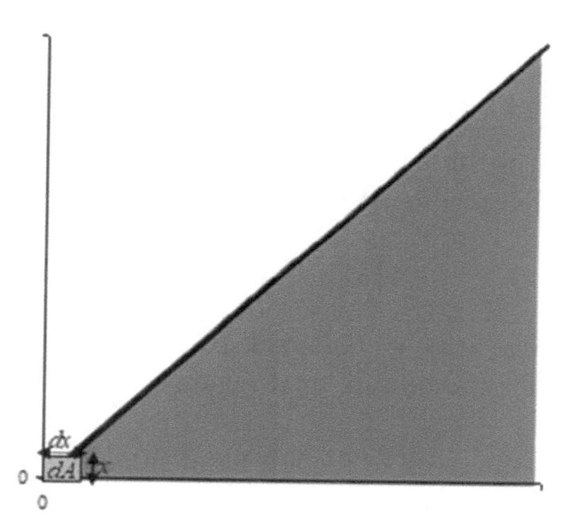

FIGURE 7.9 The triangle's area and the related differential, $dA_t = x\ dx$.

The first subinterval in the triangle corresponding to the area's differential is represented in Figure 7.9, assuming the initial point is $x = 0$ and the associated area is also zero, $A_t = 0$. Note the area's differential evaluates the initial increase within a numerical approach only, since a triangle of leg $x = 0$ would be a void concept.

A second subinterval can be added to the first one, as represented in Figure 7.10a; that is, the second differential element presents height x and differential width, dx, and the related differential of area, dA, indicates the area increasing. Then, a third subinterval corresponding to a third differential element of height x and differential width, dx, is also added to the area under analysis; note the area increase is considered by the associated differential of area, dA, as presented in Figure 7.10b. Successively, more and more subintervals of differential width, dx, are fully added to the area until the right border of the interval is reached (Figure 7.10c); in this way, the differential areas for all the subintervals are integrally added.

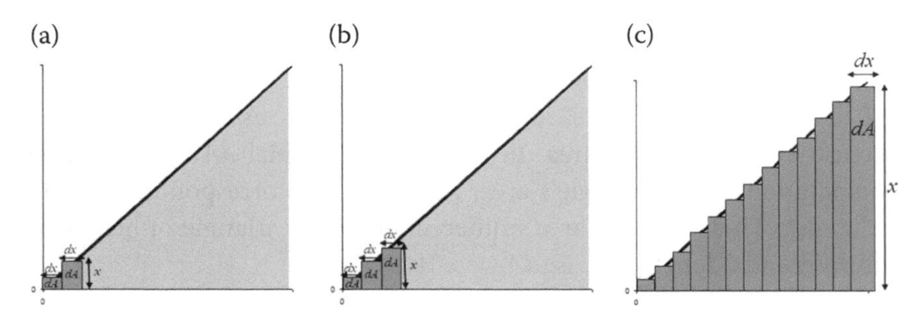

FIGURE 7.10 Adding more subintervals onto the triangle's area: (a) Two subintervals; (b) three subintervals; and (c) a large number of subintervals.

The relation for the integral sum of all the area differentials, dA_t, with height, x, and differential length, dx, follows:

$$\int dA_t = \int x\ dx$$

Note the triangle's area is obtained from the integral sum of all differential areas for the infinitesimal regions.

- That is, A_t is the result of the **left hand side** of the prior relation; in other words, the original triangle is obtained when all the infinitesimal regions are concatenated all together.
- About the **right hand side**, the triangle's area is obtained by finding the function $A_t(x)$ that becomes x, when such function is derived (again, *Why?*); the constant derivation results zero, then the area's function corresponds to a quadratic function since the derivation of x^2 is driving $2x$; thus the quadratic function is divided by two in the relation:

$$A_t(x) = \frac{x^2}{2} + c$$

In fact, the coefficient 2 in the fraction's denominator is balancing the factor 2 originated from the square power's derivation; again, the area's relation is rapidly confirmed:

$$A_t{}'(x) = \left(\frac{x^2}{2} + c\right)' = \frac{2x}{2} + 0 = x$$

Additionally, from the initial value, when the leg length is null, $x = 0$, the triangle's area is null too, and the constant term is also zero:

$$x = 0 \Rightarrow A_t = 0$$

$$A_t = 0 \Leftrightarrow 0 = \frac{0^2}{2} + c \Leftrightarrow 0 = c$$

Finally,

$$A_t(x) = \frac{x^2}{2} + 0 = \frac{x^2}{2}$$

And as a confirmation step, the triangle's area can be obtained by multiplying length, x, and height, also x, and dividing by two:

$$A_t(x) = \frac{x.x}{2} = \frac{x^2}{2}$$

7.3.3 Area of Circle

The relation for the circle's area, A_c, of radius r is also a classic result:

$$A_c(r) = \pi\, r^2$$

Again, the differential of area function, dA_c, allows us to estimate the increase in the circle's area. In fact, the differential dA_c is estimating the alteration in the circle's area, A_c, when the corresponding radius, $x = r$, changes. Therefore, the area's differential for the circle of radius, x, and differential radius, dx, corresponds to the tiny ring added to the circle:

$$dA_c = (2\pi x)\, dx$$

Note that this is equal to the circumference's perimeter with radius $x = r$ multiplied by the radius differential $dx = dr$, which makes sense when selecting polar coordinates to address the circle's area. In this way, using polar coordinates to find the circle's area becomes similar to using Cartesian coordinates to find the rectangle's area.

The first ring added to the circle is increasing the associated area and corresponds to the area's differential, as represented in Figure 7.11. Other subintervals can be added to the first one, while the related differentials of area, dA_c, indicate the area increasing. Successively, more and more subintervals of differential radius, $dx = dr$, are fully added to the area, until the final radius for the circle is reached. That is, the differential areas for all the rings are integrally added, as follows in the relations:

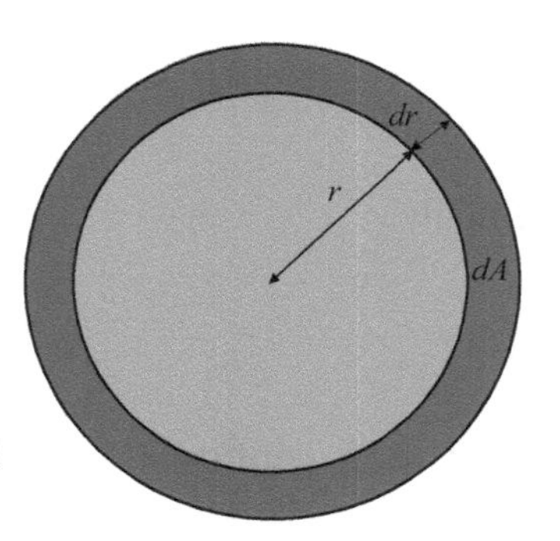

FIGURE 7.11 The circle's area and the related differential, $dA_c = (2\pi\ r)dr$.

$$\begin{aligned} \int dA_c &= \int 2\pi x\ dx \\ &= 2\pi \int x\ dx \end{aligned}$$

Note the treatment of coefficient 2π follows the linearity procedure, while the circle's area is obtained from the integral sum of all differential areas associated with the infinitesimal rings.

- That is, A_c is the result of the **left hand side** of the prior relation; in other words, the circle is obtained when all the infinitesimal rings are concatenated all together.
- Evaluating the **right hand side**, the circle's area is obtained by finding the function $A_c(x)$ that becomes x when it is derived (once again: *Why?*); since the derivative of a constant is zero, then it corresponds again to the quadratic function divided by 2:

$$A_c(x) = 2\pi\frac{x^2}{2} + c$$

And after simplifying the coefficient 2, the relation becomes

$$A_c(x) = \pi\ x^2 + c$$

Once more, the area relation can be confirmed by derivation:

$$A_c{}'(x) = (\pi\ x^2 + c)' = \pi(2x) + 0 = 2\pi\ x$$

About the initial value, when radius is zero, $x = 0$, the circle's area is null too, and the constant term also becomes zero. Once again note the numerical approach, since a circle of radius $x = 0$ would result in a void concept.

$$x = 0 \Rightarrow A_c = 0$$

$$A_c(0) = 0 \Leftrightarrow 0 = \pi 0^2 + c \Leftrightarrow 0 = c$$

Finally,

$$A_c(x) = \pi\, x^2 + 0 = \pi\, x^2$$

And it is well known the circle's area can be obtained by multiplying π and the square of radius; noting that, $r = x$, the classic expression for the circle's area is confirmed in this example:

$$A_c(r) = \pi\, r^2$$

7.4 INTEGRAL SUMS

The integral sum is calculated by partitioning the area in rectangles, for example, assuming all together those rectangles form a region similar to the area under analysis. Calculating the area for each rectangle is straightforward, by multiplying the height and width at each subinterval i; thereafter the total area is obtained by summing all of these areas together.

The Riemann's integral sum treats all the rectangles' areas, with each term representing the area of a rectangle with height $f(x_i)$ and width $(t_i - t_{i-1})$:

$$\sum_{i=1}^{n} f(x_i)\,(t_i - t_{i-1}), \text{ and } t_{i-1} \le x_i \le t_i$$

- Therefore, let $f(x)$ be a non-negative function defined on a closed interval [a, b] and P a partition of n subintervals,

$$P = \{t_0, t_1, t_2, \ldots, t_{i-1}, t_i, t_{i+1}, \ldots t_{n-1}, t_n\}$$

  that is,

$$a = t_0 < t_1 < t_2 < \ldots < t_{i-1} < t_i < t_{i+1} < \ldots < t_{n-1} < t_n = b$$

- At each subinterval i, let m_i be the function's infimum,

$$m_i = \inf\{f(x){:}\ t_{i-1} \le x \le t_i\}$$

and M_i the associated supremum:

$$M_i = \sup\{f(x): t_{i-1} \le x \le t_i\}$$

- Directly related concepts are the lower sum, L, that provides an inferior estimate for the integral sum, or by lower values,

$$L[f(x), P] = \sum_{i=1}^{n} m_i (t_i - t_{i-1})$$

and the upper sum, U, that provides a superior estimate or by excess,

$$U[f(x), P] = \sum_{i=1}^{n} M_i (t_i - t_{i-1})$$

Noting that $m_i \le M_i$, then it follows

$$L[f(x), P] \le U[f(x), P]$$

Also note that for all the partitions, P, that can be defined,

$$L[f(x), P] \le \int_a^b f(x)\, dx$$

while

$$\int_a^b f(x)\, dx \le U[f(x), P]$$

Usually, the region evaluated by the rectangles is not exactly the area to be measured, therefore the integral sum differs from the area under analysis. The error can be reduced by successively using a greater number of subintervals n; when dividing up the area, using thinner and thinner rectangles, both the lower and upper sums are nearing and converging to the exact value for the area.

- When n tends to infinite, the subintervals width approaches zero, while the lower sum and the upper sum tend to the same value:

$$U[f(x), P] - L[f(x), P] < r$$

In such a case, that value is associated with the area between the function's line and the horizontal axis, and it corresponds to the integral sum:

$$\begin{aligned}\lim_{n\to\infty} U\,[f\,(x),\,P] &= \lim_{n\to\infty} L\,[f\,(x),\,P]\\ &= \int_a^b f\,(x)\,dx\end{aligned}$$

7.4.1 Constant Function

The area for the rectangle, A_r, of height 1 and length x is simple and widely used:

$$A_r\,(x) = Height \times Width = 1.x$$

- Let $f(x) = 1$, the closed interval $[0, b]$, and P a partition of n subintervals,

$$P = \{t_0,\ t_1,\ t_2,\ \ldots,\ t_{i-1},\ t_i,\ t_{i+1},\ \ldots\ t_{n-1},\ t_n\}$$

that is,

$$0 = t_0 < t_1 < t_2 < \ldots < t_{i-1} < t_i < t_{i+1} < \ldots < t_{n-1} < t_n = b$$

A typical partition considers the use of regular subdivisions of an interval; namely, the nth regular subdivision of closed interval $[0, 1]$ considers the subinterval borders at regular pace $(1/n)$:

$$\left(0 = \frac{0}{n}\right) < \frac{1}{n} < \frac{2}{n} < \ldots < \frac{(i-1)}{n} < \frac{i}{n} < \ldots < \frac{(n-1)}{n} < \left(\frac{n}{n} = 1\right)$$

Similarly, assuming n equal subintervals for interval $[0, b]$, the border points in partition P are located at distance (b/n) from their adjacent points:

$$\left(0 = \frac{0\,b}{n}\right) < \frac{1b}{n} < \frac{2b}{n} < \ldots < \frac{(i-1)\,b}{n} < \frac{ib}{n} < \ldots < \frac{(n-1)\,b}{n} < \left(\frac{n\,b}{n} = b\right)$$

- At each subinterval i, the function's infimum, m_i, is 1 because this is the constant value for the function at hand (Figure 7.12),

$$m_i = 1$$

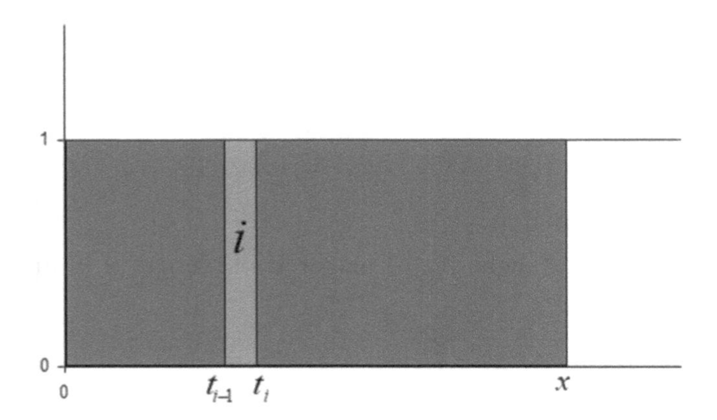

FIGURE 7.12 Rectangle with height 1 and width x; infimum and supremum are equal and both take 1 at each subinterval i.

while the associated supremum M_i takes also the constant 1:

$$M_i = 1$$

- The lower sum, L, assumes the constant height at each subinterval, 1, and the regular pace (b/n), as follows:

$$\begin{aligned} L[f(x), P] &= \sum_{i=1}^{n} m_i(t_i - t_{i-1}) \\ &= \sum_{i=1}^{n} 1\frac{b-0}{n} = \frac{b}{n}\sum_{i=1}^{n}(1) = \frac{b}{n}(n) \\ &= b \end{aligned}$$

- In the same way, the upper sum, U, assumes also the constant height at each subinterval, 1, and the regular pace (b/n):

$$\begin{aligned} U[f(x), P] &= \sum_{i=1}^{n} M_i(t_i - t_{i-1}) \\ &= \sum_{i=1}^{n} 1\frac{b-0}{n} = \frac{b}{n}\sum_{i=1}^{n}(1) = \frac{b}{n}(n) \\ &= b \end{aligned}$$

- The constant value for the lower sum, L, is not altered when the number of subintervals n tends to infinite,

$$\lim_{n\to\infty} L[f(x), P] = \lim_{n\to\infty}(b) = b$$

And the same result holds for the upper sum, U:

$$\lim_{n \to \infty} U\,[f\,(x),\, P] = \lim_{n \to \infty} (b) = b$$

That is, the value is the same for both the limits of upper sum and lower sum,

$$\lim_{n \to \infty} U\,[f\,(x),\, P] = \lim_{n \to \infty} L\,[f\,(x),\, P] = b$$

The integral sum of $f(x) = 1$ in the closed interval $[0, b]$ thus exists, and its value is:

$$\int_0^b 1 \; dx = b$$

- Complementary, when b is 1, the rectangle of height 1 and width 1 corresponds to a square of area 1: $\int_0^1 1 \; dx = 1$.

  When b is 2, the rectangle of width 2 and height 1 presents area 2: $\int_0^2 1 \; dx = 2$.

  And when b is 3, the corresponding rectangle of width 3 presents area 3: $\int_0^3 1 \; dx = 3$.

  Finally, assuming the rectangle width takes a general value, b, the area for the rectangle of height 1 is:

$$\int_0^b 1 \; dx = b = 1.b$$

- From another point of view, the area for the "slice" in the interval [1, 2] corresponds to the area in the larger interval [0, 2] minus the area for the partial region [0, 1],

$$\begin{aligned} \int_1^2 1 \; dx &= \int_0^2 1 \; dx - \int_0^1 1 \; dx \\ &= 2 - 1 = 1 \end{aligned}$$

  The area for the "slice" in the interval [2, 3] corresponds to the larger area in the interval [0, 3] minus the partial area in [0, 2],

$$\begin{aligned} \int_2^3 1 \; dx &= \int_0^3 1 \; dx - \int_0^2 1 \; dx \\ &= 3 - 2 = 1 \end{aligned}$$

Note the total area of the three "slices" corresponds to the rectangle of height 1 and width 3, with area 3; this total is confirmed by straightly adding the partial areas of the three "slices," since all of them are 1.

7.4.2 Linear Function

Just reminding the area of the isosceles triangle, A_t, of equal legs of length x is:

$$A_t(x) = \frac{Leg \times Leg}{2} = \frac{x.x}{2}$$

- Let $f(x) = x$, the closed interval $[0, b]$, and P a partition of n subintervals,

$$P = \{t_0, t_1, t_2, \ldots, t_{i-1}, t_i, t_{i+1}, \ldots t_{n-1}, t_n\}$$

  that is,

$$0 = t_0 < t_1 < t_2 < \ldots < t_{i-1} < t_i < t_{i+1} < \ldots < t_{n-1} < t_n = b$$

  Assuming n equal subintervals for the closed interval, the border points for the partition P are as follows:

$$\left(0 = \frac{0\,b}{n}\right) < \frac{1b}{n} \ldots < \frac{2b}{n} < \ldots < \frac{(i-1)b}{n} < \frac{ib}{n} < \ldots < \frac{(n-1)b}{n} < \left(\frac{n\,b}{n} = b\right)$$

  Note that as the number of subintervals n increases, the width of each subinterval diminishes; and when n tends to infinite, the subinterval width aproaches zero.

- At each subinterval i, the function's infimum, m_i, is located at the left border of the subinterval because the function is strictly increasing (Figure 7.13),

$$m_i = t_{i-1}$$

  while the associated supremum M_i is defined at the related right border:

$$M_i = t_i$$

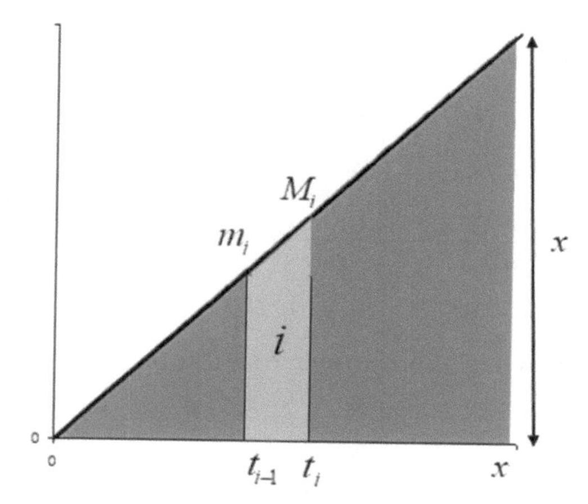

FIGURE 7.13 Triangle with equal legs of length x; infimum at the left border and supremum at the right border of each subinterval i.

- The lower sum, L, assumes the regular width (b/n) and the infimum, m_i, located at the left border at each subinterval:

$$\begin{aligned} L[f(x), P] &= \sum_{i=1}^{n} m_i (t_i - t_{i-1}) \\ &= \sum_{i=1}^{n} t_{i-1} \frac{b-0}{n} = \sum_{i=1}^{n} \left(\frac{(i-1)b}{n} \right) \frac{b}{n} = \left(\frac{b}{n} \right)^2 \sum_{i=1}^{n} (i-1) \\ &= \left(\frac{b}{n} \right)^2 \sum_{j=0}^{n-1} j \end{aligned}$$

And the upper sum, U, assumes the regular width (b/n) too, while the supremum, M_i, is located at the right border of each subinterval:

$$\begin{aligned} U[f(x), P] &= \sum_{i=1}^{n} M_i (t_i - t_{i-1}) \\ &= \sum_{i=1}^{n} t_i \frac{b-0}{n} = \sum_{i=1}^{n} \left(\frac{i\, b}{n} \right) \frac{b}{n} \\ &= \left(\frac{b}{n} \right)^2 \sum_{i=1}^{n} i \end{aligned}$$

- The summation of the first n integer numbers is well known,

$$\sum_{i=1}^{n} i = 1 + 2 + 3 + \ldots + i + \ldots + (n-1) + n = \frac{n(n+1)}{2}$$

and using this expression in the lower sum, L:

$$L[f(x), P] = \left(\frac{b}{n}\right)^2 \sum_{j=0}^{n-1} j$$

$$= \left(\frac{b}{n}\right)^2 \frac{(n-1)(n-1+1)}{2} = b^2 \frac{(n-1)n}{2n^2} = \frac{b^2}{2}\frac{n-1}{n}$$

Taking limit when the number of subintervals n tends to infinite, then

$$\lim_{n\to\infty} L[f(x), P] = \frac{b^2}{2}\lim_{n\to\infty}\left(\frac{n-1}{n}\right) = \frac{b^2}{2}(1) = \frac{b^2}{2}$$

- Similarly, using the summation expression in the upper sum, U:

$$U[f(x), P] = \left(\frac{b}{n}\right)^2 \sum_{i=1}^{n} i$$

$$= \left(\frac{b}{n}\right)^2 \frac{n(n+1)}{2} = b^2 \frac{n(n+1)}{2n^2} = \frac{b^2}{2}\frac{n+1}{n}$$

- Taking limit when the number of subintervals n tends to infinite,

$$\lim_{n\to\infty} U[f(x), P] = \frac{b^2}{2}\lim_{n\to\infty}\left(\frac{n+1}{n}\right) = \frac{b^2}{2}(1) = \frac{b^2}{2}$$

That is, the value is the same for both the limits of upper sum and lower sum,

$$\lim_{n\to\infty} U[f(x), P] = \lim_{n\to\infty} L[f(x), P] = \frac{b^2}{2}$$

Then the integral sum of $f(x) = x$ in the closed interval $[0, b]$ exists, and its value is:

$$\int_0^b x \; dx = \frac{b^2}{2}$$

- Complementary, when b is 1, $\int_0^1 x \; dx = \frac{1^2}{2} = \frac{1}{2}$.

  Also when b is 2, $\int_0^2 x \; dx = \frac{2^2}{2} = \frac{4}{2} = 2$.

And again, when b is 3, $\int_0^3 x \; dx = \frac{3^2}{2} = \frac{9}{2}$.

Finally, assuming the reference dimension takes a general value, b, the triangle's area with equal height and width, x, is:

$$\int_0^b x \; dx = \frac{b^2}{2} = \frac{b.b}{2}$$

- From another point of view, the area for the "slice" in the interval [1, 2] corresponds again to the area in the larger interval [0, 2] minus the area for the partial region [0, 1],

$$\begin{aligned}\int_1^2 x \; dx &= \int_0^2 x \; dx - \int_0^1 x \; dx \\ &= \frac{4}{2} - \frac{1}{2} = \frac{3}{2}\end{aligned}$$

Once again, the area for the "slice" in the interval [2, 3] corresponds to the larger area in the interval [0, 3] minus the partial area [0, 2],

$$\begin{aligned}\int_2^3 x \; dx &= \int_0^3 x \; dx - \int_0^2 x \; dx \\ &= \frac{9}{2} - \frac{4}{2} = \frac{5}{2}\end{aligned}$$

Note the total area of the three "slices" corresponds to the triangle of height and width 3, that is, with area 9/2; this total is confirmed by straightly adding the partial areas of the three slices, namely, 1/2, plus 3/2, and plus 5/2.

7.4.3 The Cost Function

For the barrel's cost function, $C(d)$, the instance with volume $V = 1.0 \text{ m}^3$, is presented (Figure 7.14):

$$C(d) = 100\left(\frac{\pi}{2}d^2 + \frac{4\,V}{d}\right) = 50\pi \; d^2 + \frac{400}{d}$$

For a non-negative function $y(x)$ in the closed interval $[a, b]$, the mean value $\bar{y}$ in such interval can be estimated through the area under the function's line. In fact, that area corresponds to the mean height $\bar{y}$ of the

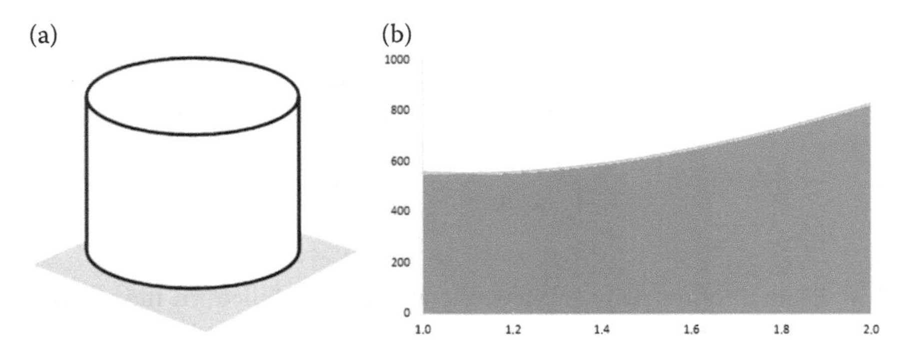

FIGURE 7.14 Cost function in the interval [1, 2].

associated function multiplied by the interval width, $(b - a)$, as indicated in the following relation:

$$Area = MeanHeight \times Width$$

and also using the integral expression,

$$\int_a^b y(x)dx = \bar{y}(b - a)$$

Then the mean height, or also, the function's mean value $\bar{y}$ in the interval $[a, b]$, can be obtained from the integral relation:

$$\bar{y} = \frac{\int_a^b y(x)dx}{(b - a)}$$

In the same way, when the diameter, $x = d$, ranges in the interval [1, 2], the mean cost $\bar{C}$ can be estimated through the mean height of the associated function $C(x)$ and the interval width, $(2 - 1)$, as follows:

$$\int_1^2 C(x)dx = \bar{C}(2 - 1)$$

Or also,

$$\bar{C} = \frac{\int_1^2 C(x)dx}{(2 - 1)} \Leftrightarrow \bar{C} = \int_1^2 C(x)dx$$

The mean value for the cost function $C(x)$ in the interval [1, 2] can be calculated through the integral sum:

$$\int_1^2 C(x)\,dx = \int_1^2 50\pi\, x^2 + \frac{400}{x}dx$$

This integral can be calculated by decomposing the two terms, the first term in x^2 and the second term in $(1/x)$, similar to working with summations and the associated linearity properties:

$$50\pi \int_1^2 x^2\,dx + 400 \int_1^2 \frac{1}{x}dx$$

The calculation for the first term using the integral sum follows, while the calculation for the second term is presented later, in two separate sub-sections. Thereafter, the results for these two terms are combined, and the final result for the mean cost is obtained.

A. The integral sum for the quadratic term, x^2

The objective is to develop the integral sum and evaluate the definite integral $\int_1^2 x^2\,dx$, that is, by addressing the area below the line of quadratic function x^2 in the closed interval [1, 2].

Let $f(x) = x^2$ and the closed interval $[0, b]$; and P is a partition of n regular subintervals,

$$P = \{t_0, t_1, t_2, \ldots, t_{i-1}, t_i, t_{i+1}, \ldots t_{n-1}, t_n\}$$

that is,

$$0 = t_0 < t_1 < t_2 < \ldots < t_{i-1} < t_i < t_{i+1} < \ldots < t_{n-1} < t_n = b$$

- At each subinterval i, the function's infimum, m_i, is located at the left border of the subinterval because the function is strictly increasing (Figure 7.15),

  $$m_i = (t_{i-1})^2$$

  while the associated supremum M_i is defined at the related right border:

  $$M_i = (t_i)^2$$

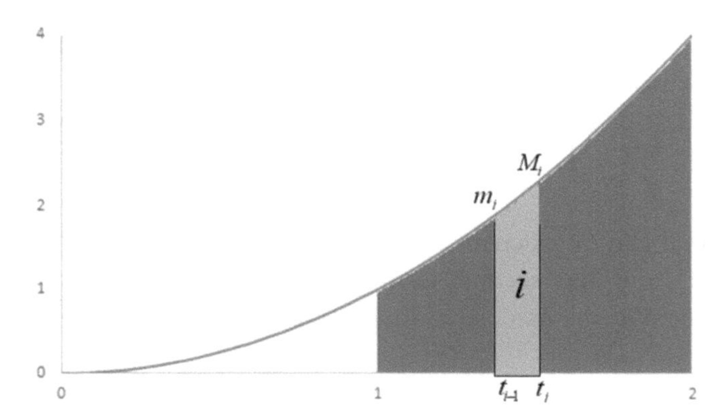

FIGURE 7.15 Increasing function $f(x) = x^2$ in interval [1, 2]; infimum at the left border and supremum at the right border of each subinterval i.

- Assuming n equal subintervals over the closed interval, the border points for the partition P are as follows:

$$\left(0 = \frac{0\,b}{n}\right) < \frac{1b}{n} < \frac{2b}{n} < \ldots < \frac{(i-1)b}{n} < \frac{ib}{n} < \ldots < \frac{(n-1)b}{n} < \left(\frac{n\,b}{n} = b\right)$$

- The lower sum, L, in the usual mode:

$$\begin{aligned} L[f(x), P] &= \sum_{i=1}^{n} m_i(t_i - t_{i-1}) \\ &= \sum_{i=1}^{n} (t_{i-1})^2 \frac{b}{n} = \sum_{i=1}^{n} \left(\frac{(i-1)\,b}{n}\right)^2 \frac{b}{n} = \left(\frac{b}{n}\right)^3 \sum_{i=1}^{n} (i-1)^2 \\ &= \left(\frac{b}{n}\right)^3 \sum_{j=0}^{n-1} j^2 \end{aligned}$$

- And also the upper sum, U:

$$\begin{aligned} U[f(x), P] &= \sum_{i=1}^{n} M_i(t_i - t_{i-1}) \\ &= \sum_{i=1}^{n} (t_i)^2 \frac{b}{n} = \sum_{i=1}^{n} \left(\frac{i\,b}{n}\right)^2 \frac{b}{n} = \left(\frac{b}{n}\right)^3 \sum_{i=1}^{n} i^2 \end{aligned}$$

- The summation of the square of the first n integer numbers is well established,

$$\sum_{i=1}^{n} i^2 = 1^2 + 2^2 + 3^2 + \ldots + i^2 + \ldots + (n-1)^2 + n^2$$
$$= \frac{n(n+1)(2n+1)}{6}$$

and using this expression in the lower sum, L:

$$L[f(x), P] = \left(\frac{b}{n}\right)^3 \sum_{j=0}^{n-1} j^2$$
$$= \left(\frac{b}{n}\right)^3 \frac{(n-1)n(2n-2+1)}{6} = b^3 \frac{(n-1)n(2n-1)}{6n^3}$$
$$= b^3 \frac{(n-1)(2n-1)}{6n^2} = \frac{b^3}{3} \frac{2n^2 - 3n + 1}{2n^2}$$

Taking limit when the number of subintervals n tends to infinite,

$$\lim_{n\to\infty} L[f(x), P] = \frac{b^3}{3} \lim_{n\to\infty} \left(\frac{2n^2 - 3n + 1}{2n^2}\right) = \frac{b^3}{3}(1) = \frac{b^3}{3}$$

- Similarly, using the expression for the square of the first n integer numbers in the upper sum, U:

$$U[f(x), P] = \left(\frac{b}{n}\right)^3 \sum_{i=1}^{n} i^2$$
$$= \left(\frac{b}{n}\right)^3 \frac{n(n+1)(2n+1)}{6} = b^3 \frac{n(n+1)(2n+1)}{6n^3}$$
$$= b^3 \frac{(n+1)(2n+1)}{6n^2} = \frac{b^3}{3} \frac{2n^2 + 3n + 1}{2n^2}$$

And taking limit when the number of subintervals n tends to infinite,

$$\lim_{n\to\infty} U[f(x), P] = \frac{b^3}{3} \lim_{n\to\infty} \left(\frac{2n^2 + 3n + 1}{2n^2}\right) = \frac{b^3}{3}(1) = \frac{b^3}{3}$$

- That is, the value is the same for both the limits of upper sum, U, and lower sum, L,

$$\lim_{n\to\infty} U\,[f\,(x),\,P] = \lim_{n\to\infty} L\,[f\,(x),\,P] = \frac{b^3}{3}$$

Then the integral sum for $f(x) = x^2$ in the closed interval $[0, b]$ exists, and its value is:

$$\int_0^b x^2\,dx = \frac{b^3}{3}$$

- When b is 1, $\int_0^1 x^2\,dx = \frac{1^3}{3} = \frac{1}{3}$.

And also when b is 2, $\int_0^2 x^2\,dx = \frac{2^3}{3} = \frac{8}{3}$.

In this way, the "slice" corresponding to the interval [1, 2] follows by subtracting the last two areas,

$$\begin{aligned}\int_1^2 x^2\,dx &= \int_0^2 x^2\,dx - \int_0^1 x^2\,dx\\ &= \frac{8}{3} - \frac{1}{3} = \frac{7}{3}\end{aligned}$$

Finally, the value for the first term of the cost function is obtained:

$$50\pi \int_1^2 x^2\,dx = 50\pi\left(\frac{7}{3}\right) = \frac{350\pi}{3}$$

B. The integral sum for the reciprocal term, 1/*x*

Now, the second term for the mean cost is addressed via the following integral sum:

$$400\int_1^2 \frac{1}{x}dx$$

The objective is to develop this integral sum and evaluate the definite integral $\int_1^2 \frac{1}{x}dx$, that is, by evaluating the area below the line of $(1/x)$ in the closed interval [1, 2], as in Figure 7.16.

Let $f(x) = 1/x$, the closed interval [1, 2], and P a partition of n regular subintervals,

$$P = \{t_0,\, t_1,\, t_2, \ldots, t_{i-1},\, t_i,\, t_{i+1},\, \ldots\, t_{n-1},\, t_n\}$$

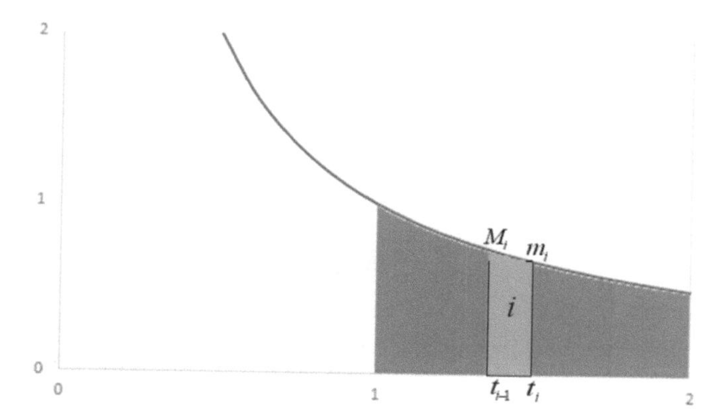

FIGURE 7.16 Decreasing function $f(x) = 1/x$ in interval [1, 2]; infimum at the right border and supremum at the left border of each subinterval i.

that is,

$$1 = t_0 < t_1 < t_2 < \ldots < t_{i-1} < t_i < t_{i+1} < \ldots < t_{n-1} < t_n = 2$$

- At each subinterval, i, the function's infimum, m_i, is located at the right border of the subinterval because the function is strictly decreasing (Figure 7.16),

$$m_i = \frac{1}{t_i}$$

while the associated supremum M_i is defined at the left border:

$$M_i = \frac{1}{t_{i-1}}$$

Assuming n equal subintervals in [1, 2], the border points for the partition P are:

$$\left(1 = 1 + \frac{0}{n}\right) < 1 + \frac{1}{n} < 1 + \frac{2}{n} < \ldots < 1 + \frac{(i-1)}{n} < 1 + \frac{i}{n} < \ldots$$
$$< 1 + \frac{(n-1)}{n} < \left(1 + \frac{n}{n} = 2\right)$$

- Therefore, by defining the regular subintervals and the infimum at each subinterval, the lower sum, L, follows:

$$
\begin{aligned}
L[f(x), P] &= \sum_{i=1}^{n} m_i (t_i - t_{i-1}) \\
&= \sum_{i=1}^{n} \frac{1}{t_i} \cdot \frac{1}{n} = \sum_{i=1}^{n} \frac{1}{\left(1 + \frac{i}{n}\right)} \cdot \frac{1}{n} = \sum_{i=1}^{n} \frac{1}{(n+i)} \\
&= \frac{1}{n+1} + \frac{1}{n+2} + \ldots + \frac{1}{n+i} + \ldots + \frac{1}{n+(n-1)} + \frac{1}{n+n} \\
&= \frac{1}{n+1} + \frac{1}{n+2} + \ldots + \frac{1}{n+i} + \ldots + \frac{1}{2n-1} + \frac{1}{2n}
\end{aligned}
$$

- Similarly, the upper sum, U, follows too:

$$
\begin{aligned}
U[f(x), P] &= \sum_{i=1}^{n} M_i (t_i - t_{i-1}) \\
&= \sum_{i=1}^{n} \frac{1}{t_{i-1}} \cdot \frac{1}{n} = \sum_{i=1}^{n} \frac{1}{\left(1 + \frac{i-1}{n}\right)} \cdot \frac{1}{n} = \sum_{i=1}^{n} \frac{1}{(n+i-1)} \\
&= \frac{1}{n} + \frac{1}{n+1} + \ldots + \frac{1}{n+i-1} + \ldots + \frac{1}{n+(n-2)} \\
&\quad + \frac{1}{n+(n-1)} \\
&= \frac{1}{n} + \frac{1}{n+1} + \ldots + \frac{1}{n+i-1} + \ldots + \frac{1}{2n-2} + \frac{1}{2n-1}
\end{aligned}
$$

- The summation of the first n harmonic numbers is well established in the literature,

$$
H_n = \sum_{i=1}^{n} \frac{1}{i} = = \frac{1}{1} + \frac{1}{2} + \frac{1}{3} + \ldots + \frac{1}{i} + \ldots + \frac{1}{n-1} + \frac{1}{n}
$$

It is also assumed an approximation using the Euler constant, $\gamma \cong 0.5772$, with an error term that approaches zero when the number of subintervals n increases, ε_n,

$$
H_n = \ln(n) + \gamma + \frac{1}{2n} - \varepsilon_n
$$

Using the harmonic relation in the lower sum, L:

$$\begin{aligned} L[f(x), P] &= \frac{1}{n+1} + \frac{1}{n+2} + \ldots + \frac{1}{n+i} + \ldots + \frac{1}{2n-1} + \frac{1}{2n} \\ &= H_{2n} - H_n \end{aligned}$$

Taking limit when the number of subintervals n tends to infinite,

$$\begin{aligned} \lim_{n\to\infty} L[f(x), P] &= \lim_{n\to\infty} (H_{2n} - H_n) \\ &= \lim_{n\to\infty} \left\{ \left[\ln(2n) + \gamma + \tfrac{1}{2(2n)} - \varepsilon_{2n} \right] \right. \\ &\quad \left. - \left[\ln(n) + \gamma + \tfrac{1}{2n} - \varepsilon_n \right] \right\} \\ &= \lim_{n\to\infty} [\ln(2n) - \ln(n)] \\ &= \lim_{n\to\infty} \ln\left(\tfrac{2n}{n}\right) \\ &= \ln\left(\lim_{n\to\infty} \tfrac{2n}{n}\right) = \ln(2) \end{aligned}$$

Similarly, using the Harmonic relation in the upper sum, U:

$$\begin{aligned} U[f(x), P] &= \frac{1}{n} + \frac{1}{n+1} + \ldots + \frac{1}{n+i-1} + \ldots + \frac{1}{2n-2} + \frac{1}{2n-1} \\ &= H_{2n-1} - H_{n-1} \end{aligned}$$

Again, taking limit when the number of subintervals n tends to infinite,

$$\begin{aligned} \lim_{n\to\infty} U[f(x), P] &= \lim_{n\to\infty} (H_{2n-1} - H_{n-1}) \\ &= \lim_{n\to\infty} \left\{ \left[\ln(2n-1) + \gamma + \frac{1}{2(2n-1)} - \varepsilon_{2n-1} \right] \right. \\ &\quad \left. - \left[\ln(n-1) + \gamma + \frac{1}{2(n-1)} - \varepsilon_{n-1} \right] \right\} \\ &= \lim_{n\to\infty} [\ln(2n-1) - \ln(n-1)] \\ &= \lim_{n\to\infty} \ln\left(\frac{2n-1}{n-1}\right) \\ &= \ln\left(\lim_{n\to\infty} \frac{2n-1}{n-1}\right) = \ln(2) \end{aligned}$$

- That is, the value is the same for both the limits of upper sum, U, and lower sum L,

$$\lim_{n\to\infty} U[f(x), P] = \lim_{n\to\infty} L[f(x), P] = \ln(2)$$

Then the integral sum for function $f(x) = 1/x$ in the closed interval $[1, 2]$ exists, and the related value is:

$$\int_1^2 \frac{1}{x}dx = \ln(2)$$

In this way, the value for the second term in the cost function is obtained too:

$$400\int_1^2 \frac{1}{x}dx = 400\ln(2)$$

Finally, conjugating the values obtained for the two terms, the mean cost $\bar{C}$ when the barrel's diameter ranges in the interval $[1, 2]$ is:

$$\begin{aligned}\bar{C} &= 50\pi\int_1^2 x^2dx + 400\int_1^2 \frac{1}{x}dx \\ &= \frac{350\pi}{3} + 400\ln(2) \\ &\cong 643.78\ €\end{aligned}$$

7.5 CONCLUDING REMARKS

The integral sum is a very important topic in applied mathematics, given the scores of theoretical approaches and computational methods that can efficiently address problems based on areas quadrature. In fact, squaring the area of a figure or the region below the function's line presents important applications. It includes, among others: the mean cost for a component when different sizes are manufactured; the accumulated probabilities in statistics, the inference tests and related measures; as well as physics relations addressing non-constant acceleration, variable velocity, and motion in space.

Integral sums are aiming at area estimates; for a non-negative function, $y(x)$, in a closed interval $[a, b]$, they evaluate the area below the function's line and above the horizontal axis, and for that:

A. A suitable partition P of the interval $[a, b]$ is developed; usually a simple figure is selected (e.g., rectangle, square, trapezoid, circle) for estimating the area elements, and the related subintervals are defined at regular pace.

B. For each partition P and at each subinterval i, the infimum and the supremum are defined; while the first value is considered in the construction of lower sums, L, the latter is considered in upper sums, U.

C. Then, both the lower sum L and the upper sum U are evaluated as inferior and superior estimates, respectively; summation results (e.g., first integer numbers; the square of first integer numbers; Harmonic numbers; geometric sums) and summation properties (e.g., telescopic, index alterations) are very often required.

D. Additionally, the sums convergence when the number of subintervals tends to infinite needs a proper demonstration, in a way to verify the existence and uniqueness of limit; upon confirmation, the limit-value for those sums is taken as the integral value.

E. Else, if the referred sums (or other sums, e.g., the "left-sum," "right-sum," or even the "mean-value sum" at each subinterval i) present different limits or the related limits do not exist, then the integral value cannot be defined.

That is, the area alteration due to the addition of a single area element (e.g., a very thin rectangle) can be evaluated by the associated differential of area. Or better, the area's differential can be estimated by the rectangle's height—the function value $y(x)$—multiplied by the differential width, dx—the infinitesimal change in variable x.

Typically, these steps A to E are difficult, long, and more effective tools that allow faster area calculations are needed. Note that when estimating the total area, the integral sum of all the area differentials is a pertinent approach, and the function's anti-derivative—the function $Y(x)$ that results in $y(x)$ upon derivation—is important to the solution.

These integral tools are presented in the next chapter, by introducing the indefinite integral that supports the solution of integral sums using the anti-derivation approach. The main aims for next chapter thus follow:

- Addressing the primitive process (anti-derivation) as the reverse process of derivation, as well as focusing on the main properties of primitives.
- Promoting the calculation of integrals (anti-derivatives, primitives) based on the derivatives of elementary functions already studied.

CHAPTER 8

Indefinite Integral

This chapter addresses the indefinite integral, the primitive function; first, basic notions and main properties are introduced, and then immediate primitives are presented. The linearity attributes for the indefinite integral are discussed too, namely, the integration of both the functions sum and the function times a constant. The general methods of integration are also addressed, including integration by decomposition, by substitution, and by parts; a primitive for the cost function is also worked out via decomposition. In addition, specific methods for the integration of rational functions are described; with specific substitutions, both the irrational and trigonometric functions can be transformed, and then treated as rational functions.

8.1 INTRODUCTION

In a general mode, the study of indefinite integral (*alias*, primitive function, antiderivative) is supported in functions derivation, an opposite attribute arising from the primitive definition: $F(x)$ is the primitive or antiderivative of function $f(x)$, if deriving $F(x)$ results in $f(x)$.

- In this way, primitivation (anti-derivation) can be seen as reverting the derivation process. The verification or confirmation of antiderivatives is thus based on the derivation of elementary functions previously studied. Simple primitives are introduced in the initial sections, and a collection of immediate primitives is also presented.
- Primitivation and related properties are also suitable for the reverse verification, and the linearity attributes of limits and derivatives also transit to primitives; namely, for the primitive of sums and for the primitive of constant times function.
- The cost function is suitable for integration by decomposition, each term is separated and its coefficients are factored out; a primitive that crosses the X-axis at point 1 is focused on, while simple instances of areas calculation (rectangle, triangle) are also featured.

DOI: 10.1201/9781003461876-8

- Specific techniques for rational functions follow systematic procedures, and several cases are presented and solved; other specific procedures exist for both the irrational and trigonometric functions, and simple instances are treated for illustration purposes only.

In this way, the chapter continues with the properties of indefinite integral while immediate primitives are presented; in addition, two main results are focused on, namely, the linearity for integration both of sum of functions and for the multiplication by a constant. After that, general methods of integration are addressed: integration by decomposition, when facing sums; by substitution, when treating a composite function that could be originated by the derivation's chain rule; by parts when focusing the product of two functions. And finally, the integration of rational functions is treated, the resulting primitives presenting quite different forms (e.g., logarithm, power, arctangent).

8.2 INDEFINITE INTEGRAL

The properties of indefinite integral and the main primitives are treated in this section; basic notions about the indefinite integral are introduced, and then important properties and immediate primitives are presented and discussed.

Primitive Function (Indefinite Integral)

Given any elementary function, $f(x)$, if there exists another elementary function, $F(x)$ such that,

$$F'(x) = \frac{dF(x)}{dx} = f(x)$$

then $F(x)$ is the primitive (antiderivative) of function, $f(x)$, and the indefinite integral is:

$$\int f(x)dx = F(x) + c$$

Both the expressions indefinite integral and primitive are commonly used. In fact, the indefinite integral represents a family of functions, since the derivative of every constant, c, is zero and thus can be neglected on verification procedures.

From the primitive definition, and with the inverse approach of derivation or differentiation, the following results follow successively:

- The derivative of an indefinite integral is equal to the function to be integrated, that is, if $F'(x) = f(x)$, then

$$\left[\int f(x)\,dx\right]' = [F(x) + c]' = f(x)$$

- The differential of an indefinite integral is equal to the expression under the integral sign:

$$d\left[\int f(x)\,dx\right] = f(x)\,dx$$

Such equality directly results from the primitive function's differential $dF = F'dx$, and then by substituting the previous result.

- The indefinite integral for the differential of a function is equal to the sum of this function and a constant:

$$\int dF = F(x) + c$$

Again, the indefinite integral represents a family of functions; typically, the integration constant, c, can be calculated either from the initial value, the final value, or other reference value for the function under study.

Immediate Primitives

Therefore, the primitiveness (anti-derivation) process can be seen as the inverse of derivation process. In this manner, simple primitives are presented, and verifying or confirming such antiderivatives is based on deriving elementary functions already studied.

- The primitive of a constant function, $f(x) = k$, corresponds to a linear function:

$$\int k\,dx = k\,x + c$$

The verification procedure is based on derivation; the aim is obtaining again the constant function:

$$(k\,x + c)' = k + 0 = k$$

As the area estimates in prior chapter, the rectangle of height 1, 2, 3, … , k, and width, x, presents an increasing area that starts at zero: $(1x)$, $(2x)$, $(3x)$, … , (kx), as in Figure 8.1.

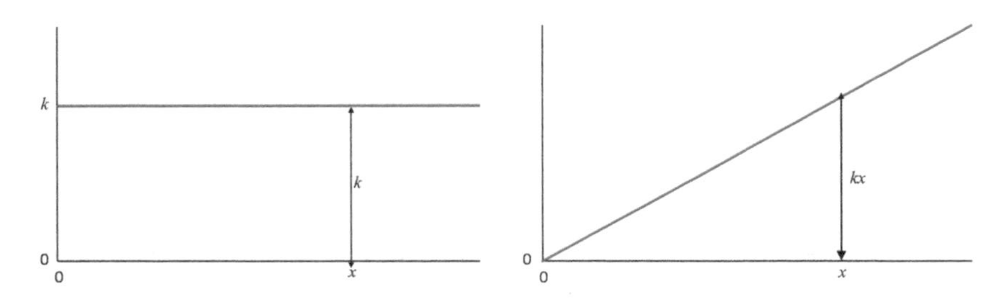

FIGURE 8.1 Revisiting the constant, $f(x) = k$, and the linear function, $f(x) = kx$.

- And the primitive of a linear function, $f(x) = kx$, corresponds to a quadratic function:

$$\int kx\, dx = k\frac{x^2}{2} + c$$

Again, the verification procedure aims to obtain the linear function:

$$\left(k\frac{x^2}{2} + c\right)' = k\frac{2x}{2} + 0 = kx$$

By analogy with the prior area estimates too, for the triangle of positive width (leg), x, and slope 1, 2, 3, … , k, then its area will correspond to the product height by width (leg times leg) divided by two: $(1x).x/2$, $(2x).x/2$, $(3x).x/2$, … , $(kx).x/2$.

- Now, the primitive of a quadratic function, $f(x) = x^2$, corresponds to a cubic function:

$$\int x^2\, dx = \frac{x^3}{3} + c$$

When verifying it, we obtain again the original function:

$$\left(\frac{x^3}{3} + c\right)' = \frac{3x^2}{3} + 0 = x^2$$

Based on prior area estimates too (Chapter 7), the integration of both quadratic and reciprocal function, $f(x) = x^2$ and $f(x) = 1/x$ as in Figure 8.2, do not correspond to regular geometric figures; by

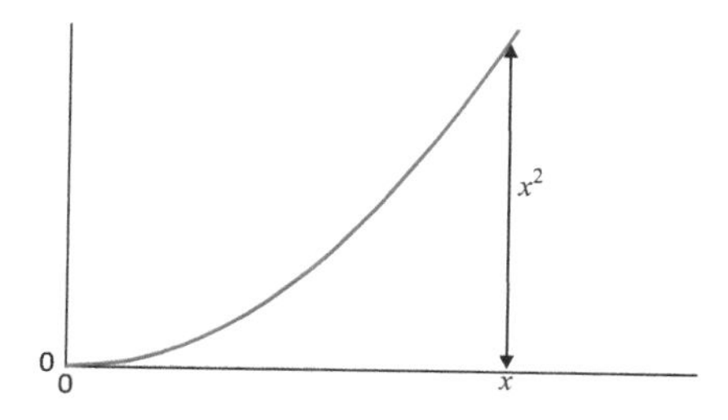

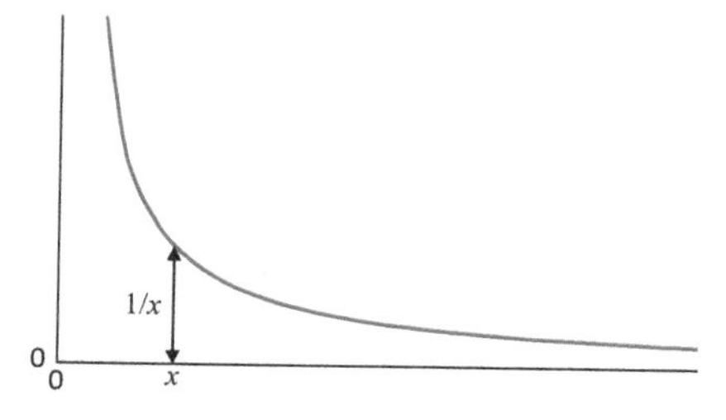

FIGURE 8.2 Revisiting the quadratic, $f(x) = x^2$, and the reciprocal function, $f(x) = 1/x$.

allowing estimates of areas defined through continuous functions, the integral calculation also overpasses such analogies with simple regular areas.

- And the primitive of a general power function of nth-order, $f(x) = x^n$, corresponds to a power function of higher order, $n + 1$, divided by such augmented order:

$$\int x^n \, dx = \frac{x^{n+1}}{n+1} + c$$

In fact, by deriving such promitive the original function is obtained, that is, by multiplying it by the augmented order, $n + 1$, and decreasing the power order by one,

$$\left(\frac{x^{n+1}}{n+1} + c\right)' = (n+1)\frac{x^{n+1-1}}{n+1} + 0 = x^n$$

Obviously the power integration formula does not hold for $n = -1$; in this case the natural logarithm occurs,

$$\int \frac{1}{x} dx = \ln(x) + c$$

The last result is directly confirmed by derivation of the logarithm in second member. Similarly, the integration of negative integer powers or fractional powers can be verified too; namely, for the primitive of power function with $n = -2$,

$$\int \frac{1}{x^2} dx = \int x^{-2}\, dx = \frac{x^{-2+1}}{-2+1} + c = \frac{x^{-1}}{-1} + c = \frac{-1}{x} + c$$

And for the fractional power, 1/2, the primitive of quadratic root function follows:

$$\int \sqrt{x}\, dx = \int x^{1/2}\, dx = \frac{x^{1/2+1}}{(1/2)+1} + c = \frac{x^{3/2}}{(3/2)} + c = \frac{2}{3} x\sqrt{x} + c$$

In general, the integration of power functions, with both integer and fractional powers, either positive or negative powers (except for $n = -1$), corresponds to a power function of superior order, $n + 1$, divided by such augmented order.

- In fact, the referred procedure inverts the power derivation, and also the steps sequence: it first considers multiplying by the power n, and then decreasing the power by one: $(x^n)' = n\ x^{n-1}$.
- In the specific instance of $n = 0$, the primitive reduces itself to the integration of a constant function, $f(x) = x^0 = 1$,

$$\int x^0\, dx = \frac{x^{0+1}}{0+1} + c = x + c$$

And the integration of exponential functions with natural base, e, or other positive base different of one, $a \neq 1$, in general corresponds to the exponential function divided by the natural logarithm of the base. In fact, the exponential's derivation results also in the exponential function,

$$\int e^x\, dx = e^x + c$$

Since the exponential's derivation supposes multiplication by the natural logarithm of the base, the following formula can be confirmed by derivation of the second member too,

$$\int a^x\, dx = \frac{a^x}{\ln(a)} + c$$

- In case of, $a = 2$ or $a = 10$, the related primitives consider the exponentials divided by the natural logarithm of the base at hand:

$$\int 2^x \, dx = \frac{2^x}{\ln(2)} + c, \text{ or also, } \int 10^x \, dx = \frac{10^x}{\ln(10)} + c$$

- In the specific instance of $a = e$, the related primitive reduces itself to the particular case of integrating the exponential of natural base, with $\ln(e) = 1$:

$$\int e^x \, dx = \frac{e^x}{\ln(e)} + c = e^x + c$$

About the integration of trigonometric functions, namely, sine or cosine, in general corresponds to the complementary function divided by the coefficient on variable, x. In fact, the sine's derivation results in the cosine function; then, in the anti-derivation sense,

$$\int \cos(x) \, dx = \sin(x) + c$$

And the following formula can be also confirmed by derivation of the second member:

$$\int \sin(x) \, dx = -\cos(x) + c$$

- If the variable's coefficient is different than 1, for instance, 2 or 3, the related primitives consider the complementary function (sine *vs.* cosine), with the proper sign (inverting the derivation sense too), and dividing by 2 or 3, respectively:

$$\int \cos(2x) \, dx = \frac{\sin(2x)}{2} + c$$

$$\int \sin(3x) \, dx = \frac{-\cos(3x)}{3} + c$$

In addition, for the integration of tangent or cotangent functions, the situation is not so simple. In fact, the tangent's derivation in the second member, as well as for the cotangent function, allows us to obtain the following equalities:

$$\begin{aligned}\int \frac{1}{\cos^2(x)}\,dx &= \tan(x) + c\\ \int \frac{1}{\sin^2(x)}\,dx &= -\cotan(x) + c\end{aligned}$$

- Recalling the tangent function is the ratio between sine and cosine functions, and the cotangent function is the ratio between cosine and sine, then it can be shown that:

$$\int \cotan(x)\,dx = \int \frac{\cos(x)}{\sin(x)}dx = \int \frac{\sin'(x)}{\sin(x)}dx = \ln|\sin(x)| + c$$

And also that,

$$\int \tan(x)\,dx = \int \frac{\sin(x)}{\cos(x)}dx = -\int \frac{\cos'(x)}{\cos(x)}dx = -\ln|\cos(x)| + c$$

Due to the importance of inverse trigonometric functions, along with their wide application in a number of integration developments, e.g., rational and irrational functions (section 8.5), the following equalities can be confirmed by deriving the right hand side:

$$\begin{aligned}\int \frac{dx}{1+x^2} &= \arctan(x) + c\\ \int \frac{dx}{\sqrt{1-x^2}} &= \arcsin(x) + c\end{aligned}$$

Finally, a simple collection of these immediate primitives follows, with some important notes about the search of antiderivatives:

$$\begin{array}{ll}\int k\,dx = kx + c & \int x^n\,dx = \frac{x^{n+1}}{n+1} + c\\ \int \frac{1}{x}\,dx = \ln(x) + c & \int \sqrt{x}\,dx = \frac{2}{3}x\sqrt{x} + c\\ \int e^x\,dx = e^x + c & \int a^x\,dx = \frac{a^x}{\ln(a)} + c\\ \int \cos(x)\,dx = \sin(x) + c & \int \sin(x)\,dx = -\cos(x) + c\\ \int \frac{1}{\cos^2(x)}\,dx = \tan(x) + c & \int \frac{1}{\sin^2(x)}\,dx = -\cotan(x) + c\\ \int \cotan(x)\,dx = \ln[\sin(x)] + c & \int \tan(x)\,dx = -\ln[\cos(x)] + c\\ \int \frac{dx}{1+x^2} = \arctan(x) + c & \int \frac{dx}{\sqrt{1-x^2}} = \arcsin(x) + c\end{array}$$

- Some primitives present the same form as the integrand function (e.g., power, exponential, sine and cosine functions); this occurs because the derivation of such functions allows to maintain the original form, with important developments in other topics (e.g., Taylor series in Chapter 10).
- However, other primitives can be very different from the integrand function, such as the logarithm, square root, tangent, or arctangent function, and their search can be difficult.
- Beyond the difficulties to obtain primitives, in certain cases it is not possible to define them through a finite number of operations and elementary functions, for instance:

$$\int e^{-x^2}\, dx, \int \cos(x^2)\, dx, \int \sin(x^2)\, dx$$

Then, numerical methods (e.g., series, trapezoid rule) are also used to evaluate integrals.

8.3 PROPERTIES OF INDEFINITE INTEGRAL

In this section, two main properties of indefinite integral are focused on; namely, the linearity attributes both for integration of sums and function times a constant. In fact, these properties about primitives (antiderivatives) are based on the properties of derivatives, and thus can be directly verified by deriving the second member.

The integral of a functions sum is equal to the sum of their integrals:

$$\int [f_1(x) + f_2(x)]\, dx = \int f_1(x)\, dx + \int f_2(x)\, dx = F_1(x) + F_2(x)$$

By taking derivative on the right hand side, and since the derivative of a sum becomes the sum of derivatives, then the original sum in the first member is obtained:

$$F_1{}'(x) + F_2{}'(x) = \left(\int f_1(x)\, dx\right)' + \left(\int f_2(x)\, dx\right)' = f_1(x) + f_2(x)$$

The integral of a function multiplied by constant corresponds to the multiplication of that constant by the function's integral:

$$\int k.f(x)\,dx = k.\int f(x)\,dx = k.F(x)$$

Again, by deriving the second member, and since the derivative of constant times function corresponds to constant times the derivative function, $(k.f)' = k.f'$, then the equality is confirmed too:

$$[k.F(x)]' = \left(k.\int f(x)\,dx\right)' = k.\left(\int f(x)\,dx\right)' = k.f(x)$$

Assuming the operations are possible and all the antiderivatives exist, recall that derivatives are defined as limits on function increments, and similar properties hold for function limits too. For example, the limit for the sum of functions corresponds to the sum of function limits, and the sequences sum corresponds to the sum of the convergent sequences.

- Similar properties also hold in different mathematical subjects; they are known as linearity properties, and are important in statistics (e.g., the average of a two distributions sum corresponds to the sum of the two average distributions), in operational calculus (e.g., the Laplacian of two functions sum corresponds to the sum of the two associate Laplacians), or in sum series (e.g., the series of a sum corresponds to the sum of the two series), obviously assuming such sums are possible and the associated objects exist.

However, the integral of functions multiplication does not correspond to the multiplication of integrals, because the derivative of functions product does not correspond to the product of derivatives. In fact, the referred derivative presents a non-linear form,

$$[f(x).g(x)]' = f'(x).g(x) + f(x).g'(x)$$

and taking antiderivatives from both sides of the equality,

$$\int [f(x).g(x)]'dx = \int f'(x).g(x)\,dx + \int f(x).g'(x)\,dx$$

$$f(x).g(x) = \int f'(x).g(x)\,dx + \int f(x).g'(x)\,dx$$

then two different expressions can be obtained, one for each integral in the second member. While this expression is preferable when the primitive of $f'(x)$ is easily treated,

$$\int f'(x).g(x)dx = f(x).g(x) - \int f(x).g'(x)dx$$

the next expression is thus preferred when the primitive of $g'(x)$ is known,

$$\int f(x).g'(x)dx = f(x).g(x) - \int f'(x).g(x)dx$$

These two expressions are commonly utilized to treat primitives of functions multiplication, the integration by parts. However, it is also very useful to transform the multiplication in sum of functions, and then treat term by term that sum: the integration by decomposition. These methods can be applied in a general manner to the symbolic search of indefinite integrals.

8.4 GENERAL METHODS OF INTEGRATION

The general methods of integration are addressed in this section, namely, integration by decomposition, by substitution, and by parts; these integration methods are preferably utilized when facing, respectively: (i) sum of functions; (ii) a composite function that could be originated by the derivation's chain rule; and (iii) the multiplication of two functions.

8.4.1 Integration by Decomposition—A Primitive for the Cost Function

From the derivation rule for a functions sum, it results that:

$$\int (f + g)(x)\, dx = F(x) + G(x) + c$$

The integration of polynomials is a good application, and a suitable instance will follow too.

Recall the cost function $y(x)$

$$y(x) = 100\left(\frac{\pi}{2}x^2 + \frac{4}{x}\right) = 50\pi\ x^2 + \frac{400}{x}$$

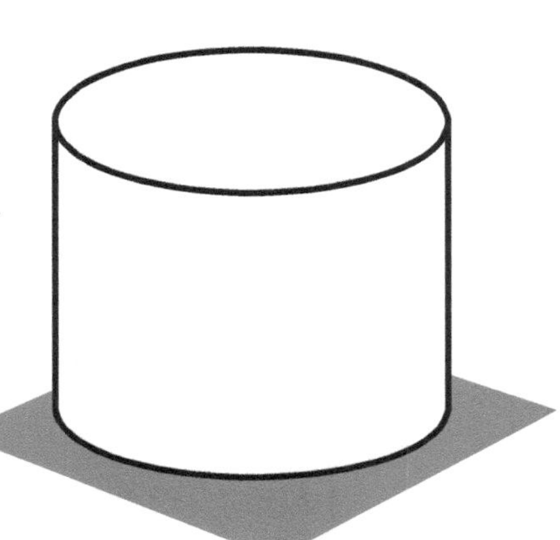

The primitive is obtained by decomposition, since the integral for the sum of functions corresponds to the sum of integrals at each term.

$$Y(x) = \int 50\pi\ x^2 + \frac{400}{x}dx = \int 50\pi\ \ x^2\ dx + \int \frac{400}{x}dx$$

In order that immediate primitives are recognized and treated, the coefficients are factored out,

$$Y(x) = 50\pi \int x^2dx + 400\ \int \frac{1}{x}dx = 50\pi\frac{x^3}{3} + 400\ \ln|x| + c$$

To estimate the integration constant, c, assume that, $Y(1) = 0$ (Figure 8.3); thereafter,

$$Y(1) = 0 \Rightarrow 50\pi\frac{(1)^3}{3} + 400\ \ln|1| + c = 0$$

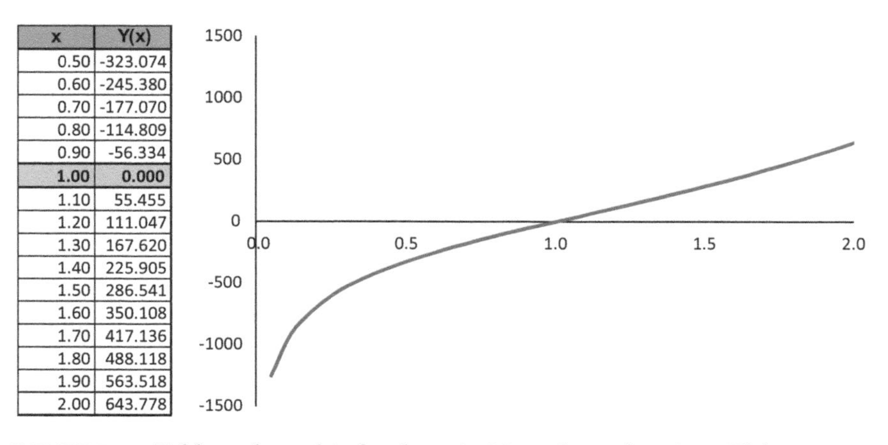

x	Y(x)
0.50	-323.074
0.60	-245.380
0.70	-177.070
0.80	-114.809
0.90	-56.334
1.00	0.000
1.10	55.455
1.20	111.047
1.30	167.620
1.40	225.905
1.50	286.541
1.60	350.108
1.70	417.136
1.80	488.118
1.90	563.518
2.00	643.778

FIGURE 8.3 Table and graphic for the primitive of cost function, $Y(x)$, crossing the X-axis at point 1.

The logarithm in the second term is zero and the result is: $c = -\frac{50\pi}{3}$; then,

$$Y(x) = 50\pi\frac{x^3}{3} + 400\ln|x| - \frac{50\pi}{3}$$

- In addition, let the fifth order polynomial be $P_5(x)$, which corresponds to the Taylor polynomial for the cost function at point, $x = 1$:

$$P_5(x) = 2400 - 6000x + (50\pi + 8000)x^2 - 6000x^3 + 2400x^4 - 400x^5$$

The primitive is obtained by decomposition too; each term is treated separately and a sixth order polynomial is then obtained:

$$\int P_5(x)\,dx = \int [2400 - 6000x + (50\pi + 8000)x^2 - 6000x^3 + 2400x^4 - 400x^5]\,dx$$

$$P_6(x) = \int 2400\,dx - \int 6000x\,dx + \int (50\pi + 8000)x^2dx - \int 6000x^3dx + \int 2400x^4dx - \int 400x^5dx$$

$$P_6(x) = 2400\int 1\,dx - 6000\int x\,dx + (50\pi + 8000)\int x^2dx - 6000\int x^3dx + 2400\int x^4dx - 400\int x^5dx$$

Treating the primitives at each term, and adding the integration constant:

$$\begin{aligned} P_6(x) &= 2400x - 6000\frac{x^2}{2} + (50\pi + 8000)\frac{x^3}{3} - 6000\frac{x^4}{4} + 2400\frac{x^5}{5} - 400\frac{x^6}{6} + c \\ &= 2400x - 3000x^2 + \frac{(50\pi + 8000)}{3}x^3 - 1500x^4 + 480x^5 - \frac{200}{3}x^6 + c \end{aligned}$$

8.4.2 Integration by Substitution

From the chain derivation rule that applies to the composite function, it can be recognized that the integrand function appears as:

$$[(f \circ g)(x).g'(x)] = f[g(x)].g'(x)$$

Then, identifying the function, $g(x)$, which is accompanying its own derivative, $g'(x)$,

$$\int f[g(x)].g'(x)dx = F[g(x)] + c$$

Or even substituting such function, $u = g(x)$, the integral will present a more friendly form:

$$\int f(u)du = F(u) + c$$

- The proposed procedure follows:

  1. Substitution

$$\begin{cases} u & = & g(x) \\ du & = & g'(x)dx \end{cases}$$

  2. The primitive calculation

$$\int f(u)du = F(u) + c$$

  3. Inversion of substitution

$$F(u) + c = F[g(x)] + c$$

Examples

- Let it be, $\int -2x \ e^{-x^2}dx$;

The function, $g(x) = -x^2$, is presented with its derivative too, $g'(x) = -2x$;

1. Substitution

$$\begin{aligned} u &= -x^2 \\ du &= -2x\,dx \end{aligned}$$

2. The primitive calculation

$$\int -2x\ e^{-x^2} dx = \int e^u\, du = e^u + c$$

3. Inversion of substitution

$$e^u + c = e^{-x^2} + c$$

The primitive finally follows; the result in second member can be confirmed by derivation:

$$\int -2x\ e^{-x^2}\, dx = e^{-x^2} + c$$

- Now, let it be, $\int 3x^2\ \cos(x^3 + 1)dx$;

The function, $g(x) = x^3 + 1$, occurs jointly with its derivative, $g'(x) = 3x^2$;

1. Substitution

$$\begin{aligned} u &= x^3 + 1 \\ du &= 3x^2\,dx \end{aligned}$$

2. The primitive calculation

$$\int 3x^2\ \cos(x^3 + 1)dx = \int \cos(u)du = \sin(u) + c$$

3. Inversion of substitution

$$\sin(u) + c = \sin(x^3 + 1) + c$$

The primitive follows, and the result can be verified by derivation too:

$$\int 3x^2 \cos(x^3 + 1)dx = \sin(x^3 + 1) + c$$

8.4.3 Integration by Parts

Let the elementary functions, $f(x)$ and $g(x)$, have antiderivatives $F(x)$ and $G(x)$; non-linearity arises from the product's derivation (Section 8.3),

$$f(x).g(x) = \int f'(x).g(x)dx + \int f(x).g'(x)dx$$

and two different expressions are obtained, by solving one for each integral in the second member. From the integration point of view, it is preferable to consider the two following equalities:

$$\begin{aligned}\int f(x).g(x)dx &= F(x).g(x) - \int F(x).g'(x)dx \\ &= f(x).G(x) - \int f'(x).G(x)dx\end{aligned}$$

Therefore, the critical use of this method is proposed, by selecting one of the functions to be integrated, obtaining or $F(x)$ or $G(x)$, and the other to be derived, respectively $g'(x)$ or $f'(x)$.

Examples

- Let it be, $\int x \cos(x)dx$;
  Applying the first relation and considering the two functions orderly, then the integration of linear function, x, originates a quadratic function and a more difficult integral arises in the second term:

$$\int f(x).g(x)dx = F(x).g(x) - \int F(x).g'(x)dx$$

$$\begin{aligned}\int x \cos(x)dx &= \frac{x^2}{2}\cos(x) - \int \frac{x^2}{2}\cos'(x)dx \\ &= \frac{x^2}{2}\cos(x) + \frac{1}{2}\int x^2 \sin(x)dx\end{aligned}$$

From the other side, such an issue is avoided by applying the second relation:

$$\int f(x).g(x)dx = f(x).\ G(x) - \int f'(x).\ G(x)dx$$

$$\int x\ \cos(x)dx = x\ \sin(x) - \int 1.\sin(x)dx = x\ \sin(x) + \cos(x) + c$$

- Let it be, $\int \arctan(x)dx = \int 1.\arctan(x)dx$;
  Applying the first relation to the second integral, then the integration of constant function, 1, originates a linear function and the issue of calculating the primitive of arctangent is solved via the respective derivative:

$$\int f(x).g(x)dx = F(x).g(x) - \int F(x).g'(x)dx$$

$$\begin{aligned}\int 1.\arctan(x)dx &= x.\arctan(x) - \int x.\arctan'(x)dx \\ &= x.\arctan(x) - \int x.\frac{1}{1+x^2}dx \\ &= x.\arctan(x) - \frac{1}{2}\int \frac{2x}{1+x^2}dx \\ &= x.\arctan(x) - \frac{1}{2}\ \ln|1 + x^2| + c\end{aligned}$$

- Let it be, $\int x^3\ dx = \int x^2.x\ dx$, and treating the power functions by parts;
  Applying the first relation to the second integral, carefully note the exchange in the third step to avoid circular reasoning when integrating the cubic power:

$$\int f(x).g(x)dx = F(x).g(x) - \int F(x).g'(x)dx$$

$$\int x^2.x\ dx = \frac{x^3}{3}.x - \int \frac{x^3}{3}.(1)\ dx$$

$$\int x^3\ dx = \frac{x^4}{3} - \frac{1}{3}\int x^3\ dx$$

$$\int x^3\,dx + \frac{1}{3}\int x^3\,dx = \frac{x^4}{3}$$

$$\frac{4}{3}\int x^3\,dx = \frac{x^4}{3}$$

$$\int x^3\,dx = \frac{x^4}{3}.\frac{3}{4}$$

$$\int x^3\,dx = \frac{x^4}{4} + c$$

In addition, treating the general relation of power functions by parts too, the exchange in the third step is avoiding circular reasoning once again:

$$\int f(x).g(x)\,dx = f(x).\ G(x) - \int f'(x).\ G(x)\,dx$$

$$\int x^n.\ x\,dx = x^n.\ \frac{x^2}{2} - \int (n\ \ x^{n-1})\frac{x^2}{2}dx$$

$$\int x^{n+1}\,dx = \frac{x^{n+2}}{2} - \frac{n}{2}\int x^{n+1}dx$$

$$\int x^{n+1}\,dx + \frac{n}{2}\int x^{n+1}dx = \frac{x^{n+2}}{2}$$

$$\frac{n+2}{2}\int x^{n+1}dx = \frac{x^{n+2}}{2}$$

$$\int x^{n+1}dx = \frac{x^{n+2}}{2} + c$$

Integration by parts allows the integration of functions product, and it provides a suitable pathway to find the primitive of inverse functions through the utilization of the respective derivatives; however, some pitfalls occur, namely, the critical selection of the best expression to apply, as well as avoiding circular thinking.

8.5 SPECIFIC METHODS OF INTEGRATION

The primitive of rational functions can be found in a systematic manner, and this section is mainly dedicated to the integration of such functions. The resulting primitives can assume very different forms (e.g., logarithms, powers, arctangents), and only quadratic functions in the fraction denominator are treated here, due both to the difficulty level and the required tasks that grow rapidly. In addition, irrational functions and trigonometric functions can be treated with specific methods too, but they can be transformed in rational functions and then addressed in a similar way as presented here.

Integration of Rational Functions

Rational functions consider the quotient between polynomials of real coefficients, with domain that excludes the roots of the denominator's polynomial. The integration of rational functions, after the due division of polynomials, comes down as a regular fraction: the polynomial degree on the numerator is lower than that on the denominator. And every polynomial on denominator can be decomposed into a sum of simple linear or quadratic elements; therefore, any regular fraction can be decomposed into such partial fractions.

- At now, linear and quadratic polynomials on denominator are focused on, together with their roots, while the decomposition of rational functions into partial fractions exceeds the text purpose. Depending on the root type, real or complex, and their multiplicity, one or more, the following types can be defined: (i) simple real root; (ii) multiple real roots; (iii) simple conjugate complex roots; and (iv) multiple conjugate complex roots.

Case (1) ***Simple Real Roots***

$$\int \frac{A}{x-r}dx = A.\ln|x-r| + c$$

The result corresponds to the immediate primitive of the natural logarithm, since one real root, r, occurs for the linear function (first order polynomial) in denominator.

- Let it be, $\int \frac{2}{x-3}dx$;
  Applying the relation presented in above, identifying the fraction parameters and the real root, $A = 2$ and $r = 3$, then the integral for this simple case is:

$$\int \frac{2}{x-3}dx = 2\ \ln|x-3| + c$$

The calculation detailing step by step follows:

$$\int \frac{2}{x-3}dx = 2\int \frac{1}{x-3}dx = 2\int \frac{(x-3)'}{x-3}dx = 2\ \ln|x-3| + c$$

Case (2) ***Multiple Real Roots***

$$\int \frac{A}{(x-r)^k}dx = \frac{A}{(1-k).(x-r)^{k-1}} + c$$

The result is the immediate primitive of the integer power of degree k, because a real root of multiplicity k occurs for the power function (k^{th}-order polynomial) in denominator.

- Let it be, $\int \frac{4}{(x-5)^2}dx$;
  Again applying the relation, identifying the fraction parameters and the real root, $A = 4$ and $r = 5$, as well as the root multiplicity, $k = 2$, then the integral follows:

$$\int \frac{4}{(x-5)^2}dx = \frac{4}{(1-2).(x-5)^{2-1}} + c = \frac{-4}{x-5} + c$$

The calculation step by step follows too:

$$\int \frac{4}{(x-5)^2}dx = 4\int (x-5)^{-2}\,dx = 4\frac{(x-5)^{-2+1}}{-2+1} = 4\frac{(x-5)^{-1}}{-1}$$

$$= \frac{-4}{x-5} + c$$

Case (3) ***Simple Conjugate Complex Roots***

$$\int \frac{Mx + N}{x^2 + B\ x + C}dx = \frac{M}{2}\ln|x^2 + B\ x + C| + \frac{2N - M\ B}{\sqrt{4C - B^2}}\arctan\left(\frac{2x + B}{\sqrt{4C - B^2}}\right) + c$$

In the first term, the logarithm function appears introducing the derivative of the denominator's polynomial into the numerator, and then the logarithm corresponds to such primitive; after that, the denominator is manipulated to obtain a sum of squares, originating the primitive of the arctangent function in the second term.

- Let it be, $\int \frac{x-6}{x^2+1}dx$;
  Applying the relation and identifying the fraction parameters ($B = 0$ and $C = 1$; $M = 1$ and $N = -6$), then the integral follows:

$$\begin{aligned}\int \frac{x - 6}{x^2 + 1}dx &= \frac{1}{2}\ln|x^2 + 0\ x + 1| + \frac{2(-6) - 1(0)}{\sqrt{4(1) - (0)^2}}\arctan\left(\frac{2x + (0)}{\sqrt{4(1 - (0)^2}}\right) + c\\ &= \frac{1}{2}\ln|x^2 + 1| + \frac{-12}{\sqrt{4}}\arctan\left(\frac{2x}{\sqrt{4}}\right) + c\\ &= \frac{1}{2}\ln|x^2 + 1| - 6\ \arctan(x) + c\end{aligned}$$

And now, detailing every step in this integration: first, introducing the derivative of the denominator into the numerator, $2x$, and the logarithm results in the first term; after that, the primitive of the arctangent function directly occurs in the second term.

$$\begin{aligned}\int \frac{x - 6}{x^2 + 1}dx &= \frac{1}{2}\int \frac{2x}{x^2 + 1}dx + \int \frac{-6}{x^2 + 1}dx\\ &= \frac{1}{2}\ln|x^2 + 1| - 6\int \frac{1}{x^2 + 1}dx\\ &= \frac{1}{2}\ln|x^2 + 1| - 6\ \arctan(x) + c\end{aligned}$$

Case (4) ***Multiple Conjugate Complex Roots***

$$\int \frac{Mx + N}{(x^2 + Bx + C)^k} dx = \frac{1}{(1 - k).\ (x^2 + Bx + C)^{k-1}} + I_k$$

where

$$I_k = \int \frac{dx}{(x^2 + Bx + C)^k} = \frac{2k - 3}{2(C - B^2/4).(k - 1)}.I_{k-1} + \frac{x + B/2}{2(C - B^2/4).(k - 1).(x^2 + Bx + C)^{k-1}}$$

In the first term, the primitive of the power function is obtained by introducing into the numerator the derivative of the denominator's polynomial, as in Case 3; likewise, the denominator is manipulated to obtain a sum of squares, originating the recurrence formula in I_k that arises from the second term.

Many expressions exist for the integration of rational functions, not to mention for irrational functions, trigonometric functions, and other functions of interest; the proposed procedure includes also the identification of parameters, roots and their multiplicities, and then using properly those tables and repositories of integrals.

Integration of Irrational functions

In the simplest cases, integrals for irrational functions can be reduced through appropriate substitutions to integrals of rational functions, and these primitives are capable of determination. Several cases can be typified, but the integration of irrational functions in terms of elementary functions is not always obtainable. A simple example follows now.

Example

- Let it be, $\int \frac{\sqrt{x+9}}{x} dx$;
  The proposed approach considers a specific variable substitution that is able to treat the square root in the numerator, namely:

$$t^2 = x + 9 \Rightarrow \begin{cases} x & = t^2 - 9 \\ dx & = 2t\ dt \end{cases}$$

Substituting both the variable and the associated differential, the integral of a rational function is obtained:

$$\int \frac{\sqrt{t^2}}{t^2 - 9} 2t \ dt = 2\int \frac{t^2}{t^2 - 9} dt$$

The systematic procedure for rational functions considers the treatment of a regular fraction, obtaining the partial fractions associated to the denominator roots, and then obtaining the related integral. In this case, the two denominator roots are real and different roots, thus the logarithm form is foreseen in this integration.

$$2\int \frac{t^2}{t^2 - 9} \ dt = 2\int \frac{t^2 - 9 + 9}{t^2 - 9} dt = 2\int 1 + \frac{9}{t^2 - 9} dt$$

$$= 2\int 1 \ dt + 18\int \frac{1}{t^2 - 9} dt = 2t + 18\int \frac{(1/6)}{t - 3} + \frac{(-1/6)}{t + 3} dt$$

$$= 2t + 3 \ln|t - 3| - 3 \ln|t + 3|$$

Finally, by inverting the substitution, $t = \sqrt{x + 9}$, the primitive follows:

$$\int \frac{\sqrt{x + 9}}{x} dx = 2\sqrt{x + 9} + 3 \ln\left|\frac{\sqrt{x + 9} - 3}{\sqrt{x + 9} + 3}\right| + c$$

Integration of Trigonometric Functions

The trigonometric functions can be also reduced to rational functions, namely, by using the well-known variable transformation:

$$t = \tan\left(\frac{x}{2}\right) \Rightarrow \begin{cases} x & = & 2\arctan(t) \\ dx & = & \frac{2}{1+t^2} dt \end{cases}$$

By the trigonometric equality:

$$\sin^2\left(\frac{x}{2}\right) + \cos^2\left(\frac{x}{2}\right) = 1$$

$$\tan^2\left(\frac{x}{2}\right) + 1 = \frac{1}{\cos^2\left(\frac{x}{2}\right)}$$

$$t^2 + 1 = \frac{1}{\cos^2\left(\frac{x}{2}\right)}, \text{ and then } \cos^2\left(\frac{x}{2}\right) = \frac{1}{t^2 + 1}$$

In addition,

$$\sin^2\left(\frac{x}{2}\right) = 1 - \cos^2\left(\frac{x}{2}\right) = 1 - \frac{1}{t^2 + 1} = \frac{t^2 + 1 - 1}{t^2 + 1} = \frac{t^2}{t^2 + 1}$$

From the relations for angles bisection, the associated expressions for sine and cosine follow:

$$\sin(x) = 2\sin\left(\frac{x}{2}\right)\cos\left(\frac{x}{2}\right) = 2\sqrt{\frac{t^2}{t^2 + 1}}\sqrt{\frac{1}{t^2 + 1}} = \frac{2t}{t^2 + 1}$$

$$\cos(x) = \cos^2\left(\frac{x}{2}\right) - \sin^2\left(\frac{x}{2}\right) = \frac{1}{t^2 + 1} - \frac{t^2}{t^2 + 1} = \frac{1 - t^2}{t^2 + 1}$$

However, the resulting rational functions often are difficult, since fractions with complex roots and different multiplicities may arise; thus trigonometric manipulations and transformations (e.g., transforming the product of sines and/or cosines in a sum) are very useful too. Such procedures are well studied, and systematic approaches to the integration of trigonometric functions are also very common.

8.6 CONCLUDING REMARKS

Important notions about the indefinite integral, with the main aim of studying the properties and tools for the function anti-derivation (primitivation), are presented and discussed.

- First, basic notions about the primitive function are introduced, and then important properties and immediate primitives are focused on. Two attributes of indefinite integral are remarked, namely, the linearity associated with both the integral of a functions sum and the product of constant by function. It is also known that the primitive of a given function will not always exist, or it will not be obtained in an obvious way.

- To obtain primitives, it is useful to manipulate the integrand function in order to recognize in it the existence of immediate primitives. Therefore, general methods of integration are preferred, namely, integration by decomposition, by substitution, and by parts. These methods are preferably utilized when facing, respectively; sum of functions; a composite function that could be originated by the derivation's chain rule; the multiplication of two functions.
- The integration of rational functions is also treated, and these primitives usually can be found. Such primitives can assume very different forms (e.g., logarithm, power, arctangent), and only quadratic functions in the fraction denominator are treated here, because both the inherent difficulty and required time grow rapidly. Also the irrational functions and trigonometric functions can be treated in a systematic manner with specific methods for such functions; alternatively, they can be transformed in rational functions and then addressed in a similar way as presented here.

Beyond the notions about primitives, the general and specific methods of integration, the next chapter is directed at the definite integral. Thus, the main properties of definite integrals are addressed, including the useful mean theorems, Barrow's Rule and the Fundamental Theorem of Calculus that associate integration and derivation. Applications to area calculations and improper integrals are detailed too, the latter is also extending integration into unbounded domains and unbounded functions.

CHAPTER 9

Definite Integral

In this chapter, the main results about definite integrals are addressed, focusing important properties and theorems, including the useful mean theorems. After that, Barrow's Rule and the Fundamental Theorem of Calculus are also introduced, and several applications are described (e.g., power and exponential functions). Applications to area calculation are detailed too, in a way to distinguish positive and negative functions, as well as positive and negative integration domains. Finally, improper integrals are addressed, including the integration of both unbounded domains (the integration range is infinite) and unbounded functions (the integration range includes a point where the function tends to infinite); convergence criteria are also indicated for both the referred types of improper integrals.

9.1 INTRODUCTION

The definite integral is a powerful tool for solving problems in contexts as, among many others, calculating areas and volumes, obtaining the pressure exerted by liquids, estimating probabilities in statistical models, calculating centers of gravity for irregular figures, or evaluating the mean value for a continuous function in a given interval.

In the preceding chapter, primitives and indefinite integrals were addressed. Now, the solution of a simple problem is revisited: the area calculation for the figure defined by the graphic of a positive continuous function and the horizontal X-axis.

- Note the possibility of calculating areas is greatly increased in this subject compared to what is previously studied; until now, only areas limited by polygons or the area of a circle have been worked, but integration makes it possible to calculate a multiplicity of areas and figures, as long as it is possible to define them from a continuous function.

DOI: 10.1201/9781003461876-9

- For a continuous and positive function, the relationship between the area under the graphic and its antiderivative is remarked; the area function, $A(x)$, is thus defined as an integral function where the variable is the upper limit, x.

  The infinitesimal alteration of area function, $A(x)$, is connecting both differential and integral subjects, and it paves the way to both the Fundamental Theorem of Calculus and the calculation of definite integrals through the primitive function (Barrow's rule).
- The calculation of areas bounded by graphics of continuous functions is proposed, with several cases being compared. The functions and primitives in such cases are already known, in this way encouraging both the results verification and graphics comparison.
- Note the similarity of function integration with the calculation of function limits, with both the topics benefiting from continuity attributes and derivation insights. Namely, in function limits (Chapter 3), the study is initiated with the limit at a point, then it also considered the limit at infinite, and the infinite limit of a function (e.g., near a vertical asymptote); this pattern can be also perceived for the definite integral, with the study initiating in a closed range, then extending into an infinite range, and even unbounded functions are treated as improper integrals.

In the continuation, important properties and theorems about definite integrals are presented, including the mean theorems for integral calculus. Then, Barrow's rule and the fundamental theorem of calculus are introduced, while several numerical instances are treated. After that, applications to area calculations are presented, for instance, in a way to distinguish positive and negative functions. Finally, improper integrals are addressed, either when the integration range is infinite or when that range includes a point where the function is unbounded.

9.2 PROPERTIES AND THEOREMS

A simple approach to the main properties and results of definite integrals is presented, by assessing integral sums, the related summations, and the graphical perception of area. Namely:

- The multiplication of function by constant and the integral of function sum correspond to the similar properties associated with derivatives, limits, or even summations; the linearity properties hold, respectively, both for the constant multiplying the function integral and the sum of integrals.

- In addition, the integral's mean value and the extended mean value are similar to the associated properties for derivatives, namely, the theorem about the derivative's mean value and its generalization to the increments relation between two functions. Of course, this holds if the operations are possible and the integrals exist.
- Two results about the existence of definite integral follow; since the main purpose in this text is focusing on applied and computational topics, they are indicated without proof.
  - If function $f(x)$ is continuous in the closed interval [a, b], then $f(x)$ is integrable in that interval; that is, the definite integral $\int_a^b f(x)\,dx$ exists.
  - Every function $f(x)$ bounded in the closed interval [a, b], and continuous except on a finite number of points of finite discontinuity, is integrable on that interval; that is, $\int_a^b f(x)\,dx$ exists.

If f(x) is an integrable function over [a, b], then function, k.f(x), is also integrable, and $\int_a^b [k.f(x)]\,dx = k.\int_a^b f(x)\,dx$

The definite integral corresponds to the limit of integral sums, and the constant factor, k, can be factored out both of summation and of function limit. In this way, k can be factored out from the definite integral too, as occurred in the calculation of primitives.

If f(x) and g(x) are integrable functions over [a, b], then (f + g)(x) is also integrable on [a, b], and $\int_a^b [(f+g)(x)]\,dx = \int_a^b f(x)\,dx + \int_a^b g(x)\,dx$

Similarly, the definite integral corresponds to the limit of integral sums, then the function sum corresponds to the sum of functions, and the limit of functions sum corresponds as well to the sum of the related limits. Then the definite integral of function sum also corresponds to the sum of definite integrals, in the same way as the primitive of sum corresponds to the sum of primitives.

Let f(x) be integrable over the closed interval [a, b], with a < b, where for all values of this interval $f(x) \geq 0$***; then*** $\int_a^b f(x)\,dx \geq 0$

This result can be seen as analogous to the limit of a non-negative function, $f(x) \geq 0$, which is also non-negative (Chapter 3). This non-negative property also holds for definite integrals: by contradiction, assuming a negative sum could exist, e.g., $L[f(x), P] < 0$, then a false expression will occur since that function cannot take negative values.

Let f(x) and g(x) be integrable on the closed interval [a, b], a < b, where $f(x) \geq g(x)$ ***for all values in that range; then*** $\int_a^b f(x)dx \geq \int_a^b g(x)dx$

Once more, this result can be seen as analogous to the superiority relationship between limits. In fact, the current result follows from the last one, for that the difference between the two functions is defined, $h(x) = f(x) - g(x) \geq 0$; a non-negative function thus occurs, and the corresponding summations and limit will be non-negative too; the integral sum of such difference is also non-negative, and then the superiority relationship holds for definite integrals.

If f(x) is integrable on the closed interval [a, b], with a < b, where $m \leq f(x) \leq M$ ***for all values; then*** $m(b-a) \leq \int_a^b f(x)dx \leq M(b-a)$

From the extreme values theorem (Chapter 4) in the closed interval, $[a, b]$, the function values are bounded between the minimum, m, and the maximum, M,

$$m \leq f(x) \leq M$$

Integrating over $[a, b]$ all the terms in the inequality, all the integral sums and the area estimate are bounded (Figure 9.1), in-between the shorter rectangle and the taller rectangle.

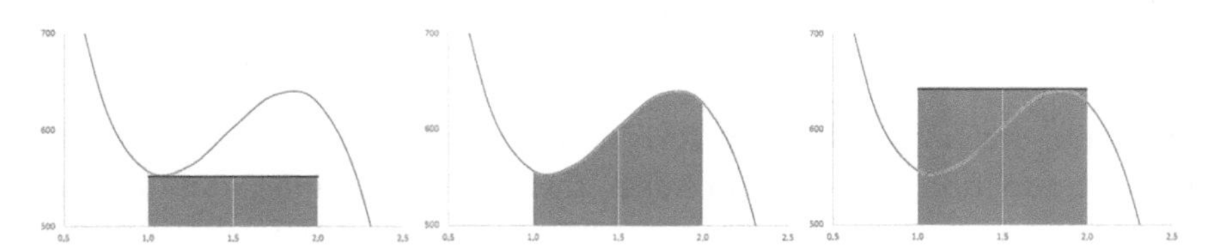

FIGURE 9.1 Areas comparison for bounded function, $m \leq f(x) \leq M$.

Mean Theorems of Integral Calculus—The two following results are also known as mean theorems for integral calculus; the first one allows us to estimate the area below the graphic of a continuous function over $[a, b]$, assuming a rectangle with the same area and width, $(b–a)$, in order to evaluate the mean height of that rectangle; the second mean theorem extends the first, in a similar mode the mean value on differential calculus (Lagrange theorem) is extended to the increments relation between two functions (Cauchy theorem).

(I) *Let f(x) be continuous in the closed interval* $[a, b]$*; then there will be a number,* $c \in [a, b]$*, such that* $\int_a^b f(x)dx = f(c)(b - a)$

In the closed interval, $[a, b]$, both the function and the definite integral are bounded,

$$m(b - a) \leq \int_a^b f(x)dx \leq M(b - a)$$

$$m \leq \frac{\int_a^b f(x)dx}{b - a} \leq M$$

From the intermediate values theorem (Chapter 4), a continuous function cannot vary from one value to another without going through all the intermediate values; then the function takes, at least once, any intermediate value between the maximum and the minimum, e.g., at point c,

$$f(c) = \frac{\int_a^b f(x)dx}{b - a}, \qquad a < c < b$$

The mean value in $[a, b]$ corresponds to the rectangle height (Figure 9.2), in a manner that the rectangle's area is equal to the area below the graphic line, which is calculated by the integral:

$$\int_a^b f(x)dx = f(c)(b - a)$$

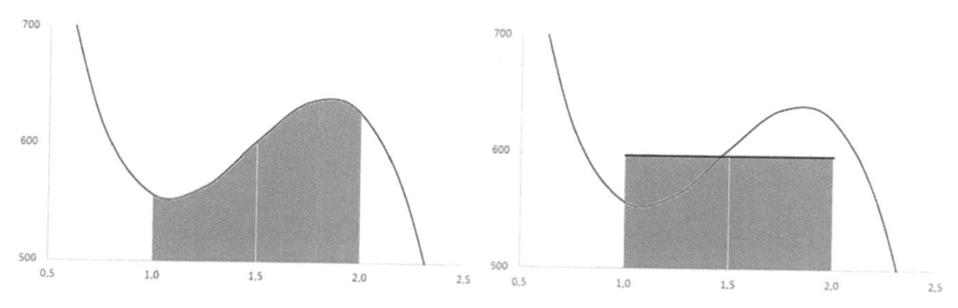

FIGURE 9.2 Mean value, $f(c)$, in the closed interval $[a, b]$.

(II) ***Let f(x) be continuous on the closed interval [a, b], and g(x) be an integrable and non-negative function on that interval; then there will be a number*** $c \in [a, b]$, ***such that*** $\int_a^b f(x).g(x)\,dx = f(c)\int_a^b g(x)\,dx$

The approach is similar to the one utilized for the first mean theorem, because in the closed interval $[a, b]$, both the function and the definite integral are bounded, and then

$$m\int_a^b g(x)\,dx \le \int_a^b f(x).g(x)\,dx \le M\int_a^b g(x)\,dx$$

$$m \le \frac{\int_a^b f(x)\,dx}{\int_a^b g(x)\,dx} \le M$$

Again from the intermediate values theorem, the continuous function takes, at least once at point c, that intermediate value that satisfies the relation,

$$f(c) = \frac{\int_a^b f(x).g(x)\,dx}{\int_a^b g(x)\,dx}, \qquad a < c < b$$

In this way, the definite integral for the functions multiplication can be evaluated through one single integral multiplied by a suitable intermediate value, $f(c)$:

$$\int_a^b f(x).g(x)\,dx = f(c)\int_a^b g(x)\,dx, \qquad a < c < b$$

In the particular case that, $g(x) = 1$, the integral in the second member reduces itself to the rectangle of height, 1, and width, $(b - a)$, and the first mean theorem of integrals is obtained.

$$\int_a^b f(x).1\,dx = f(c)\int_a^b 1\,dx \Rightarrow \int_a^b f(x)\,dx = f(c)\,(b - a)$$

9.3 THE FUNDAMENTAL THEOREMS OF CALCULUS

The study of area function's infinitesimal alteration implies very important results in calculus, since it relates both differential and integral subjects, as well as allows the calculation of definite integrals utilizing the primitive (Barrow's rule).

- In order to evaluate the area below the graphic of a continuous and non-negative function, $f(x)$, consider the area function, $A(x)$, defined as an integral where the variable is the upper limit, x, and the integration parameter is, t:

$$A(x) = \int_a^x f(t)\,dt$$

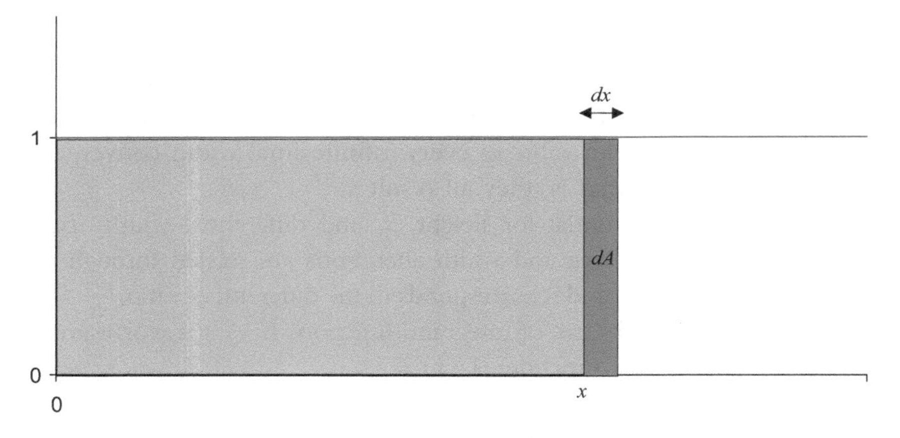

FIGURE 9.3 The area's infinitesimal alteration for constant function, $f(x) = 1$.

- The area for the rectangle of height, $f(x) = 1$, and width, x, corresponds to the product function, $A_r(x) = 1.\,x$ (Figure 9.3). And the rectangle area as a function on the integral upper limit, x, with integration parameter, t, is $A_r(x) = \int_0^x 1\,dt$. For the area increment due to width alteration, the height is constant and equal to 1; so the maximum, minimum, or the mean value in every segment results 1.

  Consequently, the width infinitesimal alteration, dx, originates the area differential, $dA_r = 1\,dx$. The derivative of area function, $A_r{}'(x)$, relates the area alterations due to width increments; in fact, it is defined by the ratio between the area and width differentials, $\frac{dA_r}{dx} = 1$. In this way, the antiderivative of integrand function, $f(x) = 1$, corresponds to the related area function: $A_r(x) = x$.

- The area for the triangle of equal height and width, $f(x) = x$, corresponds to the well-known function, $A_t(x) = \frac{x.x}{2}$ (Figure 9.4).

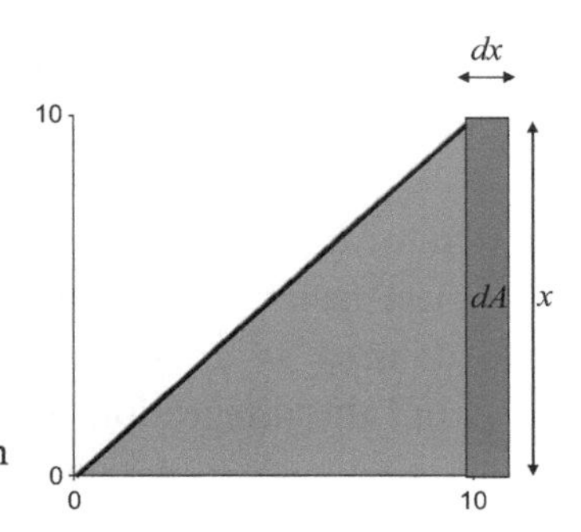

FIGURE 9.4 The area's infinitesimal alteration for linear function, $f(x) = x$.

In addition, the triangle area as a function on the integral upper limit, x, is $A_r(x) = \int_0^x t\, dt$. For the area increment, height and width are related through function, $f(x) = x$; then the maximum, minimum, or mean value in every infinitesimal width converge to the same point, that is, they all result x.

The area differential for height, x, and differential width, dx, is $dA_t = x\, dx$. The area and width alterations are related through the derivative, $A_t'(x)$, and it corresponds to the differentials ratio, $\frac{dA_t}{dx} = x$. Thus, the antiderivative of integrand function, $f(x) = x$, corresponds to the related area function: $A_t(x) = \frac{x^2}{2}$.

(I) *Let f(x) be integrable and continuous at any point, c, on the closed interval* [*a, b*]*; then function A(x) will be differentiable at that point, c, and* $A'(c) = f(c)$.

In the closed interval $[a, b]$, the derivative of continuous function, $f(x)$, at the intermediate point, c, for the area function, $A(x) = \int_a^x f(t)dt$, follows

$$A'(c) = \lim_{h\to 0} \frac{A(c+h) - A(c)}{h}$$

$$= \lim_{h\to 0} \frac{\int_a^{c+h} f(t)dt - \int_a^c f(t)dt}{h} = \lim_{h\to 0} \frac{\int_c^{c+h} f(t)dt}{h}$$

From the extreme values theorem,

$$m.h \le \int_c^{c+h} f(t)dt \le M.h$$

Taking limit when h approaches zero, either from the positive or negative side, the area increment is trapped into zero, and the $f(x)$ continuity implies also continuity for integral function, $A(x)$ (Figure 9.5).

$$\left.\begin{array}{lcl}\lim_{h\to 0}(m.h) & = & 0\\ \lim_{h\to 0}(M.h) & = & 0\end{array}\right\} \Rightarrow \lim_{h\to 0}\int_{c}^{c+h} f(t)\,dt = 0$$

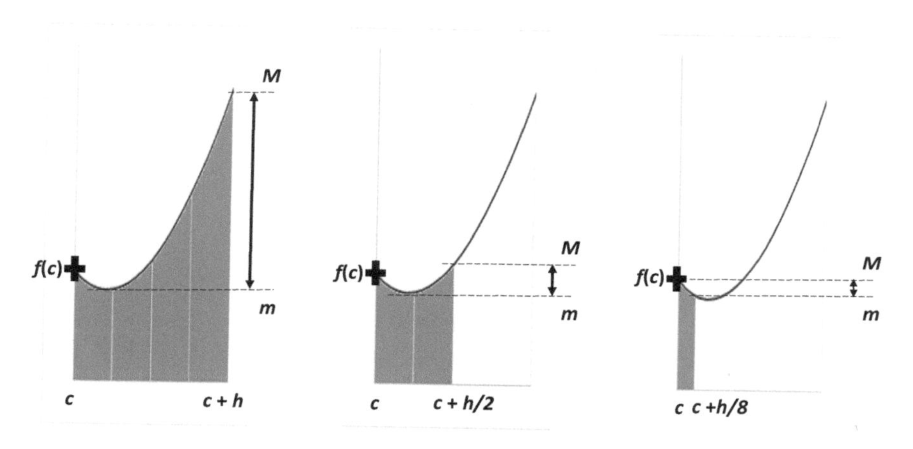

FIGURE 9.5 Area alteration for continuous function, $f(x)$, at point c.

Conjugating with the integral's mean value theorem,

$$m \leq \frac{\int_{c}^{c+h} f(t)\,dt}{h} \leq M$$

the function continuity is ensuring the following limits exist, and finally,

$$\begin{array}{lcl}\lim_{h\to 0}(m) & = & f(c)\\ \lim_{h\to 0}(M) & = & f(c)\end{array} \Rightarrow \lim_{h\to 0}\frac{\int_{c}^{c+h} f(t)\,dt}{h} = f(c)$$

$$A'(c) = f(c)$$

Finally, such a result can be generalized for all points, x, in the closed interval $[a, b]$,

$$A'(x) = \frac{d}{dx}\left(\int_a^x f(t)dt\right) = f(x), \quad \forall\, x \in [a, b]$$

(II) ***Let f(x) be a continuous function on the closed interval*** **[*a, b*];** ***if F(x) is antiderivative of function f(x), F'(x) = f(x), then***

$$\int_a^b f(x)dx = F(b) - F(a).$$

It is very useful to note that both equalities hold, $A'(x) = f(x) = F'(x)$, and then the two primitives differ from a constant, c,

$$A(x) = F(x) + c$$

Evaluating the area function at starting point, a,

$$A(a) = \int_a^a f(x)dx = 0$$

$$0 = F(a) + c$$

$$c = -F(a)$$

In this way, $A(x) = F(x) - F(a)$, and evaluating the area function at final point, b,

$$A(b) = F(b) - F(a)$$

Or better,

$$\int_a^b f(x)dx = F(b) - F(a)$$

This approach assumed the area function is zero at the segment's initial point, $A(a) = 0$, and then it starts to accumulate until reaching the final point, $A(b)$. However, the area evaluation can start at any point, *e.g.*, at $x = 0$, and the mean value in the closed interval $[a, b]$, corresponds to the difference of accumulated areas per unit of width alteration, x,

$$\begin{aligned} f(c) &= \frac{A(b) - A(a)}{b - a}, \quad a < c < b \\ &= \frac{[F(b) + c] - [F(a) + c]}{b - a} \\ &= \frac{F(b) - F(a)}{b - a} \end{aligned}$$

By comparison with the mean value definition,

$$f(c) = \frac{\int_a^b f(x)\,dx}{b - a}, \quad a < c < b$$

then,

$$\frac{\int_a^b f(x)\,dx}{b - a} = \frac{F(b) - F(a)}{b - a}$$

And the integral calculation through the difference of primitives follows again:

$$\int_a^b f(x)\,dx = F(b) - F(a)$$

9.3.1 Linear Function

Several examples based on linear (and constant) functions are presented, they complement previous developments and area estimates (Chapter 7).

- The constant function, $f(x) = 1$, corresponds to the rectangle's height and for each width step of size 1, the rectangle's area results $1 \times 1 = 1$, as in Figure 9.6.

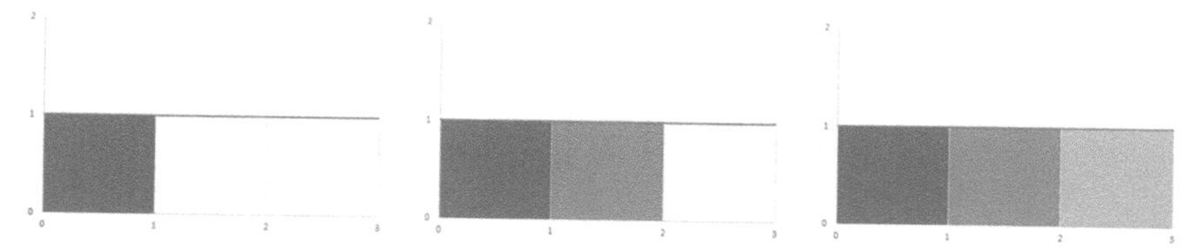

FIGURE 9.6 Area partitions for constant function, $f(x) = 1$.

Each partition corresponds to a square of width 1, and the associated area also results 1. For the general case, utilizing Barrow's rule in any interval $[i - 1, i]$, the primitive $F(x)$ is evaluated at the borders, i and $(i - 1)$, and associated difference, $F(i) - F(i - 1)$, is calculated:

$$A_i = \int_{i-1}^{i} 1 \; dx = [x]_{i-1}^{i} = i - (i - 1) = 1$$

Trivially, the related mean value also becomes 1, in concordance with the constant function.

- The linear function, $f(x) = x$, corresponds to both the triangle's legs, and the triangle's area results $x^2/2$; for unit increments (Figure 9.7), the area of each partition is evaluated too.

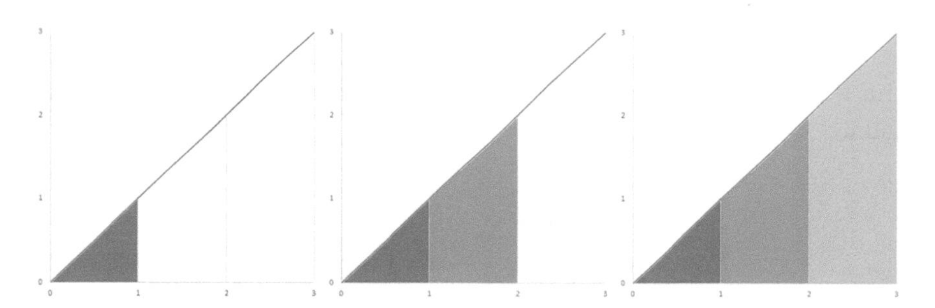

FIGURE 9.7 Area partitions for linear function, $f(x) = x$.

The area of the first triangle of leg-1 is, $(1 \times 1)/2$; the second partition corresponds to the triangle of leg-2, $(2 \times 2)/2$ minus the first triangle, $(1/2)$, that is, the area is only $3/2$:

$$A_1 = \int_0^1 x \; dx = \left[\frac{x^2}{2}\right]_0^1 = \frac{1^2}{2} - \frac{0^2}{2} = \frac{1}{2}$$

$$A_2 = \int_1^2 x \; dx = \left[\frac{x^2}{2}\right]_1^2 = \frac{2^2}{2} - \frac{1^2}{2} = \frac{3}{2}$$

In the same mode, the third partition's area corresponds to the triangle of leg-3, $(3 \times 3)/2$, minus the triangle of leg-2, $(2 \times 2)/2$:

$$A_3 = \int_2^3 x \ dx = \left[\frac{x^2}{2}\right]_2^3 = \frac{3^2}{2} - \frac{2^2}{2} = \frac{5}{2}$$

Note that the procedure can be verified, both by integral calculation and geometric approach,

$$A_T = \int_0^3 x \ dx = \left[\frac{x^2}{2}\right]_0^3 = \frac{3^2}{2} - \frac{0^2}{2} = \frac{9}{2}, \text{ and}$$

$$A_T = A_1 + A_2 + A_3 = \frac{1}{2} + \frac{3}{2} + \frac{5}{2} = \frac{9}{2}$$

For the general case, in the interval $[i - 1, i]$,

$$A_i = \int_{i-1}^{i} x \ dx = \left[\frac{x^2}{2}\right]_{i-1}^{i} = \frac{i^2}{2} - \frac{(i-1)^2}{2} = \frac{i^2 - (i^2 - 2i + 1)}{2}$$
$$= \frac{2i - 1}{2} = i - \frac{1}{2}$$

These values correspond also to the mean value at each partition, since the width increment is 1 for all partitions; thus, the arithmetic mean between two successive integers can be directly verified, namely, 0.5, 1.5, 2.5, … , $(i - 0.5)$.

9.3.2 Exponential Function

The exponential function is treated in a similar mode too, and the reasoning between the mean value at each partition and the exponential growth is envisaged.

- The exponential function, $f(x) = e^x$, presents a strong growth and a geometry-based expression is not provided; anyway, the area of each partition is evaluated, and unit increments are utilized again (Figure 9.8).

  The area of the first partition is obtained, A_1, along with the second partition:

$$A_1 = \int_0^1 e^x \ dx = [e^x]_0^1 = e^1 - e^0 = e - 1$$

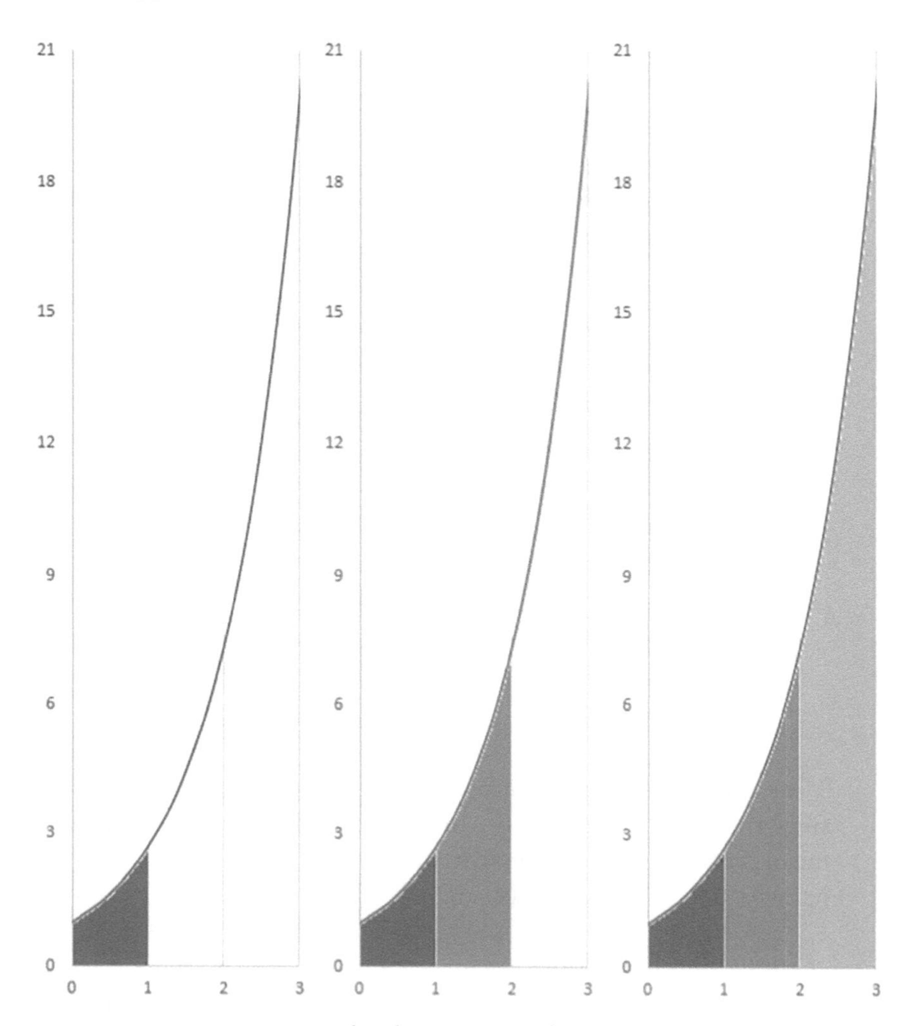

FIGURE 9.8 Area partitions for the exponential function.

$$A_2 = \int_1^2 e^x \ dx = [e^x]_1^2 = e^2 - e^1 = e(e - 1)$$

Due to the unit increments on width, these results are numerically coincident with the mean value at each partition. And also,

$$A_3 = \int_2^3 e^x \ dx = [e^x]_2^3 = e^3 - e^2 = e^2(e - 1)$$

These partial results are verified, by comparing total area once again:

$$A_T = \int_0^3 e^x \ dx = [e^x]_0^3 = e^3 - e^0 = e^3 - 1, \text{ and}$$

$$A_1 + A_2 + A_3 = (e - 1) + (e^2 - e) + (e^3 - e^2) = e^3 - 1$$

Note that the relative growth from partition to partition is constant, $A_2/A_1 = e$, and the same ratio occurs for $A_3/A_2 = e$. For the general case, in the interval $[i\text{–}1, i]$,

$$A_i = \int_{i-1}^{i} x \ dx = [e^x]_{i-1}^{i} = e^i - e^{i-1} = e^{i-1}(e - 1)$$

The area's relative growth, from partition to partition, is taking an exponential form; it corresponds to the geometric increase of mean value in every partition, i,

$$\frac{A_i}{A_{i-1}} = \frac{e^{i-1}(e - 1)}{e^{i-2}(e - 1)} = e \Rightarrow A_i = e \ A_{i-1}$$

For the area at each partition, together with the related mean value, the exponential growth is observed, namely, A_1, $(A_1 \ e)$, $(A_1 \ e^2)$, $(A_1 \ e^3)$, … , $(A_1 \ e^{i-1})$.

9.4 APPLICATIONS TO THE COST FUNCTION

Two applications addressing the mean value for the cost function are described; the first one, directly applies Barrow's rule to the diameter range [1, 2], and such unit increment on diameter allows the same numerical value to hold for the mean value; the second application focuses on the range [1, 1.2], involving the minimum cost that occurs about 1.08, and the trapezoid method is introduced.

9.4.1 The Mean of Cost Function in [1, 2]

Let it be, $y(x) = 50\pi \ x^2 + \frac{400}{x}$, and consider the area below the graphic line in the unit range [1, 2], as in Figure 9.9. Namely, if multiple production orders for barrels with diameters ranging from 1 meter to 2 meter are received during a working week, with costs spanning from about 550 € to about 800 €, then at some intermediate diameter the cost must have been 643 €.

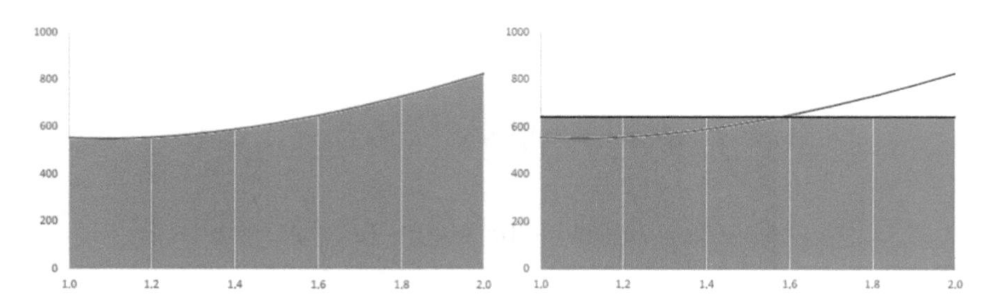

FIGURE 9.9 The mean cost over the range [1, 2].

The primitive is obtained by decomposing the sum, and calculating the primitives in each term:

$$Y(x) = \int 50\pi\, x^2 + \frac{400}{x} dx = 50\pi \int x^2\, dx + 400 \int \frac{1}{x} dx = \frac{50\pi\, x^3}{3} + 400 \ln |x|$$

By evaluating the primitive in the interval borders, the definite integral in [1, 2] is obtained too:

$$\begin{aligned}\int_1^2 50\pi\, x^2 + \frac{400}{x} dx &= Y(2) - Y(1)\\ &= \left[\frac{50\pi\, 2^3}{3} + 400\ln|2|\right] - \left[\frac{50\pi\, 1^3}{3} + 400\ln|1|\right]\\ &= \frac{50\pi\,(8-1)}{3} + 400\ln|2|\\ &= \frac{350\pi}{3} + 400\ln|2|\end{aligned}$$

In that unit range for diameter variation, the mean cost takes the same value:

$$\bar{C} = \frac{\int_1^2 50\pi\, x^2 + \frac{400}{x} dx}{2-1} = \frac{Y(2) - Y(1)}{2-1} = \frac{350\pi}{3} + 400\ln|2| \cong 643.78\ €$$

9.4.2 The Mean Cost in the Cheapest Range—The Trapezoid Rule

Since the minimum cost occurs when the diameter nears 1.08 meter, then a narrower range [1, 1.2] is focused. In addition, the trapezoid method is introduced, including both direct method and composite method to better estimates the mean in the range where cost function is lowest.

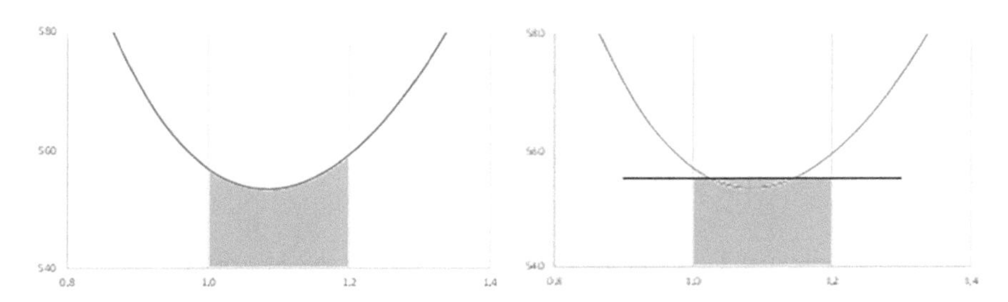

FIGURE 9.10 Mean value for the cost function in the range [1, 1.2].

Let it be again, $y(x) = 50\pi\, x^2 + \frac{400}{x}$, and consider the area below the graphic line (Figure 9.10) in the narrowed range [1, 1.2], corresponding to the definite integral:

$$\int_{1}^{1.2} 50\pi\, x^2 + \frac{400}{x} dx$$

Direct Integration Method—By the trapezoid rule, the integral is evaluated by the trapezium defined by the border points, a and b, and the corresponding function values, $f(a)$ and $f(b)$:

$$I^T \approx (b - a).\frac{f(a) + f(b)}{2}$$

By using the trapezoid of length, $(b\text{–}a)$, and height approximated by the function's arithmetic mean in that range (Figure 9.11), then the

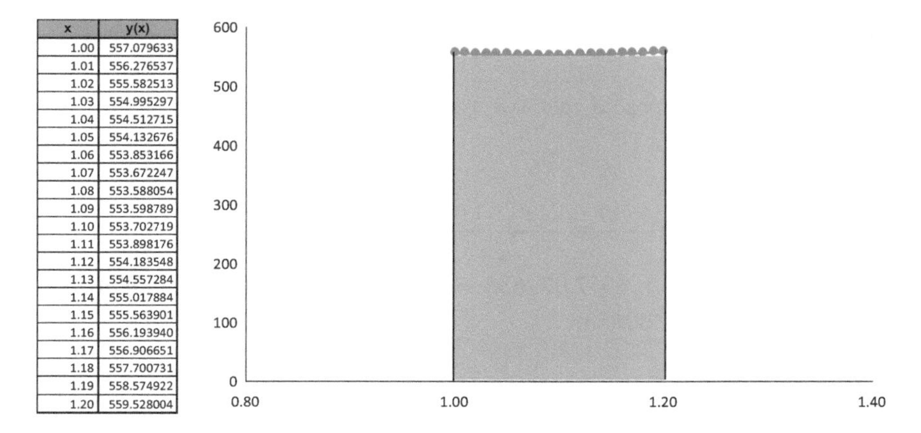

x	y(x)
1.00	557.079633
1.01	556.276537
1.02	555.582513
1.03	554.995297
1.04	554.512715
1.05	554.132676
1.06	553.853166
1.07	553.672247
1.08	553.588054
1.09	553.598789
1.10	553.702719
1.11	553.898176
1.12	554.183548
1.13	554.557284
1.14	555.017884
1.15	555.563901
1.16	556.193940
1.17	556.906651
1.18	557.700731
1.19	558.574922
1.20	559.528004

FIGURE 9.11 The cost function in the range [1, 1.2].

estimated integral is 111.660763, and the cost mean value is 558.303. The detailed calculation follows,

$$\begin{aligned} I^T &\approx (1.2-1).\frac{y(1.2)+y(1)}{2} \\ &\approx (0.2).\frac{557.079633+559.528004}{2} \approx 111.660763 \end{aligned}$$

Such estimate for the integral is satisfactory, and for the cost's mean value too, but they can be improved by increasing the number of intervals to cope with the function non-linearity.

Composite Integration Method—By the composite trapezoid rule, the integral is approximated by n subintervals with equal length, h:

$$I^{Tcomp} \approx h.\frac{f(x_0)+2.\sum_{i=1}^{n-1} f(x_i)+f(x_n)}{2}$$

In fact, with two intervals of length, $h = 0.1$, the interim value $y(1.1)$ is evaluated two times, first for the trapezoid on the left, and second for the trapezoid on the right (Figure 9.12a).

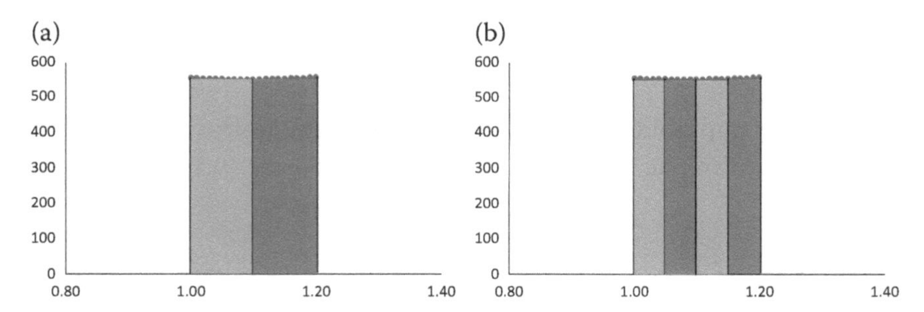

FIGURE 9.12 (a) Trapezoid rule: two partitions in the range [1, 1.2]. (b) *Ditto*, four partitions.

$$\begin{aligned} I^{Tcomp} &\approx (0.1)\frac{y(1)+2\ y(1.1)+y(1.2)}{2} \\ &\approx 0.05 \times [557.079633 + 2 \times 553.702719 + 559.528004] \\ &\approx 111.2006538 \end{aligned}$$

And with four intervals of length, $h = 0.05$ (Figure 9.12b), the interim values $y(1.05)$ and $y(1.15)$ are added and evaluated two times, since they are also bordering two trapezoids:

$$I^{Tcomp} \approx (0.05)\frac{y(1) + 2\ y(1.05) + 2\ y(1.1) + 2\ y(1.15) + f(1.2)}{2}$$

$$\approx 0.025 \times [557.079633 + 2 \times 554.132676 + 2 \times 553.702719$$

$$+ 2 \times 555.563901 + 559.528004]$$

$$\approx 111.0851557$$

n	h	Integral	Mean
1	0.2	111.6607637	558.30
2	0.1	111.2006538	556.00
4	0.05	111.0851557	555.43

A better approximation is obtained by considering 20 steps of length, $h = 0.1$; then the estimated integral is 111.048156, and the corresponding cost mean value is 555.24 €. Since the exact value for the definite integral can be verified through the function primitive, the error in such estimate is just about 1 cent, 0.01 €.

- The primitive was obtained by decomposition already: $Y(x) = \frac{50\pi\, x^3}{3} + 400\ln|x|$.

  The definite integral in the range [1, 1.2] thus follows,

$$\int_1^{1.2} 50\pi\, x^2 + \frac{400}{x}dx = Y(1.2) - Y(1)$$

$$= \left[\frac{50\pi\,(1.2)^3}{3} + 400\ln|1.2|\right] - \left[\frac{50\pi\ 1^3}{3} + 400\ln|1|\right]$$

$$= \frac{50\pi\,(1.728 - 1)}{3} + 400\ln|1.2|$$

$$= \frac{36.4}{3}\pi + 400\ln|1.2|$$

$$\cong 111.046614$$

  Finally, the mean value in the cost lowest range [1, 1.2] that involves the minimum point:

$$f(c) = \frac{\int_a^b f(x)\,dx}{b - a} = \frac{111.046614}{1.2 - 1} \cong 555.23$$

9.5 AREA CALCULATIONS

The calculation of areas defined by continuous functions correspond to scores of applications in the sciences and technological applications, including among many others, geometric representations of areas and volumes, arcs and lengths, that shall present non-negative values. As is known, for the continuous and non-negative function $f(x)$ on range $[a, b]$, the trapezoid area is given by $\int_a^b f(x)dx$; at now, simple instances on negative domains or negative integrations are focused on.

If function f(x) is non-positive in the interval [a, b], the trapezoid area is given by $\int_a^b |f(x)|dx = \left|\int_a^b f(x)dx\right|$

For instance, consider the integral of linear function $f(x) = x$, in the range $[-1, 0]$; by observation of Figure 9.13b, the triangle on the left side of Y-axis presents equal area to the triangle in the axis's right side; however, the integral calculation leads to a negative value,

$$A_{\text{Lin}} = \int_{-1}^{0} x \ dx = \left[\frac{x^2}{2}\right]_{-1}^{0} = \frac{0^2}{2} - \frac{(-1)^2}{2} = -\frac{1}{2}$$

In this way, the total in the symmetric range $[-1, 1]$ would cancel, the integral results zero:

$$A_{\text{Lin}} = \int_{-1}^{1} x \ dx = \int_{-1}^{0} x \ dx + \int_{0}^{1} x \ dx = -\frac{1}{2} + \frac{1}{2} = 0$$

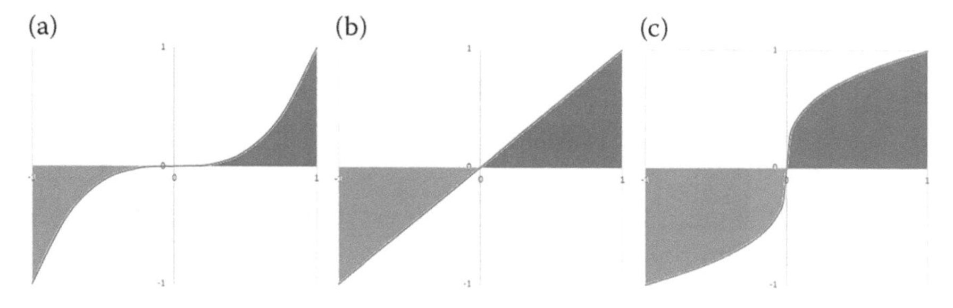

FIGURE 9.13 Area calculations for odd functions in symmetric range $[-1, 1]$. (a) Cube power function, $f(x) = x^3$. (b) Linear function, $f(x) = x$. (c) Cubic root function, $f(x) = \sqrt[3]{x}$.

Thus, the area for the triangle on the negative range is calculated in absolute value:

$$|A_{\text{Lin}}| = \left|\int_{-1}^{0} x \ dx\right| = \int_{-1}^{0} |x| dx = \int_{-1}^{0} -x \ dx = \left[\frac{-x^2}{2}\right]_{-1}^{0}$$

$$= -\frac{0^2}{2} + \frac{(-1)^2}{2} = \frac{1}{2}$$

And for the cube power function, $f(x) = x^3$, as in Figure 9.13a in the range $[-1, 0]$, the area on the left side of Y-axis is also equal to the area in the axis's right side, but with the integral calculation leading to a negative value. Thus, the absolute value is required to properly evaluate the area in the negative domain:

$$A_{Pow} = \int_{-1}^{0} x^3 \ dx = \left[\frac{x^4}{4}\right]_{-1}^{0} = \frac{0^4}{4} - \frac{(-1)^4}{4} = -\frac{1}{4}$$

$$|A_{Pow}| = \left|\int_{-1}^{0} x^3 \ dx\right| = \int_{-1}^{0} |x^3| dx = \int_{-1}^{0} -x^3 \ dx = \left[\frac{-x^4}{4}\right]_{-1}^{0}$$

$$= -\frac{0^4}{4} + \frac{(-1)^4}{4} = \frac{1}{4}$$

And also for the cubic root function, $f(x) = \sqrt[3]{x} = x^{1/3}$, the areas are symmetric in relation to the axis origin (Figure 9.13c), and again the integral calculation in $[-1, 0]$ leads to a negative value. Once more, the absolute value is utilized:

$$A_{CubRoot} = \int_{-1}^{0} \sqrt[3]{x} \ dx = \int_{-1}^{0} x^{1/3} \ dx = \left[\frac{x^{4/3}}{(4/3)}\right]_{-1}^{0}$$

$$= \frac{0^{4/3}}{(4/3)} - \frac{(-1)^{4/3}}{(4/3)} = -\frac{3}{4}$$

$$|A_{CubRoot}| = \left|\int_{-1}^{0} x^{1/3} \ dx\right| = \int_{-1}^{0} |x^{1/3}| dx = \int_{-1}^{0} -x^{1/3} \ dx = \left[\frac{-x^{4/3}}{(4/3)}\right]_{-1}^{0}$$

$$= -\frac{0^{4/3}}{(4/3)} + \frac{(-1)^{4/3}}{(4/3)} = \frac{3}{4}$$

Two notes are originated by the comparison of these instances of odd functions in the same symmetric range, $[-1, 1]$, as follows.

i. The integration in the full range, assuming both the negative and positive parts, will lead to the respective cancellation and obtaining zero in the total:

$$A_{Pow} = \int_{-1}^{1} x^3 \ dx = \int_{-1}^{0} x^3 \ dx + \int_{0}^{1} x^3 \ dx = -\frac{1}{4} + \frac{1}{4} = 0$$

$$A_{CubRoot} = \int_{-1}^{1} \sqrt[3]{x} \ dx = \int_{-1}^{0} \sqrt[3]{x} \ dx + \int_{0}^{1} \sqrt[3]{x} \ dx = -\frac{3}{4} + \frac{3}{4} = 0$$

ii. By comparison in the unit range, [0, 1], or even on [−1, 0] if absolute values are considered, both graphically (Figure 9.13a–c) and numerically, then

$$|x^3| \leq |x| \leq |\sqrt[3]{x}|$$

If function f(x) changes sign in [a, b], the area defined by the function graph, the X-axis, and the vertical lines at border points, a and b, is given by $\int_a^b |f(x)|dx$

Let it be the integral of trigonometric odd function, $f(x) = \cos(x)$, in the range $[0, \pi]$; as in Figure 9.14, the function sign changes

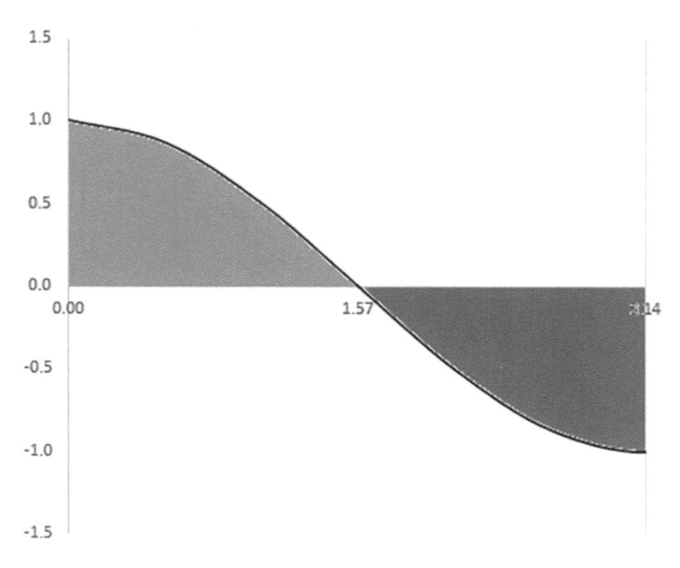

FIGURE 9.14 Cosine function in the range $[0, \pi]$.

periodically, and the second quarter is presenting a negative result that will cancel the positive integral in first quarter.

$$\begin{aligned} A_{\cos} &= \int_0^{\pi} \cos(x)\,dx = \int_0^{\pi/2} \cos(x)\,dx + \int_{\pi/2}^{\pi} \cos(x)\,dx \\ &= [\sin(x)]_0^{\pi/2} + [\sin(x)]_{\pi/2}^{\pi} \\ &= [\sin(\pi/2) - \sin(0)] + [\sin(\pi) - \sin(\pi/2)] \\ &= (1 - 0) + (0 - 1) = 1 - 1 = 0 \end{aligned}$$

In this case, neither the function at hand is odd, nor the range is symmetric; in fact, the cosine is an even function, and the domain under analysis is non-negative. Nevertheless, the sum of the two areas is canceling, and once again absolute value shall be utilized to avoid such cancellation.

$$\begin{aligned} |A_{\cos}| &= \left|\int_0^{\pi} \cos(x)\,dx\right| = \int_0^{\pi/2} \cos(x)\,dx + \int_{\pi/2}^{\pi} |\cos(x)|\,dx \\ &= [\sin(x)]_0^{\pi/2} - [\sin(x)]_{\pi/2}^{\pi} \\ &= [\sin(\pi/2) - \sin(0)] - [\sin(\pi) - \sin(\pi/2)] \\ &= (1 - 0) - (0 - 1) = 1 + 1 = 2 \end{aligned}$$

If the domain whose area is to be determined is defined by several functions, this domain must be decomposed into sub-domains without common interior points, by calculating these partial areas and adding them together.

Consider the areas defined by the graphs of two functions, as in Figure 9.15, respectively, in the range [0, 2]: (a) the area in-between the quadratic and square root functions; and (b) *ditto*, at now for the cubic and quadratic functions.

These functions are coincident in points (0, 0) and (1, 1), however the relationship between each pair under analysis changes at point (1, 1). Namely, while the quadratic function is greater than the square root for $x > 1$, $x^2 > \sqrt{x}$, as in Figure 9.15a, the opposite occurs when $x < 1$, with the square root taking greater values that the quadratic function, $x^2 < \sqrt{x}$.

In this way, the area calculations have to assess the superiority relations between the functions at hand, with the absolute value being

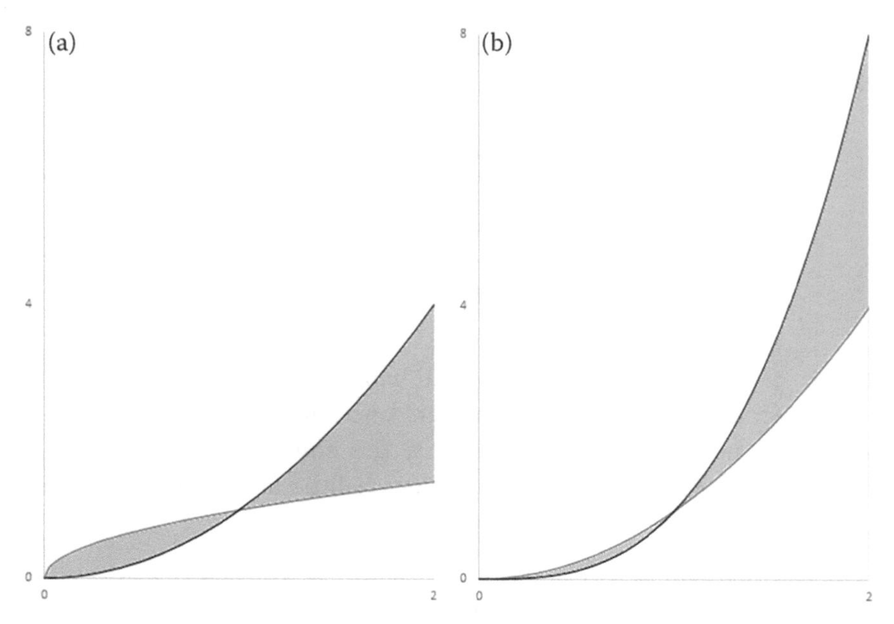

FIGURE 9.15 Calculation of areas in-between function graphs, on range [0, 2]. (a) Quadratic function and square root. (b) Cubic function and quadratic function.

useful in these type of evaluations. For the area between the quadratic and the square root function, A_A, in Figure 9.15a:

$$
\begin{aligned}
|A_A| &= \left|\int_0^2 (x^2 - \sqrt{x})\,dx\right| = \left|\int_0^2 (x^2 - x^{1/2})\,dx\right| \\
&= -\int_0^1 (x^2 - x^{1/2})\,dx + \int_1^2 (x^2 - x^{1/2})\,dx \\
&= -\left[\frac{x^3}{3} - \frac{x^{3/2}}{(3/2)}\right]_0^1 + \left[\frac{x^3}{3} - \frac{x^{3/2}}{(3/2)}\right]_1^2 \\
&= -\left(\frac{1^3}{3} - \frac{1^{3/2}}{(3/2)}\right) + \left(\frac{0^3}{3} - \frac{0^{3/2}}{(3/2)}\right) + \left(\frac{2^3}{3} - \frac{2^{3/2}}{(3/2)}\right) - \left(\frac{1^3}{3} - \frac{1^{3/2}}{(3/2)}\right) \\
&= \left(-\frac{1}{3} + \frac{2}{3}\right) + 0 + \left(\frac{8}{3} - \frac{2\sqrt{2^3}}{3}\right) - \frac{1}{3} + \frac{2}{3} = \frac{1 + 8 - 4\sqrt{2} + 1}{3} \\
&= \frac{10 - 4\sqrt{2}}{3} = \frac{\sqrt{68}}{3}
\end{aligned}
$$

For the area between the cube power and the quadratic functions, A_B, since the cube function is greater than the quadratic function for $x > 1$, and the opposite for $x < 1$ (Figure 9.15b), then:

$$
\begin{aligned}
|A_B| &= \left|\int_0^2 (x^3 - x^2)\,dx\right| = -\int_0^1 (x^3 - x^2)\,dx + \int_1^2 (x^3 - x^2)\,dx \\
&= -\left[\frac{x^4}{4} - \frac{x^3}{3}\right]_0^1 + \left[\frac{x^4}{4} - \frac{x^3}{3}\right]_1^2 \\
&= -\left(\frac{1^4}{4} - \frac{1^3}{3}\right) + \left(\frac{0^4}{4} - \frac{0^3}{3}\right) + \left(\frac{2^4}{4} - \frac{2^3}{3}\right) - \left(\frac{1^4}{4} - \frac{1^3}{3}\right) \\
&= -\left(\frac{3-4}{12}\right) + 0 + \left(\frac{16}{4} - \frac{8}{3}\right) - \left(\frac{3-4}{12}\right) = \frac{1 + 48 - 32 + 1}{12} \\
&= \frac{18}{12} = \frac{3}{2}
\end{aligned}
$$

9.6 IMPROPER INTEGRALS

The study of functions integration also includes the integration of both infinite domains, which occurs when the integration range is infinite, and unbounded functions that occurs when the integration range includes a point where the function tends to infinite. These integrals can be difficult and, in order to improve their treatment, convergence criteria are thus indicated for both the types of improper integrals.

9.6.1 Infinite Domains

For integration on infinite ranges, the following cases can be typified:

$$
\begin{aligned}
\int_a^{+\infty} f(x)\,dx &= \lim_{b\to+\infty} \int_a^b f(x)\,dx \\
\int_{-\infty}^b f(x)\,dx &= \lim_{a\to-\infty} \int_a^b f(x)\,dx \\
\int_{-\infty}^{+\infty} f(x)\,dx &= \lim_{a\to-\infty} \int_a^c f(x)\,dx + \lim_{b\to+\infty} \int_c^b f(x)\,dx
\end{aligned}
$$

The first equality focuses on the infinite domain on the interval's right side, and similarly the second equality focuses on infinite in the negative domain; the third relation applies when the unbounded domain occurs on both sides, for positive and negative domains.

- For each equality, if the limits in second member exist, the improper integrals in first member are said to converge. Else, in case those limits do not take finite value, then the integrals in the first member are said to diverge, or that they do not exist.

Examples

- Let it be, $f(x) = \frac{1}{x}$.

  The area below the function line (Figure 9.16), in the range [1, *b*], corresponds to

$$\int_1^b \frac{1}{x}dx = [\ln|x|]_1^b = \ln|b| - \ln|1| = \ln|b|$$

When *b* takes integer values, successively 2, 3, 4, … , *b*, the total area evaluated by the integral is, respectively, ln(2), ln(3), ln(4), … , ln(*b*). In the first partition, [1, 2], both the area and the mean value become ln(2); in the second partition, [2, 3], similarly, they both result ln(3) – ln(2) = ln(3/2); and in the third partition, [3, 4], they become ln(4) – ln(3) = ln(4/3).

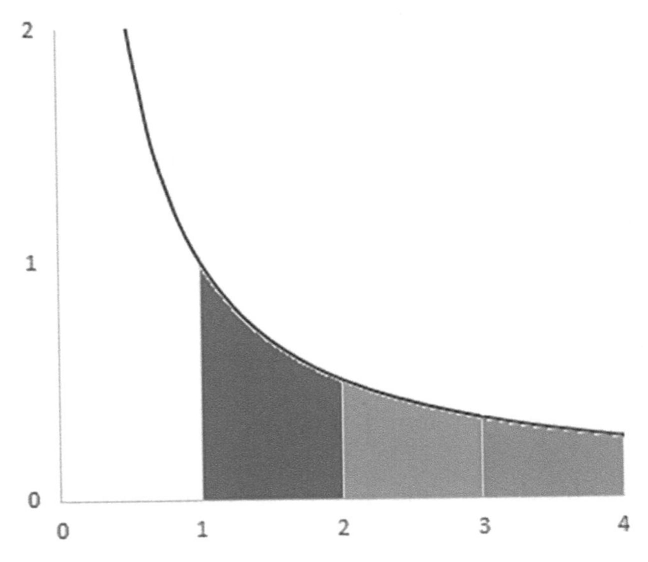

FIGURE 9.16 Area partitions for function, $f(x) = 1/x$, in the range [1, 4].

And so on, the parameter, *b*, can increase more and more, without any upper bound; by taking limit when *b* tends to infinite,

$$\int_1^{+\infty} \frac{1}{x}dx = \lim_{b\to+\infty} \int_1^b \frac{1}{x}dx = \lim_{b\to+\infty} [\ln|x|]_1^b = \lim_{b\to+\infty} [\ln|b| - \ln|1|]$$

$$= +\infty - 0 = +\infty$$

Thus, the improper integral does not converge, or also, the integral does not exist.

- At now, let it be, $f(x) = \frac{1}{x^2}$.

  In the range $[1, b]$, the area below the function line in Figure 9.17 is obtained:

$$\int_1^b \frac{1}{x^2}dx = \left[\frac{-1}{x}\right]_1^b = \left[\frac{-1}{b} + \frac{1}{1}\right] = \left(1 - \frac{1}{b}\right)$$

Again, the parameter, b, can increase without any upper bound; when b tends to infinite,

$$\int_1^{+\infty} \frac{1}{x^2}dx = \lim_{b\to+\infty} \int_1^b \frac{1}{x^2}dx = \lim_{b\to+\infty} \left[\frac{-1}{x}\right]_1^b$$

$$= \lim_{b\to+\infty} \left[\frac{-1}{b} + \frac{1}{1}\right] = -0 + 1 = 1$$

The improper integral exists, and it converges to 1. The area below the function line (Figure 9.17) is thus majored by 1.

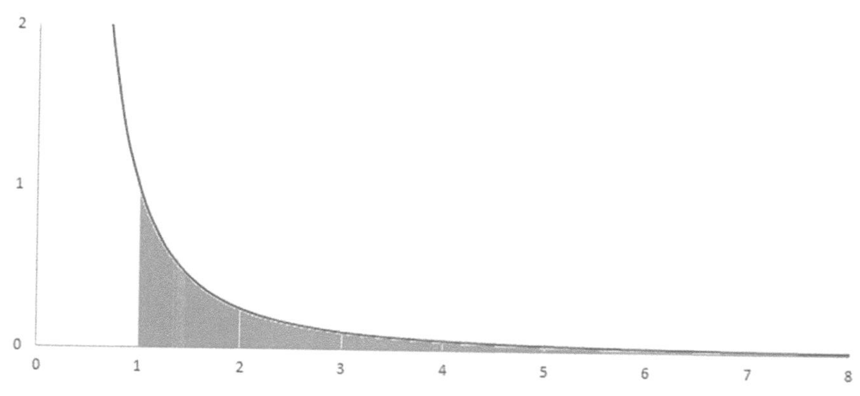

FIGURE 9.17 Majored area for function, $f(x) = 1/x^2$, in the unbounded domain, $[1, +\infty[$.

- And at now, let it be, $f(x) = \frac{1}{1+x^2}$.

  In the range $[0, b]$, the area below the function line (Figure 9.18) corresponds to:

$$\int_0^b \frac{1}{1+x^2}dx = [\arctan(x)]_0^b = \arctan(b) - \arctan(0) = \arctan(b)$$

Again, the parameter, b, can increase without any upper bound; when b tends to infinite,

$$\int_0^{+\infty} \frac{1}{1+x^2}dx = \lim_{b\to+\infty} \int_0^b \frac{1}{1+x^2}dx$$
$$= \lim_{b\to+\infty} \arctan(b) - \arctan(0) = \frac{\pi}{2} - 0 = \frac{\pi}{2}$$

The improper integral converges, it exists. The function at hand is an even function, symmetric to the Y-axis (Figure 9.18), and the negative domain is treated too. Namely,

$$\int_{-\infty}^{+\infty} \frac{1}{1+x^2}dx = \int_{-\infty}^{0} \frac{1}{1+x^2}dx + \int_0^{+\infty} \frac{1}{1+x^2}dx$$
$$= \lim_{a\to-\infty} \int_a^0 \frac{1}{1+x^2}dx + \lim_{b\to+\infty} \int_0^b \frac{1}{1+x^2}dx$$

The second term, for the positive domain was already treated; at now, the evaluation of first term is focused on when parameter a tends to negative infinite,

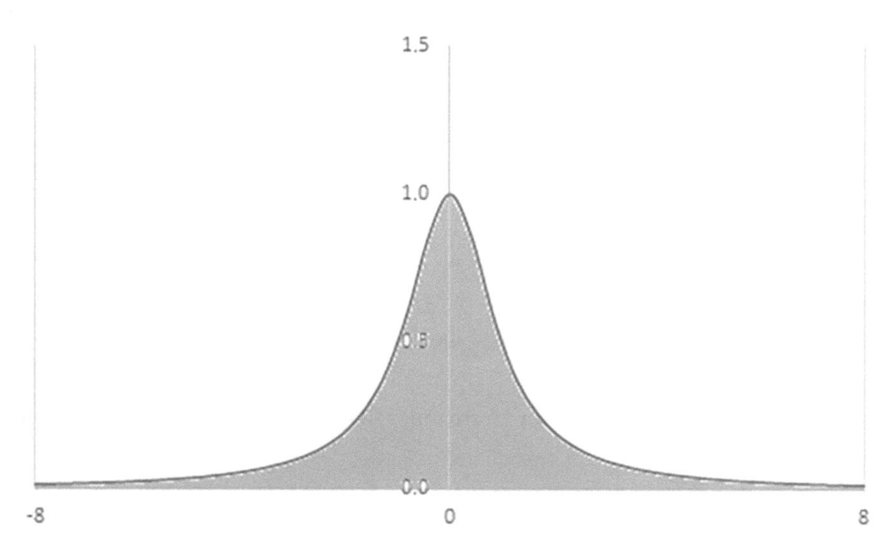

FIGURE 9.18 Majored area for function, $f(x) = 1/(1 + x^2)$, in the whole domain, $]-\infty, +\infty[$.

$$\int_{-\infty}^{0} \frac{1}{1+x^2}dx = \lim_{a\to-\infty} \int_{-\infty}^{0} \frac{1}{1+x^2}dx$$

$$= \arctan(0) - \lim_{a\to-\infty} \arctan(a) = 0 - \left(-\frac{\pi}{2}\right) = \frac{\pi}{2}$$

And finally, summing the areas for both the positive and negative domains, the improper integral exists and converges to π.

$$\int_{-\infty}^{+\infty} \frac{1}{1+x^2}dx = \frac{\pi}{2} + \frac{\pi}{2} = \pi$$

9.6.2 Unbounded Functions

For the integration of unbounded functions in a given range $[a, b]$, when a point of infinite discontinuity occurs, $\lim_{x\to c} f(x) = \pm\infty$, the following cases can be typified too:

$$\lim_{x\to b^-} f(x) = \pm\infty \Rightarrow \int_a^b f(x)\,dx = \lim_{r_b\to 0^+} \int_a^{b-r_b} f(x)\,dx$$

$$\lim_{x\to a^+} f(x) = \pm\infty \Rightarrow \int_a^b f(x)\,dx = \lim_{r_a\to 0^+} \int_{a+r_a}^{b} f(x)\,dx$$

$$\lim_{x\to c} f(x) = \pm\infty \Rightarrow \int_a^b f(x)\,dx = \lim_{r_c\to 0^+} \int_a^{c-r_c} f(x)\,dx$$

$$+ \lim_{r_c\to 0^+} \int_{c+r_c}^{b} f(x)\,dx, \qquad a < c < b$$

The first equality focuses the infinite discontinuity at the interval's right border, while the second equality similarly focuses the interval's left border; and the third relation applies when the infinite discontinuity occurs at an intermediate point, c.

- For each equality, if the limits in second member exist, the improper integrals in first member are said to converge. Else, in case those limits do not take finite value, then the integrals in first member are said to diverge, or that they do not exist.

Examples

- Let it be, $f(x) = \frac{1}{x}$.

  Note the function is not defined at, $x = 0$, an infinite discontinuity occurs in $]0, 1]$,

$$\lim_{x \to 0^+} f(x) = +\infty$$

In the range $[a, 1]$, with $a > 0$, the area below the function line corresponds to

$$\int_a^1 \frac{1}{x} dx = [\ln |x|]_a^1 = \ln |1| - \ln |a| = -\ln |a| = \ln |1/a|$$

In this way, when a takes reciprocal values, namely, 1/10, 1/100, 1/1000, ... , the areas evaluated by the integral are, respectively, ln(10), ln(100), ln(1000), ... , $\ln(1/a)$. These results are directly associated with the ranges, [0.1, 1], then [0.01, 1], and also [0.001, 1].

The first partial area, $A_1 = \ln(10)$, is colored in Figure 9.19; then the second partial area, and also the third one, can be associated with the first evaluation:

$$A_2 = \ln(100) = 2\ln(10) = 2A_1$$

$$A_3 = \ln(1000) = 3\ln(10) = 3A_1$$

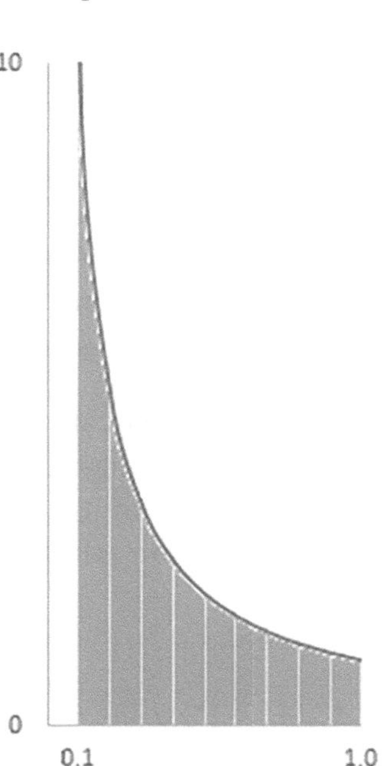

FIGURE 9.19 Area partition for function, $f(x) = 1/x$, in the range [0.1, 1].

And successively reducing parameter, a, in this geometric sequence, after n evaluations,

$$A_n = \ln(10^n) = n\ln(10) = nA_1$$

When the number of evaluations, n, tends to infinite, the area increases without upper bound too. In the opposite sense, parameter a reciprocally decreases, and decreases successively in the positive domain, nearing zero.

And this way, taking limit when a approaches zero, via the positive infinitesimal, r,

$$\lim_{a\to 0}\int_a^1 \frac{1}{x}dx = \lim_{r\to 0}\int_{0+r}^1 \frac{1}{x}dx$$

$$= \lim_{r\to 0}[\ln|x|]_r^1 = \lim_{r\to 0}[\ln|1/r|] = \ln\left[\lim_{r\to 0}|1/r|\right] = \ln|+\infty|$$

$$= +\infty$$

The improper integral $\int_0^1 (1/x)dx$ does not converge, or also, the integral does not exist.

- Now, consider $f(x) = \frac{1}{\sqrt{x}}$.
  The function is not defined at point 0, and again infinite discontinuity occurs in]0, 1]:

$$\lim_{x\to 0^+} f(x) = +\infty$$

In the range $[a, 1]$, with $a > 0$, the area below the function line corresponds to

$$\int_a^1 \frac{1}{\sqrt{x}}dx = [2\sqrt{x}]_a^1 = 2\sqrt{1} - 2\sqrt{a} = 2(1 - \sqrt{a})$$

Taking limit when a approaches zero, in a similar mode,

$$\lim_{a \to 0} \int_a^1 \frac{1}{\sqrt{x}} dx = \lim_{r \to 0} \int_{0+r}^1 \frac{1}{\sqrt{x}} dx = \lim_{r \to 0} [2\sqrt{x}]_r^1$$

$$= \lim_{r \to 0} [2 \ (1 - \sqrt{r})] = 2 \ (1 - \lim_{r \to 0} \sqrt{r})$$

$$= 2 \ (1 - 0) = 2$$

The improper integral exists, and it converges to 2. The area below the function line (Figure 9.20) is thus majored by 2.

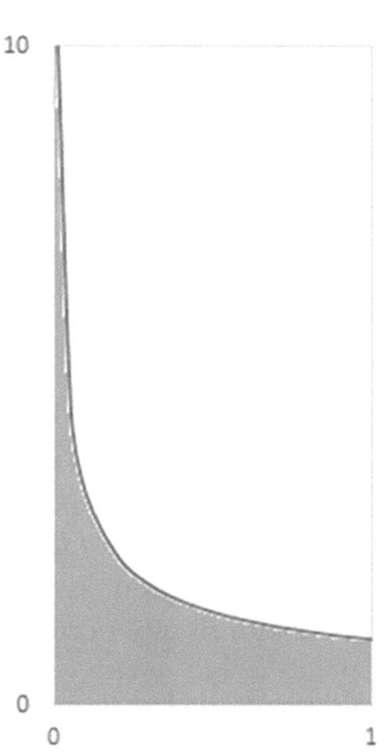

FIGURE 9.20 Majored area for function, $f(x) = 1/\sqrt{x}$, in the range]0, 1].

- Again, let $f(x) = 1/x$, but now focusing on the symmetric range [−1, 1]. Since infinite discontinuity occurs at the intermediate point, $c = 0$, the area in that range corresponds to:

$$\int_{-1}^1 \frac{1}{x} dx = \lim_{r_b \to 0^+} \int_{-1}^{0-r_b} \frac{1}{x} dx + \lim_{r_a \to 0^+} \int_{0+r_a}^1 \frac{1}{x} dx$$

The improper integral in the positive range,]0, 1], is already treated in previous instance: it is increasing without upper bound, to $+\infty$. The integral in the negative range, [−1, 0[, is symmetric to the previous one, and $\lim_{x \to 0^-} f(x) = -\infty$, so

$$\lim_{r_b \to 0} \int_{-1}^{0-r_b} \frac{1}{x} dx = \lim_{r_b \to 0} \int_{-1}^{-r_b} \frac{1}{x} dx$$
$$= \lim_{r_b \to 0} [\ln |x|]_{-1}^{r_b} = \lim_{r_b \to 0} [\ln |-r_b| - \ln |-1|]$$
$$= \ln \left[\lim_{r_b \to 0} |r_b| \right] = \ln |-\infty| = -\infty$$

The odd function, $f(x) = 1/x$, is being evaluated between symmetric borders (Figure 9.21), and the cancellation of the two improper integrals is thus expected:

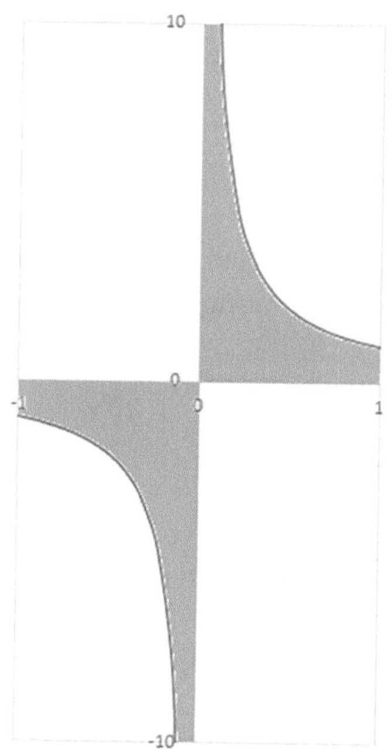

FIGURE 9.21 Odd function, $f(x) = 1/x$, in the discontinuous and symmetric range [−1, 1].

In fact, since the positive parameters nearing zero, $r_b \to 0$ and $r_a \to 0$, in the limits calculations are equivalent, then the associated calculations follow:

$$\int_{-1}^{1} \frac{1}{x} dx = \lim_{r_b \to 0^+} \int_{-1}^{0-r_b} \frac{1}{x} dx + \lim_{r_a \to 0^+} \int_{0+r_a}^{1} \frac{1}{x} dx = \lim_{r_b \to 0} [\ln |r_b|] + \lim_{r_a \to 0} [\ln |1/r_a|]$$
$$= \lim_{r_b \to 0} [\ln |r_b|] - \lim_{r_a \to 0} [\ln |r_a|] = 0$$

9.6.3 Convergence Criteria

Now, convergence criteria are indicated, for both the integrals on infinite range and integrals of unbounded functions. In this way, if the nature of the improper integral is anticipated, or a priori known, then the diverging integrals can be neglected, and the efforts associated be spared since a number of calculations are avoided.

Integration on Infinite Range

The attributes related with the comparison criterion and absolute convergence are addressed for improper integrals on positive domain, namely, $[a, +\infty[$.

Comparison Criterion—If $0 \leq f(x) \leq g(x)$ in the open interval $[a, +\infty[$, then the convergence of integral $\int_a^{+\infty} g(x)dx$ implies the convergence of integral $\int_a^{+\infty} f(x)dx$; also, the divergence of $\int_a^{+\infty} f(x)dx$, implies the divergence of $\int_a^{+\infty} g(x)dx$.

The Comparison Function: $\int_1^{+\infty} \frac{1}{x^p}dx$, with p-power.

- For $p = 1$, the improper integral is already calculated in prior instance, with the logarithm primitive originating a divergent integral. In the non-negative range,

$$\int_1^{+\infty} \frac{1}{x}dx = +\infty$$

As in Figure 9.16, by successively adding new positive partitions, each area evaluation is greater than the previous one, and such sequence is increasing monotonously: $A_n = n\ A_1$. Then the integral's convergence is associated with the upper bound's existence; in this case, the integral is diverging to infinite, $+\infty$, since such upper bound does not exist.

- For a power function, $p \neq 1$,

$$\int_1^{+\infty} f(x)dx = \lim_{b\to+\infty} \int_1^b \frac{1}{x^p}dx = \lim_{b\to+\infty} \int_1^b x^{-p}\,dx = \lim_{b\to+\infty} \left[\frac{x^{-p+1}}{-p+1}\right]_1^b =$$

$$= \lim_{b\to+\infty} \left[\frac{b^{-p+1}}{-p+1}\right] - \left[\frac{1^{-p+1}}{-p+1}\right]$$

When $p > 1$, the exponent in first term is negative, $-p + 1 < 0$, and the limit becomes zero; then the improper integral converges to the constant in the second term,

$$\int_1^{\infty} f(x)\,dx = 0 - \left[\frac{1^{-p+1}}{-p+1}\right] = \frac{1}{p-1}, \quad p > 1 \quad (\text{or } p - 1 > 0)$$

- However, when $p < 1$, the exponent in first term is positive, $-p + 1 > 0$, and the associated limit is infinite, $+\infty$; then the integral diverges.

$$\int_1^{\infty} f(x)\,dx = +\infty - \left[\frac{1^{1-p}}{1-p}\right] = +\infty, \quad p < 1 \quad (\text{or } 1 - p > 0)$$

In the positive infinite domain $[1, +\infty[$, for $p < 1$, the following relation applies,

$$x > x^p \Rightarrow \frac{1}{x} < \frac{1}{x^p}$$

and since the function at hand diverges to infinite, $+\infty$, then the second function does not allow any upper bound and also diverges to infinite, $+\infty$.

Absolute Convergence—If the integral $\int_a^{+\infty} |f(x)|\,dx$ exists, then so it does $\int_a^{+\infty} f(x)\,dx$, and the integral is said to be absolutely convergent.

- Let it be the improper integral, $\int_1^{+\infty} \frac{\cos(x)}{x^2} dx$.

  The trigonometric function, $\cos(x)$, is both positive and negative, with period changes; in addition, this function is bounded, $-1 \leq \cos(x) \leq 1 \Rightarrow |\cos(x)| \leq 1$, then

$$\frac{|\cos(x)|}{x^2} \leq \frac{1}{x^2}$$

The improper integral for the second function (Figure 9.17) exists, it converges,

$$\int_1^{+\infty} \frac{1}{x^2} dx = 1$$

then the integral under analysis is also majored and said to be absolutely convergent.

Integration of Unbounded Function

As in the prior topic, the attributes of both absolute convergence and comparison criterion also hold for unbounded functions at a given point, $\lim_{x \to c} f(x) = \pm\infty$.

Comparison Criterion—If $0 \le f(x) \le g(x)$ on the open interval, $[a, c[$, both functions being discontinuous at point, c, then the convergence of integral $\int_a^c g(x)\,dx$ implies the convergence of integral $\int_a^c f(x)\,dx$; also, the divergence of $\int_a^c f(x)\,dx$, implies the divergence of $\int_a^c g(x)\,dx$.

Comparison Function: $\int_a^c \frac{1}{(c-x)^p}dx$

- For a power function, $p \neq 1$,

$$\int_a^c \frac{1}{(c-x)^p}dx = \lim_{r_c \to 0} \int_a^{c-r_c} \frac{1}{(c-x)^p}dx = \lim_{r_c \to 0} \int_a^{c-r_c} (c-x)^{-p}\,dx$$

$$= \lim_{r_c \to 0}\left[-\frac{(c-x)^{-p+1}}{-p+1}\right]_a^{c-r_c} = \lim_{r_c \to 0}\left[\frac{(c-c+r_c)^{-p+1}}{p-1}\right]$$

$$-\left[\frac{(c-a)^{-p+1}}{p-1}\right] = \lim_{r_c \to 0}\left[\frac{r_c^{\,1-p}}{p-1}\right] - \left[\frac{(c-a)^{1-p}}{p-1}\right]$$

When $p < 1$, the exponent in first term is positive, $1 - p > 0$, and the related limit takes zero; the improper integral thus converges to the constant in second term,

$$\int_a^c \frac{1}{(c-x)^p}dx = 0 - \left[\frac{(c-a)^{1-p}}{p-1}\right] = \frac{(c-a)^{1-p}}{1-p}, \quad p < 1 \quad (\text{or } 1-p > 0)$$

For instance, revisiting the square root function (Figure 9.20), $p = 1/2$, although in the negative domain, obviously the results are coincident:

$$\int_{-1}^0 \frac{1}{(0-x)^{1/2}}dx = \int_{-1}^0 \frac{1}{\sqrt{|x|}}dx = 0 - \left[\frac{(0+1)^{1-(1/2)}}{(1/2)-1}\right] = -\frac{1^{1/2}}{-(1/2)} = 2$$

- For $p = 1$, an instance of the improper integral is also calculated at point, $c = 0$, and again the logarithm primitive will originate divergence. Either for the non-negative range or the non-positive range (Figure 9.21), divergence occurs in the two cases:

$$\int_0^1 \frac{1}{x} dx = +\infty, \text{ and also } \int_{-1}^0 \frac{1}{x} dx = -\infty$$

Note that in the positive domain]0, 1], for $p > 1$, the following relation applies,

$$x > x^p \Rightarrow \frac{1}{x} < \frac{1}{x^p}$$

and since the first function does not allow upper bound and diverges to positive infinite, $+\infty$, similarly, the second function will also diverge to infinite, $+\infty$.

- However, when $p > 1$, the exponent in first term is negative, $1 - p < 0$, and the associated limit is infinite, $+\infty$; then the improper integral diverges:

$$\begin{aligned} \int_a^c \frac{1}{(c - x)^p} dx &= \lim_{r_c \to 0} \left[\frac{r_c^{1-p}}{p - 1} \right] - \left[\frac{(c - a)^{1-p}}{p - 1} \right] \\ &= +\infty - \left[\frac{(c - a)^{1-p}}{p - 1} \right] = +\infty \end{aligned}$$

Absolute Convergence—If the integral $\int_a^b |f(x)| dx$ exists, then so does $\int_a^b f(x) dx$, and the improper integral is said to be absolutely convergent.

- Let it be the improper integral, $\int_{-1}^0 \frac{1}{\sqrt[3]{x}} dx$.
  The cubic-root function is negative in the range [−1, 0[, then the associated absolute value is evaluated and compared with the square root function. For the positive range, $0 < x < 1$, and noting that, $x^3 < x^2 < x < 1$ (e.g., multiplying successively by a positive number, x, less than 1), then in the opposite sense,

$$\sqrt[3]{|x|} \geq \sqrt{|x|} \geq x \Rightarrow \frac{1}{\sqrt[3]{|x|}} \leq \frac{1}{\sqrt{|x|}}$$

The improper integral for the square root function (Figure 9.20) converges,

$$\int_{-1}^{0} \frac{1}{\sqrt{|x|}} dx = 2$$

then the integral for the cubic root function is also majored by 2, and said absolutely convergent (in fact, the corresponding absolute value is 3/2).

9.7 CONCLUDING REMARKS

The definite integral topics are very useful in many contexts, and a glance into successive extensions is presented in this chapter. The study started with the integral application into a closed range, then extended into improper integrals where infinite domains and unbounded functions are focused. In addition, the support of continuity attributes and derivation insights about the area function, $A(x)$, allowed a similarity of procedures and practices, with enriched results from graphics comparison and qualitative approaches.

- Properties and theorems about definite integrals are presented, namely, the two mean theorems; they are also used to obtain the mean of cost function on convenient ranges. In this purpose, a computational method is detailed (the Trapezoid Rule), and numerical instances are analyzed.
- Barrow's rule and the Fundamental Theorem of Calculus are focused on, while several applications are described; namely, applying Barrow's rule to the integration of power functions and exponential functions, while geometric analogies are also discussed.
- Several applications to area calculations are detailed; the purpose is to distinguish the integration of positive and negative functions, as well as comparing the function integration in positive and negative domains.
- Improper integrals are enlarging the range of integrals application, extending integration into both unbounded domains (the integration range is infinite) and unbounded functions (the integration range includes a point where the function tends to infinite). In

addition, convergence criteria are indicated for both the types of improper integrals. This topic is also of specific interest for infinite sums, namely, the convergence criteria of non-negative series (the integral criterion, in Chapter 10).

Convergence criteria and improper integrals are also important for the study of sum series, both for the convergence criteria of non-negative series and for the absolute convergence of non-positive series, to be presented in the next chapter. Initially, notions about non-negative series are introduced, and the convergence criteria for such series are treated, including the integral criterion. After that, the next chapter also considers:

- The extension onto series with negative terms, in special alternating series, focusing applications of Leibniz criterion and the related error estimates.
- The generalization onto function series, with applications to both series derivation and series integration, not to mention the development in Taylor series of differentiable functions.

CHAPTER 10

Series

This chapter presents Series and, first, important notions about sum series are introduced in conjunction with reference series (e.g., Harmonic, Geometric, Telescopic). Then, important theorems and their applications to the analysis of series convergence are presented, followed by convergence criteria for series of non-negative terms; these topics are paving the way for the study of series with negative terms too, namely, the alternating series. The Leibniz criterion and related applications to alternating series are also focused on; in particular, applications to the barrel cost series and the related error estimates are compared. In addition, function series, the convergence domain, and applications to series derivation and series integration are presented too; namely, to the cost function's power series with center, $c = 1$. Finally, Taylor formula is revisited, and the development of Taylor series is studied, in particular for the cost function.

10.1 INTRODUCTION

Series are very useful to evaluate difficult functions because they are of relatively simple utilization, the series practice and associated computational implementation is well developed, and the associated errors can be upper bounded in a systematic manner.

This chapter starts with notions about series, and the convergence criteria for series of non-negative terms. Then the study is extended onto series with negative terms, including the Leibniz criterion for alternating series, and specific applications to error estimates. After that, the generalization from numerical series onto function series follows, with pertinent applications to both derivation and integration of function series, not to mention the development in Taylor series of differentiable functions.

DOI: 10.1201/9781003461876-10

As motivation, recall the cost function $y(x)$,

$$y(x) = 100\left(\frac{\pi}{2}x^2 + \frac{4}{x}\right) = 50\pi\, x^2 + \frac{400}{x}$$

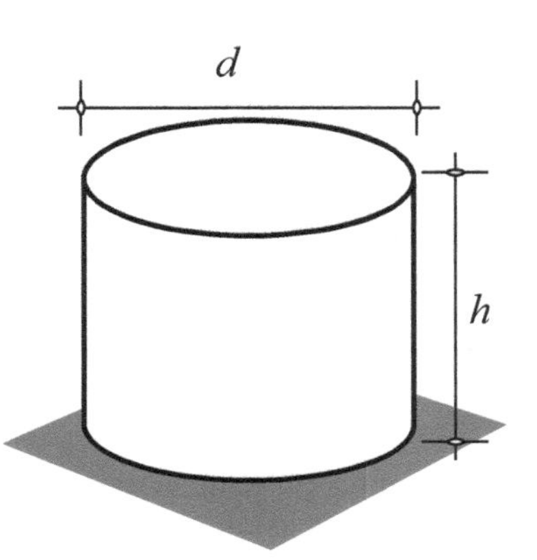

And consider also the third-order polynomial $P_3(x)$:

$$P_3(x) = 1600 - 2400x + (50\,\pi + 1600)x^2 - 400x^3$$

The function $y(x)$ and polynomial $P_3(x)$ present the same value at point $x = 1$, and the coincidence with three decimal digits occurs also in other points; they also present similar behavior in the point's neighborhood, namely inclination and curvature, as in Figure 10.1.

Let the relevant derivatives be calculated, both for function $y(x)$ and series $P_3(\mathrm{x})$, and the related values at point, $x = 1$, be compared too.

- For the function's first derivative, $y'(x)$, the first term is a quadratic term; the derivation of the negative power in the second term leads to:

x	y(x)	P3(x)
0.90	571.679	571.635
0.91	569.638	569.609
0.92	567.735	567.717
0.93	565.966	565.955
0.94	564.327	564.322
0.95	562.817	562.814
0.96	561.431	561.430
0.97	560.167	560.167
0.98	559.023	559.022
0.99	557.994	557.994
1.00	**557.080**	**557.080**
1.01	556.277	556.277
1.02	555.583	555.582
1.03	554.995	554.995
1.04	554.513	554.512
1.05	554.133	554.130
1.06	553.853	553.848
1.07	553.672	553.663
1.08	553.588	553.573
1.09	553.599	553.575
1.10	553.703	553.666

FIGURE 10.1 Coincident graphic lines for cost function, $y(x)$, and third-order polynomial, $P_3(x)$.

$$y'(x) = \left(50\pi\, x^2 + \frac{400}{x}\right)' = 50\pi\,(2x) + 400.(-1)x^{-1-1}$$
$$= 100\pi\, x - 400\, x^{-2} = 100\pi\, x - \frac{400}{x^2}$$

The first derivative, $y'(x)$, is evaluated at point 1:

$$y'(1) = 100\pi \times (1) - \frac{400}{(1)^2} = 100\pi - 400$$

- For the function's second derivative, $y''(x)$,

$$y''(x) = (100\pi\, x - 400\, x^{-2})' = 100\pi\,(1) - 400.(-2)x^{-2-1}$$
$$= 100\pi + 800x^{-3} = 100\pi + \frac{800}{x^3}$$

At the same point, $x = 1$, the second derivative:

$$y''(1) = 100\pi + \frac{800}{1^3} = 100\pi + 800$$

- The third derivative is obtained too:

$$y'''(x) = 0 + 2.400.(-3)x^{-3-1}$$
$$= -(2.3).400\, x^{-4} = \frac{-2400}{x^4}$$

And at point 1, it follows: $y'''(1) = -2400$

The derivatives associated with the third-order polynomial $P_3(x)$ are obtained too, since the derivative of a power function is also a power function, but of lower order.

$$P_3(x) = 1600 - 2400x + (50\pi + 1600)x^2 - 400x^3$$
$$P_3{}'(x) = -2400 + 2(50\pi + 1600)x - 3 \times 400x^2$$
$$P_3{}''(x) = (100\pi + 3200) - 3 \times 2 \times 400x$$
$$P_3{}'''(x) = -2400$$

- And evaluating the polynomial derivatives at, $x = 1$, the coincidence between the corresponding values of function derivatives and polynomial derivatives is confirmed:

$$\begin{aligned}
P_3(1) &= 50\pi + 400 \\
P_3'(1) &= 100\pi - 400 \\
P_3''(1) &= 100\pi + 800 \\
P_3'''(1) &= -2400
\end{aligned}$$

In addition, note the absolute error of evaluating $y(x)$ through the polynomial $P_3(x)$ is upper bounded, either at point 1.1 or at 0.9. In fact, for estimates in the neighborhood of, $x = 1$, the rounding error in the polynomial estimate occurs in the second decimal digit:

$$\begin{aligned}
|Error|_{x=1.1} &= |y(1.1) - P_3(1.1)| \\
&= |553.703 - 553.666| = 0.0337, \text{ and the error is less than } 0.0400.
\end{aligned}$$

$$\begin{aligned}
|Error|_{x=0.9} &= |y(0.9) - P_3(0.9)| \\
&= |571.679 - 571.635| = 0.0344, \text{ the error is less than } 0.0400 \text{ too.}
\end{aligned}$$

In the next section, important notions about series are addressed, in addition to the reference series due to their relevance for other topics; namely, reference series are utilized to demonstrate theorems about series and their applications to convergence analysis. Thereafter, convergence criteria for series of non-negative terms are focused on, while paving the way for the study of series with negative terms. The Leibniz criterion and related applications to alternating series are thus described, including the error estimates when using partial sums. Since series with both positive and negative terms are already studied, then function series and the convergence domain are presented, along with applications to series derivation and series integration. The Taylor formula is revisited with specific instances shown step by step, and the development of functions in Taylor series is described too.

10.2 BASIC NOTIONS ABOUT SERIES

In this section, notions about series, about sums of non-negative (positive, in latus sense) terms are presented. The Harmonic series, Geometric sums, and summation's properties are revisited, namely, the telescopic sums. Later on, these basic concepts can be enlarged to the study of series with non-positive (negative, in latus sense) terms, to the study of alternating series, as well as function series where the terms can assume either positive or negative terms.

Consider the sequence of real numbers with general term u_n; then, consider a new sequence that is summing such terms, S_n, as follows:

$$\begin{cases} S_1 = u_1 \\ S_2 = u_1 + u_2 \\ S_3 = u_1 + u_2 + u_3 \\ (\ldots) \\ S_{n-1} = u_1 + u_2 + u_3 + \ldots + u_{n-1} = \sum_{i=1}^{n-1} u_i \\ S_n = u_1 + u_2 + u_3 + \ldots + u_{n-1} + u_n = \sum_{i=1}^{n} u_i \end{cases}$$

In this way, infinite sum or infinite series of real numbers is the sequence of partial sums of general term, S_n,

$$\begin{aligned} S_n &= \sum_{i=1}^{n} u_i \\ &= S_{n-1} + u_n \end{aligned}$$

When adding a new positive term of nth order, u_n, the new sum S_n is greater than the previous one, S_{n-1}, and an increasing sequence is observed. The series convergence (or divergence) thus is related with the existence of an upper bound or majoring value (or non-existence), because a strictly increasing sequence with known upper bound is a convergent sequence.

For an infinite number of terms, n, then the limit toward which it tends is called the sum, S:

$$S = \lim S_n = \lim_{n\to\infty} \sum_{i=1}^{n} u_i = \sum_{i=1}^{\infty} u_i$$

A. *Harmonic Series*

The Harmonic series is the infinite sum of general term,

$$u_i = \left(\frac{1}{i}\right)$$

that is,

$$\begin{aligned} H &= \lim H_n = \lim_{n\to\infty} \sum_{i=1}^{n} \left(\frac{1}{i}\right) = \sum_{i=1}^{\infty} \left(\frac{1}{i}\right) \\ &= 1 + \frac{1}{2} + \frac{1}{3} + \ldots + \frac{1}{n} + \ldots \end{aligned}$$

The Harmonic series is divergent to positive infinite, $+\infty$, as follows from the integral test (in Section 10.4). In fact, adding positive terms successively, the sum is strictly increasing, and no upper bound or majoring value can be defined too.

Therefore, the Harmonic series is often used as a reference when comparing series, because every series that is equal or greater than the Harmonic series will be also diverging to infinite, $+\infty$.

The Harmonic series can be addressed as a specific instance of a general series: the P-series is reduced to the particular case of Harmonic series when power p is 1, $p = 1$,

$$\begin{aligned} P &= \lim_{n\to\infty} \sum_{i=1}^{n} \left(\frac{1}{i^p}\right) = \sum_{i=1}^{\infty} \left(\frac{1}{i^p}\right) \\ &= 1 + \frac{1}{2^p} + \frac{1}{3^p} + \ldots + \frac{1}{n^p} + \ldots \end{aligned}$$

The convergence of P-series can be confirmed for powers greater than one, $p > 1$, using the integral criterion (in Section 10.4); unbounded sums occur for P-series with power less than or equal to one, $p \leq 1$, in this way diverging to infinite, $+\infty$.

B. *Geometric Series*

The Geometric series is the infinite sum with terms in geometric sequence with ratio k, and taking one as first term, 1:

$$u_i = k^i, \quad k \in \mathrm{R}$$

That is,

$$\begin{aligned} G &= \lim G_n = \lim_{n\to\infty} \sum_{i=0}^{n} (k^i) = \sum_{i=0}^{\infty} (k^i) \\ &= 1 + k + k^2 + k^3 \ldots + k^n + \ldots \end{aligned}$$

Note the Geometric sum, G_n, with n terms in geometric sequence of ratio k, and first term, u_o, can be calculated through the relation:

$$G_n = u_0 \frac{1 - k^{n+1}}{1 - k}$$

The first term is one, $u_o = 1$, when the sum starts with the zero order's power, and the prior relation simplifies to:

$$G_n = \frac{1 - k^{n+1}}{1 - k}$$

It can be seen that the nature of a Geometric series directly depends on ratio, k, and it impacts whether the limit in the second member can be obtained:

$$\lim G_n = \lim_{n \to \infty} \frac{1 - k^{n+1}}{1 - k} = \frac{1}{1 - k} - \lim_{n \to \infty} \frac{k^{n+1}}{1 - k}$$

- Note the convergence of Geometric series is observed only when the ratio's absolute value is lower than one, $|k| < 1$; in this case, when taking limit the second term in numerator can be neglected, since it tends to zero:

$$\lim G_n = \frac{1}{1 - k} - 0 = \frac{1}{1 - k}, \quad \text{when } |k| < 1$$

- However, when the ratio's absolute value is greater than one, $|k| > 1$, the limit in the second term does not exist; for a positive ratio greater than one, $k > 1$, the series is unbounded and diverging to infinite, $+\infty$;

$$\lim G_n = \frac{1}{1 - k} + \frac{+\infty}{k - 1} = +\infty, \quad \text{when } k > 1$$

  Also note that for a negative ratio lower than -1, the series is oscillating while diverging.

- In the particular case of $k = 1$:

$$\begin{aligned} \lim G_n &= \lim_{n \to \infty} \sum_{i=1}^{n} (1^i) = \\ &= \lim_{n \to \infty} (1 + 1 + 1^2 + 1^3 \ldots + 1^i + \ldots + 1^n) \\ &= \lim_{n \to \infty} (1 + n) = +\infty \end{aligned}$$

- And for the particular case of $k = -1$,

$$\begin{aligned}\lim G_n &= \lim_{n\to\infty} \sum_{i=1}^{n} (-1)^i = \\ &= \lim_{n\to\infty} [1 + (-1) + 1^2 + (-1)^3 + 1^4 + (-1)^5 + \ldots + (-1)^n] \\ &= \lim_{n\to\infty} [1 - 1 + 1 - 1 + 1 - 1 + \ldots + 1 - 1 + \ldots + (-1)^n]\end{aligned}$$

the series is divergent. When the number of terms is even the sum is 0, since the terms can be grouped in pairs, and canceled this way; when the number of terms is odd, the sum is 1, since the terms can be also paired and canceled, except the last term that is 1. Thus, the sum is oscillating between 0 and 1, and the limit does not exist.

In the same way as the Harmonic series, Geometric series are also used in series comparison. For example, consider a Geometric series with ratio equal to or greater than one, $k \geq 1$; as shown in previous paragraphs, such Geometric series is diverging to infinite, $+\infty$; thus, every series that is equal or greater than this Geometric series will be diverging to infinite, $+\infty$, too.

From another point of view, consider a Geometric series with ratio lower than one, $k = 1/2$; as shown, such Geometric series is converging and the sum becomes 2, as follows:

$$G(1/2) = \frac{1}{1 - (1/2)} = \frac{1}{(1/2)} = 2$$

Then,

$$G(1/2) = \sum_{r=0}^{\infty} (1/2)^r = 1 + (1/2) + (1/2)^2 + (1/2)^3 \ldots + (1/2)^r + \ldots = 2$$

Thus, every series that is lower than this Geometric series will be converging too, namely, the geometric series with ratio, $k = 1/3$, is presenting a lower limit for the respective sum:

$$G(1/3) = \frac{1}{1 - (1/3)} = \frac{1}{(2/3)} = 3/2$$

In fact, neglecting the first term of zero order, all the terms of the latter series with ratio, $k = 1/3$, are lower than the corresponding terms of the prior series, then it is expected the sum will be lower too:

$$G(1/3) = \sum_{i=0}^{\infty} (1/3)^i = 1 + (1/3) + (1/3)^2 + (1/3)^3 \ldots + (1/3)^r + \ldots = 3/2$$

C. *Telescopic Series*

The Telescopic series is the infinite sum of the terms of the type,

$$(u_{i+1} - u_i)$$

where the general term considers a subtraction between two consecutive terms. The respective sum will exist, if and only if, the limit in the following relation also exists:

$$S = \lim S_n = \lim_{n\to\infty} \sum_{i=0}^{n} (u_{i+1} - u_i) = \sum_{i=0}^{\infty} (u_{i+1} - u_i)$$

Note that applying the telescopic property of summations,

$$\begin{aligned}\sum_{i=0}^{n} (u_{i+1} - u_i) &= (u_1 - u_0) + (u_2 - u_1) + (u_3 - u_2) + \ldots. + (u_{n+1} - u_n), \\ &= -u_0 + u_{n+1}\end{aligned}$$

the interim terms are all canceling, and the total includes only the first term, u_0, and the last term, u_{n+1}. When taking limit, and noting the first term is already calculated,

$$\begin{aligned}\lim_{n\to\infty} \sum_{i=0}^{n} (u_{i+1} - u_i) &= \lim_{n\to\infty} (-u_0 + u_{n+1}) \\ &= -u_0 + \lim_{n\to\infty} (u_{n+1})\end{aligned}$$

then the telescopic sum exists if and only the following limit exists too:

$$S = \lim(u_{n+1}) - u_0$$

10.3 THEOREMS AND APPLICATIONS

In this section, important theorems about series and their applications to the analysis of series nature are described. Namely, the series convergence can be studied by careful comparison with a convergent (or divergent) series, by applying linearity attributes to obtain the sum of two series or when multiplying a series by a constant, and by addressing the necessary condition (but *not-sufficient*) for convergence. For the latter, the associated corollary that indicates a non-convergent series is remarked: it directly identifies a divergent series in case the necessary condition is not satisfied, that is, when the limit of general term is not taking zero.

Theorem—Let $\sum_{i=1}^{\infty} u_i$ and $\sum_{i=1}^{\infty} v_i$ be infinite series, such that, from a certain order p, it is verified that $u_i = v_i, \ \forall \ i \geq p$; then:

i. Or both series converge;

ii. Or both series diverge.

Two examples follow using the reference series, one is showing convergence, and the other corresponds to a divergent case.

- Consider a Geometric series with ratio, $k = 1/3$, starting with term of order 2:

$$\sum_{i}^{\infty} (1/3)^i = (1/3)^2 + (1/3)^3 + (1/3)^4 + \ldots + (1/3)^i + \ldots$$

Note that this series corresponds to the prior sum, $G(1/3)$, when the two first terms are added; these two terms are also subtracted at the end, thus the sum value is not altered:

$$\sum_{i=2}^{\infty} (1/3)^i = 1 + (1/3) + (1/3)^2 + (1/3)^3 + (1/3)^4 + \ldots + (1/3)^i + \ldots -1 - (1/3)$$

The sum then results:

$$\begin{aligned} \sum_{i=2}^{\infty} (1/3)^i &= (3/2) - 1 - (1/3) \\ &= (1/2) - (1/3) = 1/6 \end{aligned}$$

This result can be confirmed with the geometric relation initiating with the order-2 term,

$$G = u_2 \frac{1}{1 - k}$$

$$G = u_2 \frac{1}{1-k} = (1/3)^2 \frac{1}{1-(1/3)} = (1/9)\frac{1}{(2/3)} = 1/6$$

- Now, consider a Geometric series with ratio $k = 3$, which clearly is a divergent series, and let it start with term of order-2 again:

$$\sum_{i=2}^{\infty} (3)^i = (3)^2 + (3)^3 + (3)^4 + \ldots + (3)^i + \ldots$$

Note that this series corresponds to the geometric sum, $G(3)$, when adding the two first terms; these two terms are subtracted in the end, thus the sum value still is the same:

$$\sum_{i=2}^{\infty} (3)^i = 1 + (3) + (3)^2 + (3)^3 + (3)^4 + \ldots + (3)^i + \ldots -1 - (3)$$

The sum then results:

$$\sum_{i=2}^{\infty} (3)^i = +\infty - 4 = +\infty$$

Once again, this result can be confirmed with the geometric relation when including the first term of order 2,

$$\begin{aligned} \lim G_n &= \lim_{n\to\infty} \left(u_2 \frac{1-k^{n+1}}{1-k} \right) \\ &= \lim_{n\to\infty} \left(3^2 \frac{1-3^{n+1}}{1-3} \right) \\ &= 9 \lim_{n\to\infty} \left(\frac{1-3^{n+1}}{-2} \right) = 9 \lim_{n\to\infty} \left(\frac{3^{n+1}-1}{2} \right) = +\infty \end{aligned}$$

Therefore, changing or deleting a finite number of terms in a series, for example the two first terms, does not change the series nature, although it may change its sum.

Theorem—If $\sum_{i=1}^{\infty} u_i$ and $\sum_{i=1}^{\infty} v_i$ are both convergent series, respectively, with sum S and T, then:

i. $\sum_{i=1}^{\infty}(u_i + v_i)$ converges, and has sum $(S + T)$.

ii. $\sum_{i=1}^{\infty}(k\ u_i)$ converges, and has sum $(k\ S)$.

Therefore, the elementary operations of summing two series or multiplying a series by a constant translate into similar operations on the respective summations (linearity properties).

- A new example follows, and again instances of the Geometric series are used. Let it be:

$$S = \sum_{i=1}^{\infty}\left(\frac{5}{2^i} - \frac{5}{3^i}\right)$$

The coefficient 5 that is multiplying the two terms is factored out of the summation,

$$S = 5\sum_{i=1}^{\infty}\left(\frac{1}{2^i} - \frac{1}{3^i}\right)$$

The summation is separated in two geometric series,

$$S = 5\left[\sum_{i=1}^{\infty}\left(\frac{1}{2^i}\right) - \sum_{i=1}^{\infty}\left(\frac{1}{3^i}\right)\right]$$

And then, both the geometric series are starting with the term of order 1,

$$\begin{aligned} S &= 5\left[(1/2)^1\frac{1}{1-(1/2)} - (1/3)^1\frac{1}{1-(1/3)}\right] \\ &= 5\left[\frac{(1/2)}{(1/2)} - \frac{(1/3)}{(2/3)}\right] \\ &= 5\left[1 - \frac{1}{2}\right] = \frac{5}{2} \end{aligned}$$

Assuming convergence, the linearity property indicates the series of a sum can be treated as the sum of two convergent series, while a multiplying constant can be factored and treated directly. Note this property is similar to the linearity attributes for limits, derivatives, or integrals.

Necessary Condition(but Not Sufficient) for Convergence—If the series $\sum_{i=1}^{\infty} u_i$ converges, then it is a necessary condition that, lim $(u_n) = 0$.

This result is important, but the condition is not sufficient; namely, the Harmonic series is divergent while the limit for the general term, $1/n$, is zero:

$$\lim_{n\to\infty}\left(\frac{1}{n}\right) = 0$$

Therefore, the following corollary allows the identification of divergent series.

Corollary—If the general term u_n of an infinite series does not approach zero when order n tends to infinite, then the series diverges.

For example, let it be: $\sum_{i=1}^{\infty}\left(\frac{i}{3i+2}\right)$;

By evaluating the limit for the general term, n, then $\lim_{n\to\infty}\left(\frac{n}{3n+2}\right) = \frac{1}{3}$.

Then the series at hand is divergent, and it can be neglected from other considerations. In fact, for a large number of terms, the related value of each term is approaching 1/3; when summing the infinite number of terms, thus the sum will be infinite too.

By analogy with limits indeterminacy, $(\infty.0)$, obtaining a finite value for a sum with infinite number of positive terms is possible when the value of each term is approaching zero. In such case, the series will deserve further attention, it may converge or diverge; reminding the Harmonic series with limit zero for the general term, or even a Geometric series with ratio lower than one, for example, $k = 1/3$:

$$\lim_{n\to\infty}\left(\frac{1}{3}\right)^n = 0$$

While this geometric series is convergent, and the related sum is 3/2, the necessary condition by itself does not allow to clarify if this series converges or diverges. Therefore, convergence criteria are developed, and they are recommended for additional study of non-negative (positive) series.

10.4 CONVERGENCE CRITERIA FOR NON-NEGATIVE SERIES

In this section, important convergence criteria for series of non-negative (positive, in latus sense) terms are focused on, namely: (A) the general comparison criterion; (B) the ratio criterion; (C) the root criterion; and (D) the integral criterion.

These criteria are simple and useful; by utilizing the absolute value to the series terms, these criteria are also suitable to the study of series with negative (in latus sense) terms, of alternating series, and function series, since the function's terms can take either positive or negative values.

A. *General Comparison Criterion*

Consider the non-negative terms, $0 \leq u_i \leq v_i$, for all values from a certain order, p. Then:

i. If series $\sum_{i=1}^{\infty} v_i$ is convergent, then series $\sum_{i=1}^{\infty} u_i$ will also be a convergent series.

ii. If series $\sum_{i=1}^{\infty} u_i$ diverges to $+\infty$, then series $\sum_{i=1}^{\infty} v_i$ will also diverge to $+\infty$.

For non-negative series and successively adding new (positive) terms, the partial sum S_n is greater than the previous one, it is an increasing sequence. Therefore, the series convergence is directly related with the existence of an upper bound; from the other side, the series is diverging to infinite, $+\infty$, in case of non-existence of such upper bound.

Two examples follow: one showing convergence by comparison with a convergent series, and aiming at lower values (comparing from below); the other corresponding to a divergent case, by comparison with a series that tends to infinite, $+\infty$, and aiming at greater values (comparing from above).

- From the first point of view, let it be:

$$\sum_{i=1}^{\infty} (1/i)^i = 1 + (1/2)^2 + (1/3)^3 + (1/4)^4 + \ldots + (1/i)^i + \ldots$$

When comparing this series with the geometric series with ratio, $k = 1/2$, term by term directly, all the terms from order 2 are equal or lower than the corresponding terms of the geometric series at hand,

$$\begin{aligned} \textstyle\sum_{i=2}^{\infty} (1/2)^i &= (1/2)^2 + (1/2)^3 + (1/2)^4 + \ldots + (1/2)^i + \ldots \\ &= (1/4)\frac{1}{1-(1/2)} = \frac{(1/4)}{(1/2)} = 1/2 \end{aligned}$$

Thus, neglecting the first term that is 1, the series under study is equal to or lower than the convergent Geometric series; then it presents an upper limit for the respective sum, 1 + 1/2 = 3/2, and the increasing sequence of non-negative sums thus is convergent.

$$\sum_{i=1}^{\infty} (1/i)^i \leq 1 + (1/2)^2 + (1/2)^3 + (1/2)^4 + \ldots + (1/2)^i + \ldots$$
$$\leq 1 + \sum_{r=2}^{\infty} (1/2)^i$$
$$\leq 3/2$$

- And from the second point of view, let it be:

$$\sum_{i=1}^{\infty} \left(\frac{1}{\sqrt{i}}\right) = 1 + \frac{1}{\sqrt{2}} + \frac{1}{\sqrt{3}} + \ldots + \frac{1}{\sqrt{i}} + \ldots$$
$$= 1 + \frac{\sqrt{2}}{2} + \frac{\sqrt{3}}{3} + \ldots + \frac{\sqrt{i}}{i} + \ldots$$

When comparing this series term by term with the Harmonic series, all the terms from order 2 are greater than the corresponding terms of the Harmonic series,

$$\sum_{i=1}^{\infty} \left(\frac{1}{i}\right) = 1 + \frac{1}{2} + \frac{1}{3} + \ldots + \frac{1}{i} + \ldots$$

The Harmonic series is diverging to infinite, $+\infty$, since no upper bound exists; the series under comparison is greater than the Harmonic series,

$$\sum_{i=1}^{\infty} \left(\frac{1}{\sqrt{i}}\right) > \sum_{i=1}^{\infty} \left(\frac{1}{i}\right)$$

then the related upper bound does not exist, and this increasing sequence of positive sums is also diverging to infinite, $+\infty$.

B. *The Ratio Criterion*

Consider the series of non-negative terms, $\sum_{i=1}^{\infty} u_i$, and obtain the finite limit of the ratio between consecutive terms, k,

$$\lim_{n\to\infty}\left(\frac{u_{n+1}}{u_n}\right) = k$$

Consequently:

i. If the ratio is lower than one, $k < 1$, then the series converges.

ii. If the ratio is greater than one, $k > 1$, then the series diverges.

iii. If the ratio is one, $k = 1$, then this ratio test is inconclusive.

The ratio criterion, also known as D'Alembert test, is very useful, since it fits well general terms with factorials, powers, quotients. However, weak points also occur, namely, the limit can be difficult to obtain, and very often this ratio test becomes inconclusive.

Three examples follow, each one addressing the key limit, respectively, a ratio limit lower than one, greater than one, and equal to one.

- For the first case, let it be:

$$\sum_{i=1}^{\infty}\left(\frac{1}{i!}\right) = 1 + \frac{1}{2!} + \frac{1}{3!} + \ldots + \frac{1}{i!} + \ldots$$

The limit of the ratio between consecutive terms,

$$\lim_{n\to\infty}\left(\frac{u_{n+1}}{u_n}\right) = \lim_{n\to\infty}\frac{\frac{1}{(n+1)!}}{\frac{1}{n!}}$$

$$= \lim_{n\to\infty}\frac{n!}{(n+1)!} = \lim_{n\to\infty}\frac{n!}{(n+1).n!} = \lim_{n\to\infty}\frac{1}{(n+1)} = 0$$

Since $k = 0 < 1$, then the series under analysis converges.

- For the divergent case, consider the geometric series:

$$\sum_{i=0}^{\infty}(3)^i = 1 + 3 + (3)^2 + (3)^3 + (3)^4 + \ldots + (3)^i + \ldots$$

The limit of the ratio between consecutive terms,

$$\lim_{n\to\infty}\left(\frac{u_{n+1}}{u_n}\right) = \lim_{n\to\infty}\frac{(3)^{n+1}}{(3)^n}$$

$$= \lim_{n\to\infty}\frac{3.(3)^n}{(3)^n} = 3.1 = 3$$

Since $k = 3 > 1$, then the series under analysis diverges and the ratio criterion is confirming the previous attributes of geometric series (in Section 10.2). In a similar mode, if the geometric ratio is lower than 1, then the non-negative series will present a ratio criterion below 1 and convergence is foreseen for such geometric series.

- For the third case, consider again the Harmonic series:

$$\sum_{i=1}^{\infty}\left(\frac{1}{i}\right) = 1 + \frac{1}{2} + \frac{1}{3} + \ldots + \frac{1}{i} + \ldots$$

The limit of the ratio between consecutive terms,

$$\lim_{n\to\infty}\left(\frac{u_{n+1}}{u_n}\right) = \lim_{n\to\infty}\frac{\frac{1}{(n+1)}}{\frac{1}{n}}$$

$$= \lim_{n\to\infty}\frac{n}{n+1} = 1$$

Since $k = 1$, then this test is inconclusive, and no further considerations can be stated here, in spite of Harmonic series be known as a divergent series.

C. *The Root Criterion*

Given the series of non-negative terms, $\sum_{i=1}^{\infty} u_i$, if there is a finite value for the root limit, k,

$$\lim_{n\to\infty}\sqrt[n]{u_r} = k$$

Then:

i. The series converges when the root limit is lower than one, $k < 1$.

ii. The series diverges when the root limit is greater than one, $k > 1$.

iii. The root test will be inconclusive if the root limit results one, $k = 1$.

The root criterion is also known as Cauchy test, it directly applies to general terms that present powers, thus it will be very useful for the study of series of power functions (in Section 10.7). However, the limit can be difficult to obtain, and the test results can be inconclusive as well.

Three examples follow again, each one addressing a value for the root limit, respectively, lower than one, greater than one, and equal to one.

- For the first case, consider the geometric series:

$$\sum_{i=0}^{\infty} (1/4)^i = 1 + (1/4) + (1/4)^2 + (1/4)^3 + (1/4)^4 + \ldots + (1/4)^i + \ldots$$

  The calculation of the root limit is straightforward, since the root is canceling the power in the general term,

$$\lim_{n \to \infty} \sqrt[n]{u_r} = \lim_{n \to \infty} \sqrt[n]{(1/4)^r} = 1/4$$

  Since $k = 1/4 < 1$, then the series converges, and the root test confirms the nature of this convergent geometric series.

- In a similar mode, if the geometric ratio is greater than 1, then the non-negative series will present a root criterion above 1 and divergence is foreseen for such geometric series. In this way, consider now the following series:

$$\sum_{i=0}^{\infty} (4)^i = 1 + 4 + (4)^2 + (4)^3 + (4)^4 + \ldots + (4)^i + \ldots$$

  The root limit on the general term applies directly, and again the root cancels the power in the general term,

$$\lim_{n \to \infty} \sqrt[n]{u_n} = \lim_{n \to \infty} \sqrt[n]{(4)^n} = 4$$

  Since $k = 4 > 1$, the series under study diverges, and the root test confirms the nature of the divergent geometric series once again.

- For the third case, consider the Harmonic series:

$$\sum_{i=1}^{\infty} \left(\frac{1}{i}\right) = 1 + \frac{1}{2} + \frac{1}{3} + \ldots + \frac{1}{i} + \ldots$$

The root limit on the general term presents some difficulties, namely, it occurs an indeterminacy, 0^0, that requires the utilization of logarithm:

$$\begin{aligned}\lim_{n\to\infty}\sqrt[n]{u_n} &= \lim_{n\to\infty}\sqrt[n]{\left(\frac{1}{n}\right)} \\ &= \lim_{n\to\infty}\left(\frac{1}{n}\right)^{\frac{1}{n}} = 0^0, \ ind.\end{aligned}$$

Taking logarithm, and exchanging logarithm and limit, the indeterminacy can be manipulated in a suitable manner:

$$\begin{aligned}\ln\left[\lim_{n\to\infty}\left(\frac{1}{n}\right)^{\frac{1}{n}}\right] &= \lim_{n\to\infty}\left[\ln\left(\frac{1}{n}\right)^{\frac{1}{n}}\right] = \lim_{n\to\infty}\left[\frac{1}{n}\ln\left(\frac{1}{n}\right)\right] \\ &= \lim_{n\to\infty}\left[\frac{\ln\left(\frac{1}{n}\right)}{n}\right] = \lim_{n\to\infty}\left[\frac{-\ln(n)}{n}\right] = -\frac{\infty}{\infty}, \ ind.\end{aligned}$$

After that, by applying L'Hôpital rule,

$$\lim_{n\to\infty}\left[\frac{-\ln'(n)}{n'}\right] = \lim_{n\to\infty}\left[\frac{-(1/n)}{1}\right] = 0$$

Finally, applying exponential to revert the logarithm,

$$\exp\left\{\ln\left[\lim_{n\to\infty}\left(\frac{1}{n}\right)^{\frac{1}{n}}\right]\right\} = \exp(0) = 1$$

Then the root test is inconclusive when addressing the Harmonic series; similar to the ratio test, no further considerations can be stated here. However, the integral criterion is very effective for functions that are easily integrated and, among others of interest, the integral approach will allow to define the nature of Harmonic series.

D. *The Integral Criterion*

Let $\sum_{i=1}^{\infty} u_i$ be a series of positive and decreasing terms, and let $f(x)$ be a positive, decreasing and integrable function, such that: $f(n) = u_n, \forall\ n \in \mathrm{N}$.

Then the series under analysis:

i. Converges, if the improper integral $\int_1^{\infty} f(x)\,dx$ is also convergent;

ii. Diverges, if this improper integral also diverges.

Note the first part of the integral criterion, in (i), declares the sufficiency of the integral convergence to identify a convergent series; the second part, in (ii), states the necessity of the integral convergence, otherwise the integral divergence corresponds to a divergent series. Thus, the integral criterion is known as a necessary and sufficient criterion for series convergence, and additional details can be found in many dedicated texts.

- With the simple purpose to address the Harmonic series, consider the positive and decreasing function, $f(x) = (1/x)$, in such a manner that:

$$f(x) = \left(\frac{1}{x}\right) \Rightarrow f(n) = \left(\frac{1}{n}\right)$$

  The improper integral follows,

$$\int_1^{\infty} f(x)\,dx = \int_1^{\infty} \frac{1}{x}\,dx$$

  Using the logarithm function, and taking limit,

$$\begin{aligned} \int_1^{\infty} f(x)\,dx &= \lim_{b\to\infty} \int_1^{b} \frac{1}{x}\,dx = \lim_{b\to\infty} [\ln(x)]_1^b \\ &= \lim_{b\to+\infty} [\ln(b)] - \ln(1) = +\infty - 0 = +\infty \end{aligned}$$

  The improper integral is diverging; the integral criterion is both necessary and sufficient condition, then the Harmonic series also diverges.

- As referred, the Harmonic series can be addressed as a particular case of P-series when $p = 1$:

$$P = \sum_{i=1}^{\infty} \left(\frac{1}{i^p}\right)$$

Through the integral criterion, the convergence of P-series can be confirmed for power greater than one, $p > 1$, and it diverges when power is less than or equal to one, $p \leq 1$.

Now, consider the positive and decreasing function, $f(x)$,

$$f(x) = \left(\frac{1}{x^p}\right) \Rightarrow f(n) = \left(\frac{1}{n^p}\right)$$

The improper integral follows,

$$\int_1^{\infty} f(x)\,dx = \int_1^{\infty} \frac{1}{x^p}\,dx$$

Using the power function, $p \neq 1$, and taking limit,

$$\begin{aligned}
\int_1^{\infty} f(x)\,dx &= \lim_{b\to\infty} \int_1^b \frac{1}{x^p}dx = \lim_{b\to\infty} \int_1^b x^{-p}\,dx \\
&= \lim_{b\to\infty} \left[\frac{x^{-p+1}}{-p+1}\right]_1^b \\
&= \lim_{b\to+\infty} \left[\frac{b^{-p+1}}{-p+1}\right] - \left[\frac{1^{-p+1}}{-p+1}\right]
\end{aligned}$$

When the power is greater than one, $p > 1$, the exponent in first term is negative, and the related limit is zero; then the improper integral converges to the value in second term,

$$\int_1^{\infty} f(x)\,dx = 0 - \left[\frac{1^{-p+1}}{-p+1}\right] = \frac{1}{p-1}, \quad p > 1$$

However, when the power is lower than one, $p < 1$, the exponent in first term is positive, and the associated limit is infinite, $+\infty$; then the integral diverges, and the P-series too:

$$\int_1^{\infty} f(x)\,dx = +\infty - \left[\frac{1^{-p+1}}{-p+1}\right] = +\infty, \quad p < 1$$

10.5 NON-POSITIVE AND ALTERNATING SERIES

In this section, series with non-positive (or negative, in latus sense) terms and alternating series are studied, and used for analyzing the absolute value of the series terms. Thereafter, two types of convergence can be defined: (a) the absolute convergence, when the series of absolute values is converging; and (b) the conditional convergence, when the absolute series diverges, but the series with positive and negative terms converges. About alternating series, the Leibniz criterion and applications to error bounding are described too; namely, the estimates for evaluating the barrel's cost using a third-order polynomial, $P_3(x)$, and a fifth-order polynomial, $P_5(x)$, are compared.

A. *Series with Non-Positive Terms*

For series with non-positive terms, the methodology already studied for series with non-negative terms can be used, since

$$u_i \leq 0 \Rightarrow -u_i \geq 0$$
$$\sum_{i=1}^{\infty} u_i = -\sum_{i=1}^{\infty} (-u_i)$$

For series whose terms have any signs, both positive or negative, the above criteria can no longer be directly applied. In that case, studying the absolute value of the series terms, then the convergence criteria and other important properties for series with non-negative (positive) terms continue to be valid options. Additional details about the approaches using the absolute value to non-positive terms follow, namely, addressing absolute convergence and conditional convergence.

- ***Absolute Convergence***—If the series of absolute values $\sum_{i=1}^{\infty} |u_i|$ is convergent, then the original series $\sum_{i=1}^{\infty} u_i$, with positive and negative terms, is also convergent; in this case, the series is said to be absolutely convergent.
- ***Conditional Convergence***—A series is conditionally convergent when its limit exists, but it is not absolutely convergent; that is, the series of absolute values diverges, but the series with any sign terms converges.

The main difference between series with absolute and conditional convergence is related with the corresponding sum, that is, it depends on the nature both on the total presented and in the mode the terms are added. In this way, the sum of an absolutely convergent series does not depend on the ordering, since any rearrangement of its terms also converges absolutely. For a conditionally convergent series, if the terms are properly commuted then different results are obtained, either the result can be a selected number or even it can result on a divergent series.

B. *Alternating Series and the Leibniz Criterion*

Alternating series are series whose signs of terms are oscillating positive and negative; the Leibniz criterion plays an important role, both to study the convergence of such series and to provide an insight to the error estimates when evaluating it numerically.

Leibniz Criterion—If u_n is a sequence of decreasing non-negative terms, $u_1 \geq u_2 \geq u_3 \geq \ldots \geq 0$, and case $\lim_{n \to \infty} (u_n) = 0$, then the following alternating series are both convergent:

$$\sum_{i=1}^{\infty} [(-1)^i \, u_i] \quad \text{and} \quad \sum_{i=1}^{\infty} [(-1)^{i+1} \, u_i]$$

Consider an alternating series that satisfies both the conditions required by Leibniz criterion, namely, the terms are decreasing, $u_{n+1} \leq u_n$, and $\lim_{n \to \infty} (u_n) = 0$. Thus, the alternating series will be bounded and monotone, and these attributes indicate convergence for such infinite sum.

Consider a series with even number of terms, S_{2n}; without loss of generality, assume the first term is positive. Note the terms present alternating signs, namely, negative sign for even terms, while the odd terms present positive sign.

$$S_{2n} = u_1 - u_2 + u_3 - u_4 + u_5 - u_6 + u_7 \ldots - u_{2n-2} + u_{2n-1} - u_{2n}$$

Grouping the series terms in pairs, the difference between two consecutive terms is always positive, since the terms are decreasing (in latus sense) in absolute value,

$$S_{2n} = (u_1 - u_2) + (u_3 - u_4) + (u_5 - u_6) + \ldots + (u_{2n-1} - u_{2n})$$

Therefore, only positive quantities (in latus sense too) are added, and the total is also positive:

$$S_{2n} \geq 0$$

Separating the first term and pairing the other terms, two at a time,

$$S_{2n} = u_1 - (u_2 - u_3) - (u_4 - u_5) - (u_6 - u_7) \dots -(u_{2n-2} - u_{2n-1}) - u_{2n}$$

Again, all the subtractions inside parentheses result positive quantities, because the terms are decreasing, $(u_{2i-2} - u_{2i-1}) \geq 0$; in this way, the alternating series is upper-bounded by the first term, since it is being successively subtracted by positive quantities:

$$S_{2n} \leq u_1$$

When addressing the next series with even number of terms, S_{2n+2}, the two added terms are decreasing too; the difference between the two terms inside parentheses is positive once more,

$$S_{2n+2} = S_{2n} + (u_{2n+1} - u_{2n+2})$$

The new partial sum, S_{2n+2}, is greater than the previous one, S_{2n},

$$S_{2n+2} \geq S_{2n}$$

Finally, the alternating series is convergent, because it is simultaneously upper-bounded by the first term, u_1, and the positive sequence of partial sums is increasing monotonous.

- Note that these results are similar when addressing an alternating series with odd number of terms, S_{2n+1}; this series is both upper-bounded and lower-bounded, and it is decreasing monotonous, as follows from the relations:

$$S_{2n+1} = (u_1 - u_2) + (u_3 - u_4) + (u_5 - u_6) + \dots + (u_{2n-1} - u_{2n}) + u_{2n+1}$$
$$\Rightarrow S_{2n+1} \geq 0$$

$$S_{2n+1} = u_1 - (u_2 - u_3) - (u_4 - u_5) - (u_6 - u_7) \dots -(u_{2n} - u_{2n+1})$$
$$\Rightarrow S_{2n+1} \leq u_1$$

$$S_{2n+3} = S_{2n+1} - (u_{2n+2} - u_{2n+3}) \Rightarrow S_{2n+3} \leq S_{2n+1}, \quad \text{decreasing}$$

This series thus is convergent too, and the limit for the series with odd-number of terms, S_{2n+1}, and the series with even-number of terms, S_{2n}, is equal. When taking limits, note that $\lim_{n\to\infty} (u_{2n+1}) = 0$, and consequently

$$S_{2n+1} = S_{2n} + u_{2n+1} \Rightarrow \lim_{n\to\infty} S_{2n+1} = \lim_{n\to\infty} S_{2n}$$

- As an example, it can be verified that the alternating Harmonic series satisfies the conditions of the Leibniz criterion, and as such it is convergent. Since the Harmonic series with absolute terms is divergent, the alternating Harmonic series is conditionally convergent. The non-negative terms in the Harmonic series are decreasing,

$$1 \geq \frac{1}{2} \geq \frac{1}{3} \geq \frac{1}{4} \geq \frac{1}{5} \geq \frac{1}{6} \geq \frac{1}{7} \geq \dots \geq \frac{1}{n} \geq \dots \geq 0$$

and the general term satisfies the necessary condition (but not sufficient) for convergence of non-negative series:

$$\lim_{n\to\infty} \left(\frac{1}{n}\right) = 0$$

Then the Harmonic alternating series is satisfying the two conditions required in the Leibniz Criterion, and both the series are (conditionally) convergent. Namely the Harmonic alternating series with first term positive is converging to ln(2), while the symmetric series with first term negative tends to ln(1/2):

$$\sum_{i=1}^{\infty} \frac{(-1)^{i+1}}{i} = 1 - \frac{1}{2} + \frac{1}{3} - \frac{1}{4} + \frac{1}{5} - \frac{1}{6} + \frac{1}{7} - \dots$$
$$+ \frac{(-1)^{i+1}}{i} + \dots = \ln(2)$$

$$\sum_{i=1}^{\infty} \frac{(-1)^i}{i} = -1 + \frac{1}{2} - \frac{1}{3} + \frac{1}{4} - \frac{1}{5} + \frac{1}{6} - \frac{1}{7} + \ldots + \frac{(-1)^i}{i}$$

$$+ \ldots = -\ln(2) = \ln\left(\frac{1}{2}\right)$$

Error Majoring—The Leibniz criterion also provides an insight to the error estimates when evaluating an alternating series numerically. In fact, replacing the sum, S, by a partial sum, S_n, is equivalent to neglecting all terms from order $(n + 1)$.

$$S = S_n + (-1)^{n+1}u_{n+1} + (-1)^{n+2}\, u_{n+2} + (-1)^{n+3}u_{n+3} + \ldots$$

That is, the error is:

$$Error = \sum_{i=n+1}^{\infty} [(-1)^i\, u_i]$$

Without loss of generality, assume positive sign for the first neglected term, or better, assume that $(n + 1)$ is even:

$$\begin{aligned} Error &= u_{n+1} - u_{n+2} + u_{n+3} - u_{n+4} + u_{n+5} - u_{n+6} + u_{n+7} \ldots \\ &= u_{n+1} - (u_{n+2} - u_{n+3}) - (u_{n+4} - u_{n+5}) - (u_{n+6} - u_{n+7}) \ldots \end{aligned}$$

Once more, all the subtractions inside parentheses result positive quantities, since the subtracted terms are presented in decreasing order. Therefore, the error committed that way will be upper-bounded by the absolute value of the first neglected term:

$$Error \leq |u_{n+1}|$$

10.5.1 The Cost Alternating Series and Error Estimates

Once again, recall the cost function: $y(x) = 50\pi\, x^2 + \frac{400}{x}$.

The second term, $400/x$, corresponds to the alternating series:

$$P(x) = 50\,\pi\, x^2 + 400 \sum_{i=0}^{\infty} [(-1)^i\, (x - 1)^i]$$

Note the power series corresponding to the first part of $y(x)$, $50\pi\, x^2$, remains the same part by cancellation of various terms (Section 10.8).

Error Comparison in P_3 and in P_5

- When using the third order's power series, $P_3(x)$, the first term to be neglected is the fourth one; then the error's upper bound can be defined:

$$Error \leq |u_4|$$

The cost estimate, $y(1.1)$, through the evaluation of series $P_3(x)$ at, $x = 1.1$, is associated with an error that is equal to or less than:

$$Error \leq 400|(x - 1)^4|$$

And at point, $x = 1.1$, the upper bound for the error is:

$$\begin{aligned} Error &\leq 400|(1.1 - 1)^4| \\ &\leq 400|(0.1)^4| \\ &\leq 0.0400 \end{aligned}$$

Then, the rounding error corresponds to the second decimal digit.

- For other side, when using the fifth order's power series, $P_5(x)$, the first term to be neglected is the sixth one; then

$$Error \leq |u_6|$$

The estimate of cost, $y(1.1)$, through series $P_5(x)$ is associated with an error that is equal to or less than:

$$Error \leq 400|(x - 1)^6|$$

And at the same point, $x = 1.1$, the error majoring for the cost estimate, $y(1.1)$, is

$$\begin{aligned} Error &\leq 400|(1.1 - 1)^6| \\ &\leq 400|(0.1)^6| \\ &\leq 0.000400 \end{aligned}$$

In this way, the rounding error corresponds to the fourth decimal digit.

The Error and the Number of Terms

And what if the function estimate, $y(1.1)$, needs a specific accuracy level? For example, let the rounding digit be the 10th decimal digit, as it occurs in a number of scientific, technologic, or finance topics; then the absolute error is majored,

$$Error \leq 0.5 \times 10^{-10}$$

- Therefore, assuming $y(x)$ is estimated through the alternating series with n terms, $P_n(x)$, the associated error must satisfy:

$$Error \leq 400|(x-1)^{n+1}| \leq 0.5 \times 10^{-10}$$

And at the point at hand, $x = 1.1$,

$$400|(1.1-1)^{n+1}| \leq 0.5 \times 10^{-10}$$

$$400|(0.1)^{n+1}| \leq 0.5 \times 10^{-10}$$

The error relation is satisfied by a power series with twelve terms, $n = 12$, and neglecting the 13th term; in fact,

$$\begin{aligned} 0.4 \times 10^3|(0.1)^{n+1}| &\leq 0.5 \times 10^{-10} \\ |10^{-(n+1)}| &\leq \frac{0.5 \times 10^{-10}}{0.4 \times 10^3} \\ 10^{-(n+1)} &\leq \frac{5}{4} \times 10^{-13} \end{aligned}$$

And for equal powers of 10, the inequality is satisfied when the exponents are equal too:

$$\begin{aligned} 10^{-(n+1)} &= 10^{-13} \\ -(n+1) &= -13 \\ n+1 &= 13 \\ n &= 12 \end{aligned}$$

That is, 12 terms are needed to obtain an estimate of $y(1.1)$ with a rounding error in the 10th decimal digit, and then a P_{12} alternating series is to be calculated.

10.6 FUNCTION SERIES

Important notions about function series are presented in this section, including the convergence test for function series, together with derivation and integration of function series. In addition, these topics of series derivation and integration are illustrated for the power series of fifth order, $P_5(x)$, associated with the cost function.

- Series of functions is any series in which the general term is a function of a variable x:

$$S(x) = \lim S_n(x) = \lim_{n \to \infty} \sum_{i=1}^{n} f_i(x) = \sum_{i=1}^{\infty} f_i(x)$$

  And the convergence domain for a function series is the set of values, x, for which the infinite sum of functions converges.

Convergence Test (Weierstrass)—Let $f_n(x)$ be a sequence of functions defined over the domain, D, and consider a numerical sequence, M_n, such that,

$$\forall\ n,\ \forall\ x_0 \in D,\ |f_n(x_0)| \leq M_n$$

So if the number series $\Sigma_{i=1}^{\infty} M_i$ converges:

i. The numerical series $\Sigma_{i=1}^{\infty} f_i(x_0)$ will converge absolutely on domain D;

ii. The function series $\Sigma_{i=1}^{\infty} f_i(x)$ will converge uniformly on D for a limit function $S(x)$.

Note the uniform convergence assumes the function series is within the radius of convergence of limit function, $S(x)$, along the entire domain for the function; it implies pointwise convergence, in the sense the convergence radius is satisfied at each point in the domain, however convergence at a point does not imply uniform convergence.

Two results of great interest are presented next, one addressing the integration of function series, the other the derivation. The application to the function series for the cost function follows too, illustrating this way the empowerment it allows. Namely, beyond the treatment of complex or difficult functions that require significant work and time (e.g., statistics estimators, integral functions, numerical calculations), the two first derivatives of cost function both fit well. Such derivatives emulate well the growth and the curvature of cost function, and the same applies to the integral function, with a good fit to the cumulative value of cost function too.

- ***Integration Theorem***—Let $\Sigma_{i=1}^{\infty} f_i(x)$ be a function series that uniformly converges to the function $S(x)$ on the closed interval $[a, b]$. Then:

    i. If each of the functions $f_i(x)$ is continuous on the interval $[a, b]$, the series function $S(x)$ will also be continuous on $[a, b]$;

    ii. If $S(x)$ and each of the functions $f_i(x)$ are integrable in $[a, b]$, then

$$\int_a^b S(x)\, dx = \sum_{i=1}^{\infty} \left[\int_a^b f_i(x)\, dx \right]$$

- ***Derivation Theorem***—Let $\Sigma_{i=1}^{\infty} f_i(x)$ be a series of functions with pointwise convergence to the function $S(x)$ on the closed interval $[a, b]$.

    If all functions $f_i(x)$ are differentiable on $[a, b]$, and if the series of derivative functions $\Sigma_{i=1}^{\infty} f_i'(x)$ converge uniformly on $[a, b]$ for any continuous function, then

$$S'(x) = \sum_{i=1}^{\infty} f_i'(x)$$

10.6.1 Derivation of Cost Function Series

Again, consider the cost function, $y(x) = 50\pi\, x^2 + \frac{400}{x}$, and the fifth-order polynomial, $P_5(x)$:

$$P_5(x) = 2400 - 6000x + (50\,\pi + 8000)x^2 - 6000x^3 + 2400x^4 - 400x^5$$

- The first derivative both for cost function, $y'(x)$, and for function series, $P_5'(x)$, follows:

$$y'(x) = 100\pi\, x - \frac{400}{x^2}$$

$$\begin{aligned} P_5{}'(x) &= -6000 + 2 \times (50\,\pi + 8000)\,x - 3 \times 6000x^2 \\ &\quad + 4 \times 2400x^3 - 5 \times 400x^4 \\ &= -6000 + (100\,\pi + 16000)\,x - 18000x^2 + 9600x^3 \\ &\quad - 2000x^4 \end{aligned}$$

The table of numerical values in the neighborhood of $x = 1$ and the related graphic lines are presented in Figure 10.2. At that point the derivatives of first order for the cost function and the function series are equal, $y'(1) = P_5'(x) = -85.841$, with the accuracy provided. The derivative's zero, both for the cost function and the function series, is bounded in the range]1.08, 1.09[, where their sign changes from negative to positive. Thus, the minimum point for the cost function is confirmed, since the decreasing trend turns on an increasing trend in that range, in both cases.

In the range at hand [0.9, 1.10], the values calculated for the first-order derivatives of cost function, $y'(x)$, and function series, $P_5'(x)$, are coincident until the first decimal digit; the estimates deviation

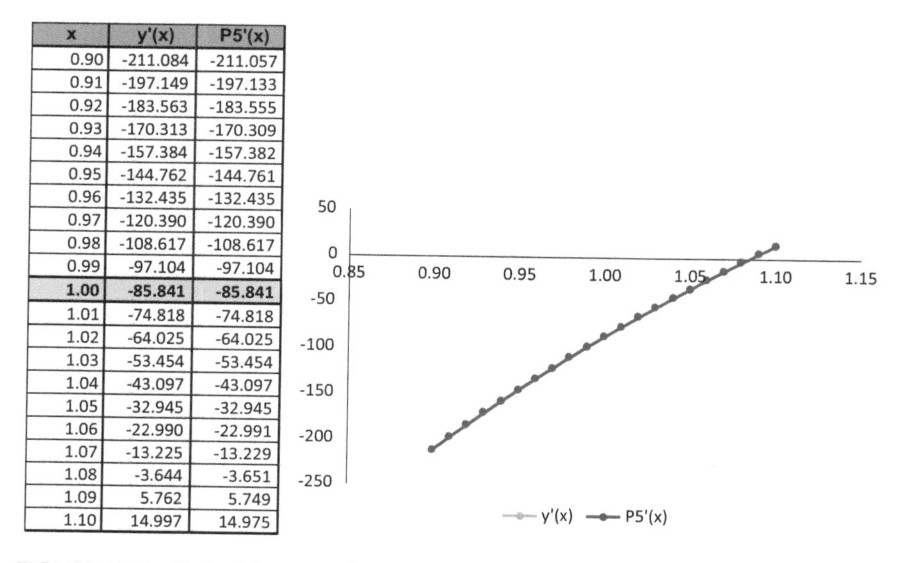

x	y'(x)	P5'(x)
0.90	-211.084	-211.057
0.91	-197.149	-197.133
0.92	-183.563	-183.555
0.93	-170.313	-170.309
0.94	-157.384	-157.382
0.95	-144.762	-144.761
0.96	-132.435	-132.435
0.97	-120.390	-120.390
0.98	-108.617	-108.617
0.99	-97.104	-97.104
1.00	**-85.841**	**-85.841**
1.01	-74.818	-74.818
1.02	-64.025	-64.025
1.03	-53.454	-53.454
1.04	-43.097	-43.097
1.05	-32.945	-32.945
1.06	-22.990	-22.991
1.07	-13.225	-13.229
1.08	-3.644	-3.651
1.09	5.762	5.749
1.10	14.997	14.975

FIGURE 10.2 Coincident graphic lines of first derivative for cost function, $y'(x)$, and function series, $P_5'(x)$.

occur only in the second, or even third decimal digit. That is, the absolute error is about 10^{-2} or 10^{-3}; if additional accuracy is required, more terms can be added to the polynomial, and then reducing the value for the first term to be neglected.

- Now, the derivatives of second order both for function, $y''(x)$, and power series, $P_5''(x)$, are focused on.

$$y''(x) = 100\pi + \frac{800}{x^3}$$

$$\begin{aligned} P_5''(x) &= (100\,\pi + 16000) - 2 \times 18000x + 3 \times 9600x^2 - 4 \times 2000x^3 \\ &= (100\,\pi + 16000) - 36000x + 28800x^2 - 8000x^3 \end{aligned}$$

In the neighborhood of, $x = 1$, table and graphics are presented in Figure 10.3.

And again, at that point, the values of second-order derivatives for the cost function and the function series are equal, $y''(1) = P_5''(1) = 1114.159$. These derivatives are positive and the values are mostly coincident, then the curvature of cost function is upwards and adequately estimated. In fact, the values calculated for the second

x	y''(x)	P5''(x)
0.90	1411.553	1410.159
0.91	1375.771	1374.871
0.92	1341.528	1340.975
0.93	1308.743	1308.423
0.94	1277.337	1277.167
0.95	1247.240	1247.159
0.96	1218.384	1218.351
0.97	1190.705	1190.695
0.98	1164.145	1164.143
0.99	1138.647	1138.647
1.00	**1114.159**	**1114.159**
1.01	1090.631	1090.631
1.02	1068.017	1068.015
1.03	1046.273	1046.263
1.04	1025.356	1025.327
1.05	1005.229	1005.159
1.06	985.855	985.711
1.07	967.198	966.935
1.08	949.225	948.783
1.09	931.906	931.207
1.10	915.211	914.159

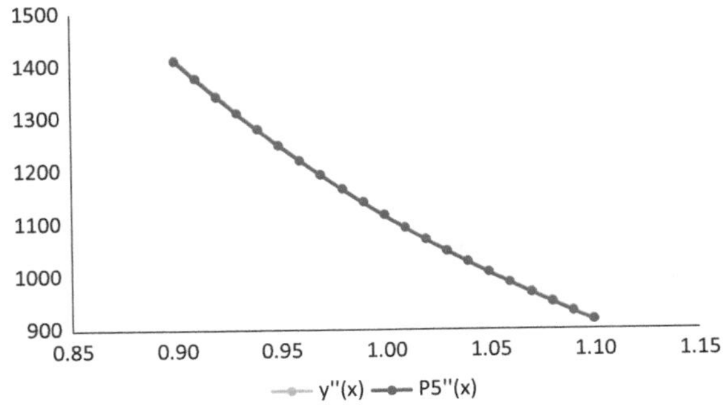

FIGURE 10.3 Coincident graphic lines of second derivative for cost function, $y''(x)$, and function series, $P_5''(x)$.

derivatives of cost function, $y''(x)$, and function series, $P_5''(x)$, are coincident until the unit's digit, the deviation on estimates mostly occur in the first decimal, or even second decimal digit. While the absolute error is about unit, the relative error is about 10^{-3} or 10^{-4}; once again, in case of additional accuracy, more terms can be added to the polynomial.

10.6.2 Integration of Cost Function Series

Consider the cost function, $y(x)$, and the fifth-order polynomial, $P_5(x)$; then, calculating the integrals of both functions, $Y(x)$ and $P_6(x)$,

$$Y(x) = \int 50\pi\, x^2 + \frac{400}{x} dx$$

$$P_6(x) = \int 2400 - 6000x + (50\,\pi + 8000)\, x^2 - 6000x^3 + 2400x^4 - 400x^5 dx$$

and including the integration constants:

$$Y(x) = 50\pi\,\frac{x^3}{3} + 400\ln|x| + c$$

$$\begin{aligned} P_6(x) &= 2400x - 6000\tfrac{x^2}{2} + (50\,\pi + 8000)\tfrac{x^3}{3} - 6000\tfrac{x^4}{4} + 2400\tfrac{x^5}{5} - 400\tfrac{x^6}{6} + c \\ &= 2400x - 3000\,x^2 + \tfrac{(50\,\pi + 8000)}{3}x^3 - 1500\,x^4 + 480\,x^5 - \tfrac{200}{3}x^6 + c \end{aligned}$$

- Assuming, $Y(1) = 0$, then the integration constant for $Y(x)$ can be calculated:

$$50\pi\frac{(1)^3}{3} + 400\ln|1| + c = 0$$

Since the logarithm in the second term becomes zero, it directly results in $c = -\frac{50\pi}{3}$; then

$$Y(x) = 50\pi\,\frac{x^3}{3} + 400\ln|x| - \frac{50\pi}{3}$$

In the same way, assuming that $P_6(1) = 0$, then the integration constant for integral series, $P_6(x)$, is calculated too:

$$2400 \times 1 - 3000 \times (1)^2 + \frac{(50\,\pi + 8000)}{3} \times (1)^3 - 1500 \times (1)^4 + 480 \times (1)^5 - \frac{200}{3} \times (1)^6 + c = 0$$

It results in $c = -980 - \frac{50\pi}{3}$, and finally,

$$P_6(x) = 2400x - 3000\,x^2 + \frac{(50\,\pi + 8000)}{3}x^3 - 1500\,x^4 + 480\,x^5 - \frac{200}{3}x^6 - 980 - \frac{50\pi}{3}$$

The associated values in the neighborhood of, $x = 1$, and the graphic lines are presented in Figure 10.4. At that point, since the integration constant is defined as zero, both the integrals for the cost function and the function series become zero, $Y(1) = P_6(1) = 0.0$.

Throughout all the range [0.9, 1.10], the values obtained from the integration of cost function, $Y(x)$, and function series, $P_6(x)$, are

x	Y(x)	P6(x)
0.90	-56.334	-56.334
0.91	-50.627	-50.627
0.92	-44.941	-44.941
0.93	-39.272	-39.272
0.94	-33.621	-33.621
0.95	-27.985	-27.985
0.96	-22.364	-22.364
0.97	-16.756	-16.756
0.98	-11.160	-11.160
0.99	-5.575	-5.575
1.00	**0.000**	**0. 000**
1.01	5.567	5.567
1.02	11.126	11.126
1.03	16.679	16.679
1.04	22.226	22.226
1.05	27.769	27.769
1.06	33.309	33.309
1.07	38.847	38.847
1.08	44.383	44.383
1.09	49.919	49.919
1.10	55.455	55.455

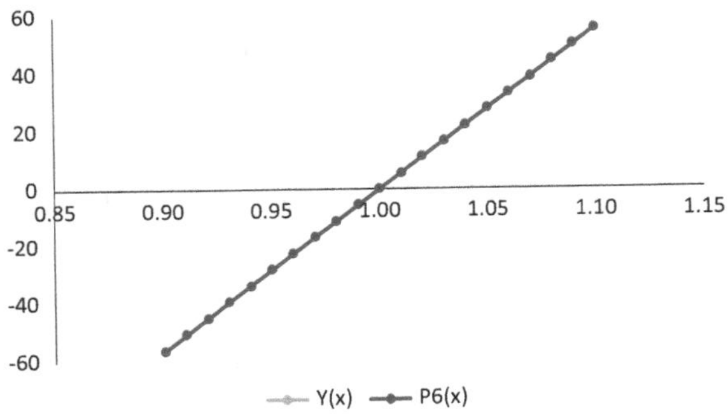

FIGURE 10.4 Coincident graphic lines (and values) for integral function, $Y(x)$, and integral series, $P_6(x)$.

coincident until the third decimal digit. Within the accuracy provided in the table, it means the deviation on estimates would occur only in the fourth decimal digit. Then, the absolute error is about 10^{-4} or 10^{-5}; since this polynomial of sixth order presents one additional term, the value for the first term to be neglected occurs in the seventh-order term, and the estimates error are being reduced.

10.7 POWER SERIES

In this section, power series are focused on, including Maclaurin series that are a particular type of power series with the center point at zero, $c = 0$. The convergence intervals are treated too, and a specific instance of power series with center, $c = 1$, is presented for the cost function.

- Power series centered at point, c, is a function series of the type:

$$a_0 + a_1(x - c) + a_2(x - c)^2 + \ldots + a_i(x - c)^i + \ldots$$
$$= \sum_{i=0}^{\infty} a_i(x - c)^i$$

where power functions are summed at each term, and
$a_0, a_1, a_2, \ldots, a_n$—the series coefficients, a sequence of real numbers;
c—the series center, a real number.

A particular case occurs when the center point is zero, $c = 0$, which is called Maclaurin series:

$$a_0 + a_1x + a_2x^2 + \ldots + a_ix^i + \ldots = \sum_{i=0}^{\infty} a_ix^i$$

The domain of convergence for a Maclaurin series is an interval centered in zero, $c = 0$, and with radius of convergence, r:

$$|x| < r$$

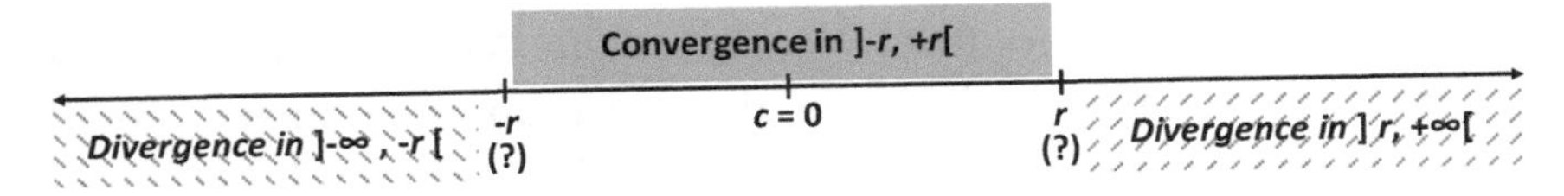

FIGURE 10.5 Convergence interval for the Maclaurin series.

That is, as presented in Figure 10.5, the Maclaurin series $\sum_{i=1}^{\infty} a_i x^i$:

i. Converges uniformly on any closed interval contained in $]-r, +r[$.

ii. Diverges outside the interval, $[-r, +r]$, for values larger than the convergence radius.

iii. At the border points of that interval, $x = -r$ and $x = +r$, the series may converge or diverge, for this purpose requiring a particular analysis.

In addition, three types of convergence intervals can be described for the power series centered at point c. In fact, as presented in Figure 10.6, only one of the following situations will occur:

i. The series converges only at the central point, $x = c$.

ii. There is a positive radius, r, such that the series is absolutely convergent if $|x - c| < r$, and is divergent if $|x - c| > r$.

iii. The series is absolutely convergent in the whole domain.

10.7.1 Convergence Interval for the Cost Power Series

The power series with center, $c = 1$, represents the cost function (more details in Section 10.8) for diameter about 1 meter, $x = 1$, as follows:

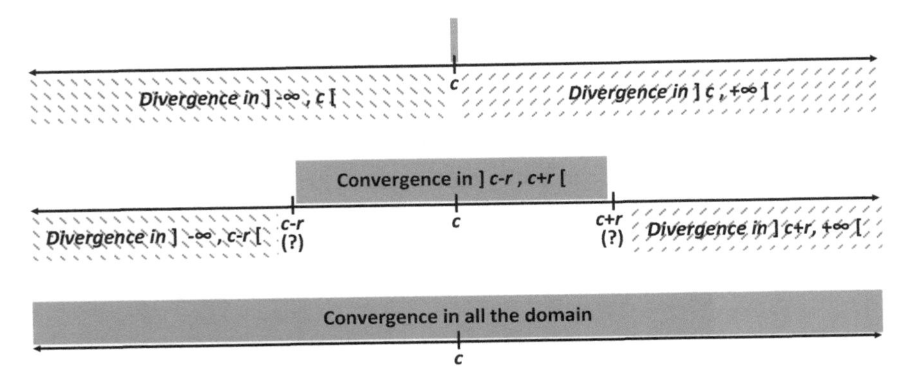

FIGURE 10.6 Types of convergence intervals for power series with center, c.

$$P(x) = 50\,\pi\,x^2 + 400 \sum_{i=0}^{\infty} [(-1)^i\,(x-1)^i]$$

Addressing only the summation in the second part that corresponds to an alternating series, $S(x)$:

$$\begin{aligned} S(x) &= \textstyle\sum_{i=0}^{\infty} [(-1)^i\,(x-1)^i] \\ &= 1 - (x-1) + (x-1)^2 - (x-1)^3 + (x-1)^4 \\ &\quad - (x-1)^5 + \ldots + (-1)^n\,(x-1)^n + \ldots \end{aligned}$$

The convergence analysis includes three approaches, namely: (a) The comparison as geometric series; (b) the ratio test; and (c) the root test.

A. *Comparison as Geometric Series*

Note the terms of series $S(x)$ correspond to the terms of a geometric progression of ratio $k_g = -(x-1)$, then the series convergence requires the ratio's absolute value be less than 1:

$$|k_g| = |-(x-1)| < 1$$

And then,

$$\begin{gathered} |x-1| < 1 \\ -1 < x - 1 < 1 \\ 0 < x < 2 \end{gathered}$$

In this range, the relation for geometric sums applies,

$$|k_g| < 1 \Rightarrow S = \frac{1}{1-k_g}$$

Assuming $0 < x < 2$ for the barrel diameter, the (absolute) geometric ratio is lower than 1, $k_g = 1 - x$, and the function series $S(x)$ absolutely converges to function $(1/x)$:

$$S(x) = \frac{1}{1-(1-x)} = \frac{1}{x}, \quad x \neq 0$$

B. *The Ratio Test*

Applying the ratio test,

$$\lim_{n\to\infty}\left(\frac{u_{n+1}}{u_n}\right) = k$$

to the absolute values of alternating series, $S(x)$, and simplifying the power factors in numerator and denominator,

$$\lim_{n\to\infty}\frac{\left|(-1)^{n+1}(x-1)^{n+1}\right|}{\left|(-1)^{n}(x-1)^{n}\right|} = \left|(-1)\,(x-1)\right| = |x-1|$$

Again, the series convergence requires the absolute value for the ratio test is less than 1:

$$|x-1| < 1, \text{ and also } 0 < x < 2$$

C. *The Root Test*

Applying the root test,

$$\lim_{n\to\infty}\sqrt[n]{u_n} = k$$

to the absolute values of alternating series $S(x)$, and canceling power and root,

$$\lim_{n\to\infty}\sqrt[n]{\left|(-1)^{n}(x-1)^{n}\right|} = \left|(-1)\,(x-1)\right| = |x-1|$$

Once again, the series convergence requires the absolute value for the ratio test is less than 1, and then $0 < x < 2$.

In addition, note that both the ratio test and root test are not driving a conclusion, neither convergence nor divergence, when the test result is $k = 1$. That is, a specific analysis is to be performed when

$$|x-1| = 1$$

$$x-1 = 1 \quad \lor\ x-1 = -1$$

$$x = 2 \quad \lor\ x = 0$$

- In this way, when the alternating series

$$S(x) = 1 - (x - 1) + (x - 1)^2 - (x - 1)^3 + (x - 1)^4 - (x - 1)^5 + \ldots + (-1)^n (x - 1)^n + \ldots$$

is evaluated at, $x = 2$, then an oscillating divergent series is obtained:

$$\begin{aligned} S(2) &= 1 - (2 - 1) + (2 - 1)^2 - (2 - 1)^3 + (2 - 1)^4 - (2 - 1)^5 \\ &\quad + \ldots + (-1)^n (2 - 1)^n + \ldots \\ &= 1 - 1 + 1 - 1 + 1 - 1 + \ldots + (-1)^n + \ldots \end{aligned}$$

- And when it is evaluated at, $x = 0$, then a series that properly diverges to infinite arises:

$$\begin{aligned} S(0) &= 1 - (0 - 1) + (0 - 1)^2 - (0 - 1)^3 + (0 - 1)^4 - (0 - 1)^5 \\ &\quad + \ldots + (-1)^n (0 - 1)^n + \ldots \\ &= 1 + 1 + 1 + 1 + 1 + 1 + \ldots + (-1)^{2n} + \ldots \end{aligned}$$

Finally, the convergence interval is]0, 2[, and the power series $S(x)$ diverges when variable, x, takes values outside that interval, as presented in Figure 10.7.

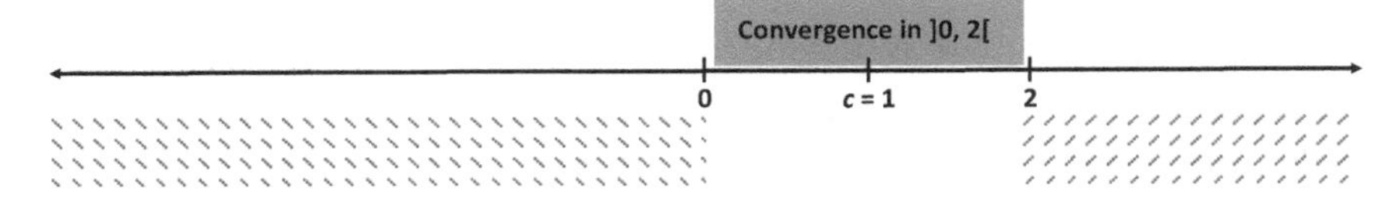

FIGURE 10.7 Convergence interval for the cost's alternating power series.

10.8 TAYLOR SERIES

The Taylor formula is revisited in this section, as well as the development of functions in Taylor series. Thereafter, the development of cost function in Taylor series with center, $c = 1$, is presented, while the steps to obtain Taylor polynomials with three and five terms—the power series $P_3(x)$ and $P_5(x)$—are detailed.

Taylor Polynomial—Consider a function $f(x)$, defined at an arbitrary point c, such that its derivatives exist up to the order $(n + 1)$, in a given neighborhood of c.

The Taylor polynomial of degree n is defined in such way that its n derivatives evaluated at point c, take the same value of the corresponding derivatives of function $f(x)$, that is,

$$\begin{cases} P_n(x) & = f(c) \\ P_n{}'(x) & = f'(c) \\ P_n{}''(x) & = f''(c) \\ P_n{}'''(x) & = f'''(c) \\ \quad (\ldots) & \\ P_n{}^{(n)}(x) & = f^{(n)}(c) \end{cases}$$

The Taylor polynomial of degree n, for function $f(x)$ at point c, thus follows:

$$P_n(x) = f(c) + \frac{(x-c)}{1!}.f'(c) + \frac{(x-c)^2}{2!}.f''(c) + \ldots + \\ + \frac{(x-c)^n}{n!}.\, f^{(n)}(c) + R_n(x)$$

with the Lagrange form for the rest term:

$$R_n(x) = \frac{(x-c)^{n+1}}{(n+1)!}.f^{(n+1)}(t), \text{ with } c < t < x$$

The Taylor polynomial is extended to a power series with center c, assuming all the n terms as powers of $(x{-}c)$, and taking limit when n tends to infinite; this development is said to be Taylor series.

Taylor Series—Let $f(x)$ be an indefinitely differentiable function in the neighborhood of point c, and represented by a power series on an open interval containing c:

$$\begin{aligned} f(x) &= \sum_{i=0}^{\infty} a_i (x-c)^i \\ &= a_0 + a_1(x-c) + a_2(x-c)^2 + \ldots + a_n(x-c)^n + \ldots \end{aligned}$$

Note the series coefficients are defined in such way the corresponding derivatives of both function, $f(x)$, and the series are equal. The series derivatives thus follow:

$$f'(x) = 0 + a_1 + 2\,a_2(x - c) + 3\,a_3(x - c)^2 + 4\,a_4(x - c)^3 \ldots$$
$$+\, n\,a_n(x - c)^{n-1} + \ldots$$

$$f''(x) = 0 + 2\,a_2 + 3.2\,a_3(x - c) + 4.3\,a_4(x - c)^2 \ldots$$
$$+\, n(n - 1)\,a_n(x - c)^{n-2} + \ldots$$

$$f'''(x) = 0 + 3.2\,a_3 + 4.3.2\,a_4(x - c) \ldots + n\,(n - 1)\,(n - 2)\,a_n(x - c)^{n-3} + \ldots$$

$$f^{(IV)}(x) = 0 + 4.3.2\,a_4 \ldots + n\,(n - 1)\,(n - 2)\,(n - 3)\,a_n(x - c)^{n-4} + \ldots$$

$$(\ldots)$$

$$f^{(n)}(x) = 0 + n\,(n - 1)\,(n - 2)\,(n - 3) \ldots 3.2.a_n(x - c)^{n-n} + \ldots$$

And evaluating these derivatives at point $x = c$, the series coefficients can be directly calculated:

$$\begin{aligned} f'(c) &= 0 + a_1 + 2\,a_2(c - c) + 3\,a_3(c - c)^2 + 4\,a_4(c - c)^3 \ldots + n\,a_n(c - c)^{n-1} + \ldots \\ &= 0 + a_1 + 0 + 0 + 0 + \ldots \\ &= a_1 \end{aligned}$$

The coefficient takes the same value as the first derivative at point c; for the second derivative,

$$\begin{aligned} f''(c) &= 0 + 2\,a_2 + 3.2\,a_3(c - c) + 4.3\,a_4(c - c)^2 \ldots \\ &\quad + n(n - 1)\,a_n(c - c)^{n-2} + \ldots \\ &= 0 + 2\,a_2 + 0 + 0 + 0 + \ldots \\ &= 2\,a_2 \end{aligned}$$

$$\text{or better, } a_2 = \frac{f''(c)}{2}$$

For the third derivative,

$$\begin{aligned} f'''(c) &= 0 + 3.2\, a_3 + 4.3.2\, a_4 (c - c) \ldots \\ &\quad + n\,(n-1)\,(n-2)\, a_n (c - c)^{n-3} + \ldots \\ &= 0 + 3.2\, a_3 + 0 + 0 + \ldots \\ &= 0 + 3.2\, a_3 \end{aligned}$$

$$\text{and again, } a_3 = \frac{f'''(c)}{3.2}$$

And the fourth derivative,

$$\begin{aligned} f^{(IV)}(c) &= 0 + 4.3.2\, a_4 \ldots \\ &\quad + n\,(n-1)\,(n-2)\,(n-3)\, a_n (c - c)^{n-4} + \ldots \\ &= 0 + 4.3.2\, a_4 + 0 + \ldots \\ &= 4.3.2\, a_4 \end{aligned}$$

$$\text{once again, } a_4 = \frac{f^{(IV)}(c)}{4.3.2}$$

(...)

In a general manner, for the *n*th-order derivative

$$\begin{aligned} f^{(n)}(c) &= 0 + n\,(n-1)\,(n-2)\,(n-3) \ldots 3.2.1\, a_n + 0 + \ldots \\ &= n\,(n-1)\,(n-2)\,(n-3) \ldots 3.2.1\, a_n \end{aligned}$$

$$\text{and finally, } a_n = \frac{f^{(n)}(c)}{n\,(n-1)\,(n-2)\,(n-3) \ldots 3.2.1}$$

Therefore, the development in Taylor series of function $f(x)$ at point c takes the following form:

$$f(x) = f(c) + \frac{f'(c)}{1!}.(x - c) + \frac{f''(c)}{2!}.(x - c)^2 + \frac{f^{(3)}(c)}{3!}.(x - c)^3 + \ldots$$

That is, the coefficients can be defined through the associated derivatives,

$$f(x) = \sum_{i=0}^{\infty} \frac{f^{(i)}(c)}{i!}.(x - c)^i,$$

$$\text{with } a_i = \frac{f^{(i)}(c)}{i!}$$

10.8.1 Development of Taylor Series for the Cost Function

Again, recall the cost function:

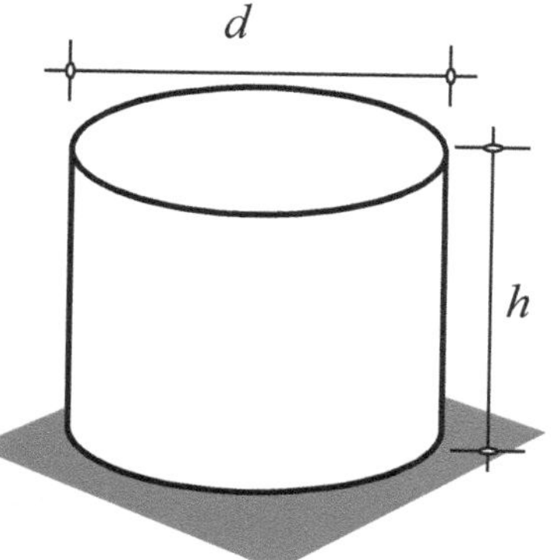

$$y(x) = 100\left(\frac{\pi}{2}x^2 + \frac{4}{x}\right) = 50\pi\, x^2 + \frac{400}{x}$$

When x is taking the center point, $c = 1$, then function $y(x)$ becomes:

$$y(1) = 100\left(\frac{\pi}{2}1^2 + \frac{4}{1}\right) = 50\pi + 400$$

- The function's first-order derivative, $y'(x)$, is already calculated in Section 10.1,

$$y'(x) = 100\pi\, x - \frac{400}{x^2}$$

  and at point $x = 1$, it takes $y'(1) = 100\pi - 400$.

- Similarly, for the second-order derivative, $y''(x)$,

$$y''(x) = 100\pi + \frac{800}{x^3}$$

  at the same point, $x = 1$, it results $y''(1) = 100\pi + 800$.

- Following with the derivation procedures, the first term in the second-order derivative $y''(x)$ is constant and can be neglected; hereafter, the second term is treated as a negative power and the derivatives of higher order can be obtained as follows:

$$y'''(x) = 0 + 2.400.(-3)x^{-3-1} = -(2.3).400\, x^{-4} = \frac{-400 \times 3!}{x^4}$$

$$y^{IV}(x) = -400.3!\,(-4)x^{-4-1} = +400.4!\, x^{-5} = \frac{400 \times 4!}{x^5}$$

$$y^{V}(x) = 400.4!\,(-5)\,x^{-5-1} = -400.5!\,x^{-6} = \frac{-400 \times 5!}{x^{6}}$$

and the derivative of nth order,

$$y^{(n)}(x) = (-1)^{n} 400\, n!\, x^{-n-1} = \frac{(-1)^{n} 400\, n!}{x^{n+1}}$$

- Then, the related values at the point, $x = 1$, are:

$$y'''(1) = -400 \times 3!$$

$$y^{IV}(1) = +400 \times 4!$$

$$y^{V}(1) = -400 \times 5!$$

and for the general case,

$$y^{(n)}(1) = (-1)^{n} 400\, n!$$

By substitution in the power series with center, $c = 1$,

$$y(x) = y(1) + \frac{y'(1)}{1!}(x-1) + \frac{y''(1)}{2!}(x-1)^2 + \frac{y^{(3)}(1)}{3!}(x-1)^3 + \ldots$$

or

$$y(x) = \sum_{i=0}^{\infty} \frac{y^{(i)}(1)}{i!}(x-1)^i, \quad \text{with} \quad a_i = \frac{y^{(i)}(1)}{i!}$$

The Taylor series for the cost function thus is:

$$\begin{aligned} y(x) = (50\pi + 400) &+ \frac{(100\pi - 400)}{1!}(x-1) + \frac{(100\pi + 800)}{2!}(x-1)^2 \\ &+ \frac{-400 \times 3!}{3!}(x-1)^3 + \frac{400 \times 4!}{4!}(x-1)^4 \\ &+ \frac{-400 \times 5!}{5!}(x-1)^5 + \frac{400 \times 6!}{6!}(x-1)^6 + \ldots \\ &+ \frac{(-1)^n\, 400 \times n!}{n!}(x-1)^n \end{aligned}$$

The factorials at each term can be simplified, and the terms with coefficients π and 400 are grouped and separated in two parts:

$$\begin{aligned} y(x) = {} & 50\pi + 100\pi(x-1) + 50\pi(x-1)^2 \\ & + 400 - 400(x-1) + 400(x-1)^2 - 400(x-1)^3 \\ & + 400(x-1)^4 - 400(x-1)^5 + \ldots + (-1)^n\, 400(x-1)^n \end{aligned}$$

Again simplifying the terms with coefficient 50π, the first part turns similar to the first term in function $y(x)$; thereafter, an alternating series with coefficient 400 is obtained in the second part, as follows:

$$\begin{aligned} y(x) = {} & 50\,\pi\,[1 + 2(x-1) + (x-1)^2] \\ & + 400[1 - (x-1) + (x-1)^2 - (x-1)^3 + (x-1)^4 \\ & - (x-1)^5 + \ldots + (-1)^n (x-1)^n + \ldots] \end{aligned}$$

Finally, the development in Taylor series for the cost function $y(x)$ with center $c = 1$ is:

$$\begin{aligned} y(x) = {} & 50\,\pi\,x^2 \\ & + 400 \sum_{i=0}^{\infty} [(-1)^i (x-1)^i] \end{aligned}$$

- From this general result, the third-order polynomial, $P_3(x)$, can be directly obtained from the series's first three terms:

$$\begin{aligned} P_3(x) = {} & 50\,\pi\,x^2 \\ & + 400[1 - (x-1) + (x-1)^2 - (x-1)^3] \end{aligned}$$

Developing the powers in the binome $(x - 1)$, namely, using the combinatorial coefficients or recalling the Pascal Triangle's coefficients for the third power, then

$$\begin{aligned} P_3(x) = {} & 50\,\pi\,x^2 \\ & + 400[1 - (x-1) + (x^2 - 2x + 1) - (x^3 - 3x^2 + 3x - 1)] \end{aligned}$$

Grouping the series in terms with the same power in variable x,

$$P_3(x) = 50\,\pi\,x^2 + 400\begin{bmatrix} 1 - (-1) + 1 - (-1) \\ + x(-1 - 2 - 3) \\ + x^2(1 + 3) \\ + x^3(-1) \end{bmatrix}$$

After simplifying the coefficients associated with that terms, it results in

$$P_3(x) = 50\,\pi\,x^2 + 400[4 - 6x + 4x^2 - x^3]$$

and then the third-order polynomial, $P_3(x)$, follows:

$$P_3(x) = 1600 - 2400x + (50\,\pi + 1600)x^2 - 400x^3$$

- To improve accuracy, the power series shall include additional terms; for example, the fifth-order series also includes the fourth and fifth powers of $(x - 1)$,

$$P_5(x) = 50\,\pi\,x^2 + 400[1 - (x - 1) + (x - 1)^2 - (x - 1)^3 + (x - 1)^4 - (x - 1)^5]$$

Developing again the powers in the binome $(x - 1)$ in the same way as for $P_3(x)$, either using the combinatorial coefficients or the Pascal Triangle's coefficients, then

$$P_5(x) = 50\,\pi\,x^2 + 400\begin{bmatrix} 1 - (x - 1) + (x^2 - 2x + 1) - (x^3 - 3x^2 + 3x - 1) + \\ (x^4 - 4x^3 + 6x^2 - 4x + 1) - (x^5 - 5x^4 + 10x^3 - 10x^2 + 5x - 1) \end{bmatrix}$$

After simplifying the terms with equal powers, by directly conjugating the corresponding coefficients, it results in

$$P_5(x) = 50\,\pi\,x^2 + 400\begin{bmatrix} 1 - (-1) + 1 - (-1) + 1 - (-1) \\ + x(-1 - 2 - 3 - 4 - 5) \\ + x^2(1 + 3 + 6 + 10) \\ + x^3(-1 - 4 - 10) \\ + x^4(1 + 5) \\ + x^5(-1) \end{bmatrix}$$

That is,

$$P_5(x) = 50\,\pi\,x^2 + 400\left[6 - 15x + 20x^2 - 15x^3 + 6x^4 - x^5\right]$$

Finally, the polynomial of fifth order, $P_5(x)$, corresponding to the power series with center, $c = 1$, for the cost function $y(x)$ is:

$$P_5(x) = 2400 - 6000x + (50\,\pi + 8000)x^2 - 6000x^3 + 2400x^4 - 400x^5$$

10.9 CONCLUDING REMARKS

Important notions about sum series, or infinite sums, with the main purpose to evaluate and emulate the cost function using partial sums are described. In this way, polynomials and sum series are useful to evaluate difficult functions; they are of relative simple utilization, the practice of computational implementation is well developed, and the associated errors are upper bounded.

- Due to their relevance for the following topics, reference series (Harmonic, Geometric, and Telescopic) are introduced; in addition, theorems about series and their applications to convergence analysis are presented too.
- Convergence criteria for series of non-negative terms are treated, while paving the way for the study of series with negative terms, in particular alternating series; the Leibniz criterion and related

applications are described, namely, the error estimates when evaluating the cost function with different alternating series.

- After that, function series are focused on, the related convergence domain is addressed, and applications to series derivation and series integration are presented; the polynomial $P_5(x)$ corresponds to a power series that emulates the cost function well, and its convergence interval is obtained too.
- Finally, Taylor formula is revisited, and the development of functions in Taylor series is studied; namely, the development of a power series with center, $c = 1$, for the cost function, while the steps to obtain partial sums with three and five terms—the polynomials $P_3(x)$ and $P_5(x)$—are detailed.

In addition, tables summarizing the main subjects described in the book are presented, including both the differential methods described until Chapter 6 and the integral topics in Chapters 7–10.

A real world problem, sizing a cylinder metal container (barrel), motivated a new visit to important Calculus results, which feature reverse mathematical applications and computational approaches (e.g., generalization/reduction, error evolution). These attributes contribute to enhancing problem solving, critical thinking, the methods' validation, with the support of both numerical and graphic approaches. Beyond automated operations and routine-basis procedures, complex problems within rich-data environments are arising, and also cascading from climate change, environment, food industry, pharmaceutics, health, onto other societal fields, while comprehensive approaches bordering Applied Mathematics are crucial to face such challenges.

Summary tables showing the main approaches and methods in this book are presented too (Tables 10.1 and 10.2).

TABLE 10.1 Main Mathematics Topics and Computational Approaches (I)

Approaches/ Methods	Mathematical Topics	Computational Approaches
First Notes	Introductory tools. Proofs and mathematical reasoning. The inverse rationale. Discussion of results.	First notes on computational reasoning. Representation in table and graphic, and close observation. Trial-and-error procedures and computational support.
Sequence of Real Numbers	Convergence radius, and sequence limits. Theorems and properties. Geometric sequences, base-2 and base-10 instances.	Simple computational methods, and verification as necessary condition. Proofs and contradiction procedures. Computational approximations and representation of real numbers.
Function Limits	Function limit as enlarged concept of two convergent variables. Composite function limit as three convergent variables in cascade. Lateral limits and remarkable limits.	Limits verification is both qualitative and quantitative, with graphical and numerical views. Treatment of important instances (trapped function, numerical conjecture, remarkable limit).
Continuity	Continuity is complementing and enlarging function limits. Generalization approach, e.g., continuity at a point, then extended onto an open and a closed range. Attributes of continuous function in closed range (extreme values, intermediate values).	A direct method holds for the limits calculation of continuous function. Graphical and numerical views, and treatment of important instances (elementary functions, cost function). Equations roots (I): the bisection method.
Derivation	Derivatives and geometric representation. Derivation rules and derivatives of important functions. Derivative of inverse function and derivatives of different orders.	Numerical instances of tangent to the function graphic at the point at hand. Obtaining derivatives using definition; then verifying via operating rules and derivation expressions. The cost function and simple instances that allow direct verification.
Sketching Functions and Important Theorems	Generalization: from derivative roots, extending to derivative (non-zero) mean value, and to the increments relation theorem. The first derivative and extreme values (max. and min.). Curvature and inflection points and the second derivative. Asymptotes (vertical, oblique, horizontal) and calculation of indeterminate limits.	Reduction: defining $g(x) = x$ in the increments relation, obtaining slope at intermediate point c, and $f'(c) = 0$ when $f(x)$ takes two times the same value. Sketching the cost function, and both the first and second derivatives. Equation roots (II): the Newton-Raphson method and error bounding.

TABLE 10.2 Main Mathematics Topics and Computational Approaches (II)

Approaches/ Methods	Mathematical Topics	Computational Approaches
Integral Sums	Integral sum and geometric interpretation. Calculation of convenient areas (rectangle, triangle, circle). Integral sums of simple functions (constant, linear, cost case).	Graphical and numerical insights; treatment of pertinent instances. Anti-derivative approach (*Why?*). Systematic verification and appreciation of circular thinking.
Indefinite Integral	Indefinite integral and related properties. General methods (decomposition, substitution, by parts). Specific methods, e.g., for rational functions.	Anti-derivative approach: primitives can be difficult or unable to obtain. Reverse rationale: derivation allows verification, and also solution improvements. A reduction approach is used very often, even for difficult integrals.
Definite Integral	Definite integral and its properties enlarge the integral sums' scopes of application. Barrow's Rule and associated instances. Improper integrals: extension to both unbounded domains and unbounded functions.	Numerical and graphic views of important functions (e.g., powers, exponential, cost function). Applications to area calculations and the Trapezoid method. Complementary convergence criteria for improper integrals.
Series	Basics about series, including main theorems and applications. Comparison of function series, the associated derivative and integral. Convergence intervals. The Taylor Series.	Enlarging the convergence criteria: from non-negative series, to non-positive, and to alternating series. Error estimates when replacing a function by its power series. Development of Taylor Series and the cost function.

Selected References

Apostol, T., *Calculus*, Vol. I, Wiley (2019). ISBN: 978-0-471-00005-1

Belien, J., Ittmann, H.W., Laumanns, M., Miranda, J.L., Pato, M.V., and Teixeira, A.P. (Eds.), *Advances in Operations Research Education – European Studies*, Springer International Series (2018). ISBN: 978-3-319-74103-1

Chapra, S.C., Canale, R.P., *Numerical Methods for Engineers*, McGraw Hill (8th Edition, 2020). ISBN: 9781260232073

Cochran, J.J., Cox, L.A., Keskinocak, P., Kharoufeh, J.P., Smith, J.C., & Ruckmann, J.-J., *Wiley Encyclopedia of Operations Research and Management Science*, John Wiley & Sons (2011). https://doi.org/10.1002/9780470400531

Conte, S.D., De Boor, C., *Elementary Numerical Analysis – An Algorithmic Approach*, SIAM (2017). ISBN: 978-1-61197-519-2

Kreyszig, E., *Advanced Engineering Mathematics*, John Wiley & Sons (10th Edition, 2020). ISBN: 9781119455929

Miranda, J.L., *Introduction to Optimization-Based Decision Making*, CRC PRESS (2021). ISBN: 9781315200323

Murty, K.G. (Ed.), *Case Studies in Operations Research*, Springer International Series, Springer (2015). ISBN: 978-1-493-91007-6

Ostrowski, A., *Lições de Cálculo Diferencial e Integral I*, Fundação Calouste Gulbenkian (1990). ISBN: 9789723102048

Pina, H., *Métodos Numéricos*, McGraw Hill Portugal (2003). ISBN: 972-8298-04-8

Spivak, M., *Calculus*, Cambridge University Press (2006). ISBN: 9780521867443

Swokovski, E.W., *Calculus*, Brooks/Cole (2000). ISBN: 9780534435387

Index

Note: Page numerals in *italic* and **bold** refer to figures and tables, respectively.